ADVANCE PRAISE FOR

Calm Clarity

"*Calm Clarity* tells Due Quach's story of transformation, weaving her rich life experience with in-depth scientific insights from across diverse fields. Due has created practical and accessible tools for readers to not just be inspired by her transformation, but to set off and experience their own profound changes as well." —SHARON SALZBERG, *New York Times* bestselling author of *Real Love* and *Real Happiness*

"*Calm Clarity* is refreshingly readable, easily accessible, and understandable; a very valuable addition to contemporary books on transforming our minds and creating the foundation of a happy life." —SYLVIA BOORSTEIN, PhD, author of *Happiness Is an Inside Job*

"Quach offers an extensive exploration of neuroscience and trauma resolution that led her from a struggling beginning with her impoverished Vietnamese family to Harvard, a Wharton MBA, and business success. She delves into the importance of meditation, ancient wisdom teachings, and the latest scientific research in transforming 'outer success' into a deep inner peace and calm clarity. This is an extraordinary work by an extraordinary woman." —ALLAN LOKOS, author of *Patience*

"Our world is in need of healing because there is so much pain and negativity being transmitted. *Calm Clarity* provides a powerful, practical, science-based approach to healing and transforming our pain and cultivating compassion to enable others to do so also. The compelling story of Due Quach's personal modern-day metamorphosis from a skeptic into a mystic will touch the hearts and minds of seekers everywhere, and her ongoing efforts to bring these tools to low-income communities to end the transmission of toxic stress and trauma is inspiring." —RICHARD ROHR, author of *Falling Upward*

"Imagine leading an organization where taking personal responsibility, having positive interpersonal relationships, performing at the highest level, and thinking creatively are commonplace. This is the future: *Calm Clarity*'s mixture of cognitive science, spirituality, and enlightened thinking are already transforming everything we know about life and work."

—HOWARD BLUMENTHAL, visiting scholar at the University of Pennsylvania

"*Calm Clarity* is an essential guide for the masses, to not only understand how the brain works in a simple but scientific way, but also to find a pathway to inner peace and a joyful life. Thanks, Due, for revealing something so complex, so simply."　　　　　—TOM CRONIN, founder of the Stillness Project and producer of *The Portal* documentary film

"One of the best books I have ever read. Yes, this book is about rewiring your brain for greater wisdom and joy, but it also shares Due's compelling story of resilience and grit, a very personal account of shifting from entanglement to enlightenment."　　　　　—RICK BELLINGHAM, EdD, CEO of iobility

"*Calm Clarity* captivates not only the mind, but also the heart, soul, and imagination. Due Quach skillfully interweaves scientific understanding with spiritual wisdom and grounds these important insights in everyday life by sharing how they enabled her inspiring personal journey of healing and transformation. This book can help anyone looking for effective tools to manage anxiety and their inner critic."

—SCOTT BARRY KAUFMAN, PHD, coauthor of *Wired to Create*

"*Calm Clarity* is the powerful story of a woman's journey from early chaos to a life of profound purpose. Combining the latest research on the brain with a skeptical dive into ancient spiritual wisdom, Due Quach develops an impressive program of 'brain hacks.' These tools can help all of us, especially those who have traumatic histories, to fashion a better life by thinking more clearly and more calmly."　　　　　—ROBERT KREIDER, former president and CEO of Devereux Advanced Behavioral Health

"Accessible yet deep, *Calm Clarity* helps readers achieve inner transformation, untangle emotional knots, and find clarity. This book is a revelatory and inspiring guide on the spiritual path that also helps people develop the leader inside them."　　　　　—PATRICIO BARRIGA, CEO and president of Fagor America, Inc.

"Due's rich personal narrative is interwoven with deep research into the way our brains work, offering lessons and tools that all of us should embrace for our own good and for the good of the world we live in."
—PHILIP W. LOVEJOY, executive director
of the Harvard Alumni Association

"*Calm Clarity* is both inspiring and practical. Due Quach combines her incredible personal journey with a deep scientific exploration and practical tools to help people investigate their own paths. This book is for seekers, pragmatists, and all those who are looking to live a deep, meaningful life."
—LARRY SCHWARTZ, founding chair
of the Institute for Jewish Spirituality

"Due Quach has given us an exceptional gift—she's taken research on neuroscience and mindfulness and made it easy to understand. But more than that, she's used this research to build a set of practices that will help you to optimize your nervous system and rise to the challenges you face. And she's done this with a keen awareness of the social realities of our time. Highly recommended!"
—PAUL ZELIZER, founder of Awarepreneurs

"Quach has broken down complex neuroscience into an easily understandable three-stage framework, making it possible for anyone interested in personal fulfillment and joy to follow a simple path and attain success."
—KRISHNA PENDYALA, president of the Mindful Nation
Foundation, Chief Empowerment Officer of ChoiceLadder Institute,
and author of *Beyond the Pig and the Ape*

"Following Due on her journey is so inspiring that you can't help but think about your own path and life's purpose. Her fascinating insights into current brain science and ancient Buddhist teachings provide the foundation for her Calm Clarity program, which all of us can use to live happier, more fulfilling lives."
—ROBIN WHITE OWEN, creative producer of MediaCombo, Inc.

"The book is a tour de force—really well written, clear, easy to follow, and inspiring."
—TOM TRITTON, former president of Haverford College
and the Chemical Heritage Foundation

Calm Clarity

How to Use Science to Rewire Your Brain for Greater Wisdom, Fulfillment, and Joy

Due Quach

A TarcherPerigee Book

tarcherperigee

An imprint of Penguin Random House LLC
375 Hudson Street
New York, New York 10014

Calm Clarity® is a registered trademark of Calm Clarity Co.

TarcherPerigee with tp colophon is a registered trademark
of Penguin Random House LLC.

Most TarcherPerigee books are available at special quantity discounts for bulk purchase
for sales promotions, premiums, fund-raising, and educational needs. Special books or
book excerpts also can be created to fit specific needs. For details, write:
SpecialMarkets@penguinrandomhouse.com.

Library of Congress Cataloging-in-Publication Data

Names: Quach, Due, author.
Title: Calm clarity : how to use science to rewire your brain for greater
    wisdom, fulfillment, and joy / Due Quach.
Description: New York : TarcherPerigee, [2018] | Includes bibliographical
    references and index.
Identifiers: LCCN 2017057371 (print) | LCCN 2017059179 (ebook) |
    ISBN 9781524704803 | ISBN 9780143130970
Subjects: LCSH: Brain—Popular works. | Neuroplasticity. | Stress
    (Physiology) | Stress (Psychology) | Calmness. | Meditation.
Classification: LCC QP376 (ebook) | LCC QP376 .Q25 2018 (print) |
    DDC 612.8/2—dc23
LC record available at https://lccn.loc.gov/2017057371

Printed in the United States of America
10 9 8 7 6 5 4 3 2 1

Book design by Katy Riegel

*This book is dedicated to my parents, who taught me
through their sacrifice, perseverance, humanity, compassion,
and kindness, lessons that no elite university ever could.*

*This book is for all who wonder: is this all there is?
May reading this help you appreciate that there is
much, much more.*

Contents

Preface

Why I Became a Mind-Hacker

I DEFINE "MIND-HACKING" as using science to enhance the best qualities of being human by proactively steering brain development in a way that physiologically supports greater physical, emotional, mental, and spiritual well-being, alignment, and integration. The saying that "necessity is the mother of invention" captures why I first created the mind-hacking techniques that eventually evolved into the Calm Clarity Program.

I had a particularly urgent need for tools that could help me stay calm and think clearly because, at the age of seventeen, I had to learn on my own how to navigate between two polar ends of the socioeconomic spectrum. In the beginning, the stress of moving between such social extremes literally drove me to the verge of insanity and compelled me to find my own way to make sense of the world and my place in it. By becoming a mind-hacker, I eventually built the inner strength and resilience needed to become a bridge between these two spheres. Today, I teach mind-hacking techniques to people across the socioeconomic spectrum, from inner-city teens to executives of leading corporations.

As a refugee from Vietnam who grew up in a poor and violent area of Philadelphia, my life changed dramatically when I enrolled at Harvard College. It was the first time I ever lived away from my family and I was not prepared in any way to handle the culture shock. It was like being dropped down alone onto another planet. Having rarely been exposed to the world outside my inner-city community and not having had the privilege of attending a private high school, I felt completely alienated with no support system and no mentorship. Soon enough, the toxic stress I felt from the competitive intensity and social isolation I experienced at Harvard caused the traumatic ordeals I had experienced and witnessed during my childhood, but never got treated for, to explode into complex post-traumatic stress disorder (PTSD). As PTSD took over my brain, my inner world turned into a living hell.

At first I had no idea what was going on. I had no idea why, after arriving at Harvard, I started to dissociate and have panic attacks, uncontrollable crying spells, and recurring nightmares that kept me from sleeping through the night. I had no idea why a voice in my head was obsessively telling me to bring my life to an end. I had no idea why all I wanted to do was stay in bed the entire day and hide from people. I privately struggled with these symptoms until I reached the point where I could hardly function. When it became almost impossible to pull myself together to attend classes or do my schoolwork, when it became clear that I was a serious danger to myself, I finally forced myself to seek help.

I was both relieved and devastated as the psychiatrist explained that my condition was linked to the traumatic ordeals I had survived as a baby when my family escaped from Vietnam and started life over in poverty. On one hand, this made it clear it wasn't my fault. I had no control over what happened to me as a baby. In fact, it made me realize how unusual and incredible it was for someone in my situation even to make it into an elite college like Harvard. On the other hand, what he said implied that early childhood trauma could be a life sentence because it affected brain development in ways that could result in lifelong challenges. He explained that there was no guarantee of a cure for my condition, which meant it was possible that I would have to struggle with these symptoms the rest of my life. This was horrible news to let sink in. It made me feel like all the hard work I'd done to break through so many barriers was all in vain. It felt so pointless to get into Harvard only to have my brain turn against me there!

The fact that I was no stranger to adversity hurt before it helped. PTSD added fuel to the anger I already felt about how unfair life seemed to be. After I arrived at campus, I felt the floor cave in under my feet as I learned that most of my classmates had been brought up in what I considered to be extreme wealth and privilege. How was I supposed to up my game to their level when I did not have their level of family support or resources? I couldn't stop fixating on how economically disadvantaged I was in comparison to my classmates and saw constant reminders of this—in the sense of entitlement, extravagant tastes, and prominent family connections that they seemed to unintentionally flaunt every day.

During my time at Harvard, it had a sink-or-swim culture and there was no support program for students in my situation. I was completely on my own to make this transition, but the problem was, I had no idea what I didn't know. Further, throughout my childhood, the only person in my life that I could 100 percent rely upon was me. Now that my brain was malfunctioning and I couldn't even rely on myself anymore, what was I supposed to do? I felt helpless, lost, and doomed.

As my world came crashing down, wallowing in self-pity only made things worse. It didn't help that PTSD made negative emotions feel even more catastrophic. I really thought my life was over. I didn't know if I could even make it to graduation. My inner demons kept telling me my situation was hopeless and now that I had lost control over my life, I could at least control how it ended. Then somehow, in the midst of my despair, something inside me said, "Hell no!" At that point, I knew I was not going to break. I was not going to let this be how my story ended. I was going to scrape together whatever willpower was left in me to find a path out of the darkness of my own mind.

I started taking the medications prescribed by the psychiatrist to alter my brain chemistry and soften the panic attacks. I also started therapy at his suggestion. In a few months, I began to stabilize, but I knew I could not become dependent on medication or therapy because I did not have reliable health insurance growing up, and once my student health insurance expired, I would no longer be able to afford these treatments. Therefore, I naturally saw my graduation as a deadline for getting myself off the drugs and therapy without relapsing.

As I searched for a solution, it dawned upon me that losing control of my brain meant that I had to stop taking it for granted. I had to take ownership of how I was affecting my own brain every day and start appreciating what my brain did for me. So I decided to learn what science could tell me about the brain and then use this knowledge to experiment with different approaches to help mine function better. The more I learned about the brain, the more fascinated I became with my own. From my research, I realized that the very same self-defense mechanisms I had built as a child to protect me from danger were now the cause of negative spirals that made me fall apart. Once I understood how my symptoms could be traced to the neural wiring inside my head, I had a very stark choice to make: change my neural wiring or self-destruct.

A very important lesson I learned from taking medication like Zoloft was that altering my neurotransmitter levels could change which neural circuits in my brain readily fired and that this could change how I felt, thought, and acted.* As I connected my thoughts and feelings to the firing of neural circuits, I realized that whatever neural circuits were activated did not necessarily provide an accurate view of reality. If the wiring was faulty, the distortion could be extreme. Since "neurons that fire together wire together," it dawned on me that I could actively think specific thoughts, do specific activities, or self-generate specific feelings to trigger specific neural pathways to fire and wire. This meant I could intentionally activate different parts of my brain, like pressing the keys of a piano to strike different chords. This was how I began to experiment with mind-hacking techniques to bring my brain back to a high level of functioning. By actively breaking my mental patterns and building up healthier mental and behavioral habits, I rewired my brain. Eventually, I managed to graduate with honors and wean myself off medication without relapsing.

During that time, there was such a stigma attached to mental health issues that I instinctively knew it was best not to broadcast what I was going through. After Harvard, as I built a career in very competitive fields such as management consulting and private equity investments, I continued to quietly use mind-hacking techniques to perform at a very high level. Yet, even though I was a passionate brain geek, I didn't talk openly about this topic because I didn't want anyone to use this information against me.

It took many years for me to eventually see my having had PTSD as a blessing in disguise because it forced me to learn about the brain and to develop a unique set of techniques to take care of and enhance the functioning

---

* Neurotransmitters are the chemicals that a neuron releases to activate another neuron to fire or to inhibit another neuron from firing. For the most part, medications for mental health conditions work by increasing or decreasing levels of certain neurotransmitters in the brain. Zoloft is a selective serotonin reuptake inhibitor, which scientists believe has the effect of increasing the activity of serotonin-driven neural pathways in a patient's brain.

of my brain—things I probably would never have done if I had a normal brain. In hindsight, early childhood trauma put me on a unique life journey that would eventually lead to my embarking on a life-changing adventure in 2012 that would in turn lead to my founding Calm Clarity. After that transformation, I finally developed the fortitude to embrace my story and summon the courage to share it openly.

In 2012, about twelve years after my college graduation, after rising to a senior investment position that far exceeded my childhood expectations for career success, I decided to take time off to go on a soul-searching mission to understand my life purpose. I had paid off all my student loans and felt it was time to pursue my aspiration to pay it forward by helping young people growing up in similar situations overcome the obstacles to realizing their potential. The challenge for me was to come up with a way to move the needle in a system that was so broken and dysfunctional that it routinely wore even the best people down and caused them to lose hope and become cynical.

Around that time, new research on the health benefits of meditation, in particular on the brain, caught my attention. I became so intrigued that I decided to gain a more direct understanding of meditation practices by going to India to learn and experience these practices directly at the "source" where they were developed. Learning how to dive more deeply into my inner world changed my worldview. I was blown away by how the ancient teachings on mind training lined up with discoveries from neuroscience. What I learned and experienced enabled me to take mind-hacking to a whole new level.

In deep states of meditation, I saw patterns for how I shifted among three primary emotional states that correspond to archetypal patterns of brain activation, which I simply label: Brain 1.0, Brain 2.0, and Brain 3.0. In Brain 1.0, I act out of fear and self-preservation. The activation of Brain 1.0 causes me to see threats everywhere and be on the lookout for the worst possible things to happen. In this state of mind, a primal set of survival instincts unleashes a persona that I call my "Inner Godzilla" to hijack my mind. I give it this name because in this state, I can get so upset that I want to either smash things or withdraw and disappear completely, just like Godzilla. Whenever the Inner Godzilla takes over, I feel like the world, and everyone in it, is out to get me and I can't trust anyone. I often just want to be left alone. If I act on these urges, I push people away and burn bridges.

In Brain 2.0, I fiercely compete for rewards and impulsively satisfy cravings and urges. The activation of Brain 2.0 causes me to get fixated on pursuing short-term rewards, to the point of developing tunnel vision. Getting the reward can feel so urgent and all-consuming that I don't care about any negative long-term consequences. In this state of mind, a different primal set of instincts unleashes a persona that I call my "Inner Teen Wolf" to hijack my mind. I give it this name because it activates animalistic urges to have status, dominance, and control. When I'm in Brain 2.0, I can get so obsessed with getting the things I want, I impulsively strong-arm people to do things my way. This often leads me to act in ways that make others resent me. In this state, it's also much harder to resist immediate gratification. I regularly binge with eating, drinking, shopping, playing video games, surfing the Internet and social media, or watching television, often to regret it later.

In Brain 3.0, I am centered and act from a space of enlightened awareness, wisdom, and creativity—a state of mind I call "Calm Clarity." In Brain 3.0, a persona I call my "Inner Sage" takes the driver's seat and acts for the greater good. I give it this name because it guides me to tune in to my inner compass and embody my inner wisdom. When my Inner Sage is at the helm, I feel aligned with my core values and connected to a sense of purpose much greater than myself. I feel a strong sense of connection to the people around me, whether I know them or not. I naturally want to make a positive impact on others in a sincerely altruistic way. I can see the bigger picture and foresee the consequences of my choices and actions in a given situation. In Brain 3.0, I experience a deep and lasting sense of contentment, appreciation, and awe for being alive.

It's only when I'm in Brain 3.0 that I feel like my real self and aspire to develop my full potential as a human being. I believe that Brain 3.0 corresponds to what the ancient Greek philosophers referred to as "eudaimonia," a state of "good spirit" that emerges from living in alignment with virtue and the highest human good and that is often translated today as "human flourishing." The ancient Greeks made an active distinction between eudaimonia, as a deeper and more enduring state of intrinsic happiness and well-being, and hedonia, as a fleeting instance of happiness that comes from enjoying pleasurable sensations. Because of this association, I often depict the Inner Sage using the image of Plato from the painting titled *The School of Athens* by Raphael when I teach on Brain 3.0.

During my inward journey, I saw that by using mind-hacking techniques to build up Brain 3.0 over the course of my life, I had increased the ability of my Inner Sage to tame my inner demons (the Inner Godzilla and Inner Teen Wolf). This gave me an innate ability to shift into Brain 3.0 when I encountered a crisis, so that, instead of panicking and making the situation worse, I could calm myself, create space to look at things from a higher perspective, and choose a course of action that would lead to better outcomes over the long term. Further, once these insights crystallized, I realized that self-actualization is tied to the development of Brain 3.0 structures.* I came to see meditation practices as tools for mind-hacking because they exercise, strengthen, and grow Brain 3.0, thus enabling my Inner Sage to come more and more to the forefront of my life.

The intense period of learning and practicing meditation accelerated the development of Brain 3.0, giving me increasingly greater access to my inner compass. I also began to feel more integrated and whole. One day, during one of my meditation retreats, I woke up with the realization that the void I had felt inside me for so many years was somehow filled. In its place was a deep sense of joy, contentment, and fulfillment. As I meditated, I became better aligned with my Inner Sage and felt more deeply connected with my life purpose. In my meditations, I started to see how all the dots in my life—dots that represented critical experiences such as struggles, failures, triumphs, and lessons learned—connected. This led me to see that my purpose was actually interwoven into my own story, arising from the hardships I'd faced and the tools I'd developed to heal and reintegrate. I eventually came to see that sharing the insights from my journey could help more people not only overcome adversity but also appreciate and more fully embrace our shared humanity. This made me realize that part of my calling would be to help foster a larger collective shift into Brain 3.0.

When I had these realizations, I also became aware that I would have to dedicate my life to bringing this vision to reality. It would be a tremendous undertaking. To begin, I would have to draw from my own life experience, from science, and from ancient wisdom to synthesize new insights and build an evidence-based training that would help people understand what it means to be human, how the way our brains are wired can help or hurt us, and how we

---

* Dictionary.com describes "self-actualization" as "the achievement of one's full potential through creativity, independence, spontaneity, and a grasp of the real world."

can take ownership of our own brain development to become the people we aspire to be—our best selves. This vision is what inspired me to build the Calm Clarity Program.

When I returned to the United States at the very end of 2012, I moved back into my parents' house in inner-city Philadelphia and grounded myself once again in the harsh realities of life in my neighborhood, a place where the teenage pregnancy and incarceration rates are higher than the college graduation rate. During this period of time when I developed and pilot tested the Calm Clarity Program, a young man was shot more than twenty times only a block away from our home, my father was held up at gunpoint by a masked robber (and luckily was not harmed), and my family went through a number of health scares and crises. Without a doubt, I consider myself the biggest beneficiary of the Calm Clarity techniques and tools. I don't know how I could have faced these challenges without them.

I intentionally designed the Calm Clarity Program to be accessible to people across diverse socioeconomic and cultural backgrounds. My vision was to build a program with universal relevance that could resonate with inner-city teenagers and executives of large corporations. Once I'd put together the first draft of the program, I began pilot testing it with students in high schools in Philadelphia to validate that what I had to share could make a difference in their lives. Some of the challenges that the students in our pilots faced included having a parent incarcerated, having a loved one become a victim of homicide, living in foster care, having parents who were addicted to drugs and were HIV positive, and becoming temporarily homeless because their family couldn't afford to pay rent.

To set the stage, I opened the first class in each pilot by sharing my own life story. Then I saw, to my own astonishment, how much of a chord it struck and how it helped the participants to embrace their own experiences and see that their own transformation was possible. At the end of each pilot, we asked students to describe the impact that the program had had on their lives. The testimonials and stories they shared were so incredibly moving, I couldn't help crying tears of joy and satisfaction as I read them. I knew I had to keep going, no matter how difficult the path might be.

From there, I began building a social enterprise business model by delivering the Calm Clarity Program as a neuroscience-based mindful leadership training to corporate executives and universities. Since then, the numerous testimonials and positive feedback from participants of widely diverse backgrounds have shown that anyone and everyone can benefit from a user manual on how the brain functions and how to master their mind so they can be their best self. The purpose of this book is to share the insights I gained with a wider audience in the hopes that it will make a positive impact on readers.

Introduction

Why I Love Neuroscience

I LOVE LEARNING about the brain because it gives me a structured, matter-of-fact approach to making sense of my emotions and inner life in a way that takes my ego out of the equation—an approach that liberates me from notions of blame, guilt, shame, and weakness. After my inner life shattered into chaos because of trauma, learning about the brain gave me the scaffolding to put the pieces back together without beating myself up for breaking down and dissociating. Understanding the brain opened up a path to self-honesty, self-acceptance, and self-compassion that just wasn't possible through any other approach I'd tried.

I know too well from firsthand experience that one of the most terrible effects of early childhood trauma is that it disrupts the development of the neural pathways that process and regulate emotions. This has since been confirmed by brain scans showing that trauma can cause the brain to dissociate by impairing the neural pathways for executive functioning, self-awareness, and language.[1] As a result, when a traumatic memory is triggered, people tend to lose the sense of being consciously in control of what is unfolding inside their minds and have a hard time describing what they are going through in words. In that way, a PTSD episode is not that different from sleepwalking or sleep terrors. It's like involuntarily time traveling to the past in a nightmare and not being able to wake up.

Understanding the brain enabled me to see that my healing and recovery essentially involved restoring and strengthening the impaired connections between Brain 3.0 (where executive functioning and positive emotions reside) and Brain 2.0 (where craving, impulses, and habits reside) and Brain 1.0 (where fear, anger, and depression reside). Understanding the neural mechanisms of emotions enabled me to no longer run from, repress, or blow up negative emotions. Instead, I can use Brain 3.0 to create a space to hold them gently, listen to them, and release them when they are ready to be let go.

Further, learning about the brain reminds me that universal aspects of the human experience are grounded in our shared biology. There is no point in beating myself up for being human. Yet there is so much to be gained by understanding and tapping into the intrinsic resilience of the human brain, body, and mind. What I've seen firsthand is that by intentionally turning on the mechanisms for healing and reintegration, all of us can get closer to realizing our full potential.

As a self-taught brain geek without formal training in neuroscience, it should be no surprise that I have an unconventional and unique perspective on the brain that comes from my personal quest to merge scientific findings with life experience to move further toward healing, thriving, and self-actualization. Through active, self-directed learning over many years, I hungrily devoured and pieced together the insights that many scientists across diverse fields such as neurobiology, psychology, physiology, immunology, and behavioral economics have captured in books, articles, and lectures.

I have also been very lucky to have had a few opportunities to spend time with researchers and scientists. For instance, during my time working in consulting, I helped develop an innovation strategy for the neuroscience division of a top pharmaceutical company and spent many hours brainstorming with the research team about how to cure diseases affecting the brain (about twelve years later, this company would serendipitously become my first corporate training client when I started offering the Calm Clarity Mindful Leadership Program to corporations). I'm also very grateful that leading neuroscientists such as David Cox at Harvard University, Sara Lazar at Massachusetts General Hospital, and Richard Davidson at the Center for Healthy Minds at the University of Wisconsin–Madison made time to talk to me and to learn about Calm Clarity.

A Quick Initial Primer on Neuroscience

Before we dive in, I want to take a moment to put in context where science currently stands in our understanding of the brain, so you have a better sense of what we know and what we don't know or are really only guessing about.

We generally think of the brain as the seat of consciousness in the human body. It mysteriously coordinates and regulates all of our critical life processes, including voluntary movements that we consciously control as well as involuntary activities like respiration, digestion, and blood circulation, which take place in the background without needing our direct control or even our conscious awareness. Scientists tend to refer to the brain dryly as a complex biological system whose primary job is information processing—but exactly how the brain processes information remains a mystery that science has not yet solved. For instance, scientists often say memories are "stored" in the brain, but they don't know how memories are stored and where exactly they are held. They know from brain imaging technology that brain activity corresponds to the firing of neural pathways, but how exactly the firing of specific neural pathways translates into emotions, like happiness, boredom, or nostalgia, and activities, like solving a problem, writing a story, or playing an instrument, is still unknown.

Anyone who wants to learn about the brain needs to keep in mind that the language scientists use to talk about the brain tends to be analogical and metaphorical. Throughout history, people tended to describe the brain using the most advanced technologies of their time; for example, the ancient Greeks described the brain as a hydraulic pumping system, and European Enlightenment philosophers described it as an intricate machine such as a clock. In the twentieth century, people started describing the brain as a computer, and in the twenty-first century, people started talking about the brain as an information processing network. These metaphors are helpful but should not be taken literally because there are many aspects of the brain that are much more sophisticated than any technologies created by human beings so far.

I also want to emphasize that science is only at the very beginning stages of understanding how the brain works. All that we know right now is only scratching the surface. This is because for most of human history, we didn't have any tools to see inside the skull of living human beings without cutting them open. Before the invention of brain imaging technologies in the 1990s, our knowledge of the brain was built from animal research and studying people with brain damage.

Studies on animals usually involved opening the skulls of monkeys and mice and using electrodes to stimulate specific areas, removing specific parts, or cutting connections between specific areas of the brain. Researchers would then observe how the animal's behavior changed to figure out what function the part they were studying controlled. Nonetheless, they could never be 100 percent sure that the analogous part of the human brain performed the same function.

The studies involving people usually fell into two categories: people who'd had brain surgery so the researchers already had an idea which parts of their brain were abnormal (and in many cases, altered or removed from their brain during the surgery) and people whose brains were not examined until after they died. Typically, scientists tied whatever behavioral abnormalities the person demonstrated to any abnormality they found in the person's brain. Then they concluded that the part of the brain that was abnormal controlled what they considered to be the function underlying the abnormal behavior.

Over the past twenty years, thanks to the development of brain imaging technologies such as functional magnetic resonance imaging (fMRI), there have been an exponentially increasing number of studies on the brain. To be clear, fMRI scans do not actually tell us what the brain is doing. They only allow researchers to take readings of blood flow and oxygen consumption while asking subjects to focus on specific mental tasks. The assumption is that areas getting more blood flow or consuming more oxygen are "activated" by the task. Then researchers look at patterns of activated neural pathways and attempt to draw conclusions based on what story they think can be reasonably told with the data. It is not foolproof. Researchers don't always agree on how to analyze and interpret the data. The reality is: everything we think we know about how the brain functions is a collection of best guesses based on available evidence.

We are about to enter into a new era of neuroscience because cutting-edge technologies now allow scientists to safely and temporarily deactivate or hyperactivate specific areas of the brain. In these new studies, scientists typically ask subjects to perform the same task, like playing a piano, under three conditions: their brain in a normal condition, their brain with a specific part deactivated, and their brain with that same part hyperactivated. These types of studies are so new that it will take scientists time to design them, run them, and digest and understand the findings and implications. In a few decades, the neurobiological information in this book could become totally outdated and obsolete. This is especially true for technical details about the functions of specific areas of the brain, the effects of various chemicals, and the mechanisms for how cognition takes place. Nonetheless, I hope that the general concepts behind the Calm Clarity framework for connecting our emotional states to three archetypal patterns of brain activation will stand the test of time.

The key thing to keep in mind is that the brain and nervous system are made of nerve cells called neurons, which essentially serve as living wires, or "biowires," that carry electrical impulses and "plug into" each other at sites called synapses, where their branches connect. Scientists estimate that the brain has somewhere between 80 billion and 100 billion neurons, with each neuron capable of connecting to about ten thousand other neurons, thereby creating a nearly infinite number of combinations. When neurons connect, they form neural pathways along which electrical impulses flow from one part of the brain to another. Activities like thinking, feeling, sensing, communicating, and flexing muscles correspond to the "firing," or "activation," of specific neural pathways connecting different areas of the brain. Many activities, like executive functioning and impulse control, involve the activation of specific neural pathways that inhibit many other neural pathways. A group of neural pathways that perform and regulate a discrete function is referred to as a neural circuit. Often, multiple neural circuits firing in unique combinations throughout the brain are involved in sophisticated functions and are thus referred to as a neuronal network, such as the imagination network and the language network.

Since I started learning about neuroscience, several breakthroughs have radically overturned the way we think about the brain. For instance, scientists used to believe that the brain could generate new neurons only during childhood and that once adulthood was reached, the brain would have a fixed number of neurons for the rest of its life. Therefore, they concluded that if any neurons were damaged, say by a stroke, the brain could not replace them by generating new neurons or rewire neural circuits to replace or repair the lost function. Then in the 1970s, several studies done with animals showed that neural pathways in the adult brain are constantly rewiring in response to experience (a quality those researchers called "neuroplasticity") and that the adult brain continuously forms new neurons (a process the researchers called "neurogenesis"). However, instead of embracing these findings, the scientific establishment initially balked and insisted something was wrong with those studies. It took decades of building consistent, irrefutable evidence before the establishment finally acknowledged that the model of the unchangeable brain needed to be revised.[2]

Today, the concept of neuroplasticity is central to our understanding of brain functioning. As explained by Richard Davidson, founder and head of the Center for Healthy Minds at the University of Wisconsin–Madison, the brain has "the ability to change its structure and patterns of activity in significant ways not only in childhood, but also in adulthood and throughout life.... That change can come about as a result of experiences we have as well as purely internal mental activity—our thoughts."[3] What this means is that the brain is continuously adapting to and being remodeled by the lives we lead. The types of experiences we had while growing up, our daily routine, and what we think about each day can make a big difference in how our brain develops going forward. For instance, right-handed violin players have more neurons assigned to the left hand to carry out finer motor control than the average right-handed person.* Brain imaging studies have also shown that cabdrivers in London have much larger than average hippocampi, an area of the brain in each hemisphere that is involved in spatial navigation; in storing detailed visual representations like maps; and in the consolidation of information from short-term to long-term memory.[4]

While we now know the brain is constantly changing, exactly how neurons organize, interconnect, and then change how they are connected is not clear. Therefore, scientists such as Daniel Siegel have come to think of the brain as a "self-organizing system" because there seems to be no "conductor" telling the neurons what to do. In biology, self-organization is a process by which a complex adaptive system composed of many elements spontaneously arranges itself in a purposeful manner in response to changes in the environment

---

* I learned from an exhibit on Einstein's brain at the Mütter Museum of the College of Physicians of Philadelphia that Einstein, as a lifelong violin player, had a larger area of his brain assigned to his left hand.

without the control or direction of an external agent. In that way, neural connections seem to automatically adapt to how they are being used.[5] Like muscles, the more we "fire" neural connections, the stronger they become; if we don't fire them, they weaken and atrophy. These aspects of neuroplasticity are often captured by two phrases, "neurons that fire together wire together" and "use it or lose it." To clarify, neural connections weaken from disuse, yet they don't actually disappear. It's more like they become dormant with the capacity to reactivate suddenly out of nowhere if triggered by strongly associated sensory cues in the external environment or by strongly associated emotional and behavioral cues.

Advances in brain imaging technologies, by providing a new understanding of neuroplasticity and neural pathways, have overturned the once dominant model of the brain as a machine made of clearly delineated parts that have fixed specific functions. Instead, we now think of the brain as a dynamic supernetwork of intelligent neurons forming many interacting and overlapping neural circuits that, when activated together in certain patterns, create specialized networks enabling sophisticated functions. This new understanding of neural networks has also shown that the widespread notion of a rational brain versus an emotional brain is a misconception that should be laid to rest. Instances of emotions, rather than being isolated to one area of the brain, involve the large-scale activation of neural networks across the brain, including areas earlier demarcated as the centers for reason.[6]

It also turns out that emotions guide decision making and that the capacity to reason is really the capacity for our conscious minds to rationalize and justify intuitive gut feelings and emotions that may not fully bubble up into conscious awareness. In fact, when the neural pathways for processing emotions and feelings are damaged, the ability to make wise and sound decisions becomes so impaired that people become "rudderless." Lisa Feldman Barrett, one of the leading researchers in affective neuroscience ("affect" is the scientific word for emotion), insists, "Affect is in the driver's seat and rationality is a passenger." She adds, "Affect is not just necessary for wisdom, it's irrevocably woven into the fabric of every decision."[7]

Today, one of the most exciting areas of research is mapping how all the neurons that make up the brain are interconnected. The formal term for this concept, the "connectome," was coined in 2005 by scientists to refer to the organization of neuronal interactions within the brains of many different species.[8] A connectome is usually rendered as an elaborate multicolored diagram of neural pathways that is stunning to look at just as a work of art. Like the Human Genome Project, the Human Connectome Project is a vast multiyear, worldwide collaborative research project to "map" the human brain as a comprehensive visual representation of the neuronal pathways and key structural elements common across 1,200 healthy volunteers. The hope is that the map will enable scientists to identify more precisely the neural pathways that underlie specific brain functions and behaviors, which could provide key clues for curing diseases and disorders.[9]

The data collection phase of the project was completed in 2015 and the researchers are now in the process of analyzing and synthesizing the data, which will likely also take many years. In July 2016, researchers released what they called "version 1.0" of a new map of the cerebral cortex, the outer layer of the human brain, which is made of roughly 20 billion neurons. This preliminary mapping identified 180 distinct areas in each hemisphere, but a lot more work still needs to be done to understand how these areas interact to enable human beings to perform so many different activities.[10] (See www.human -connectome.org for more information.)

Over the last decade neuroscience has captivated mainstream attention thanks to fascinating new discoveries like the ones I've listed. In 2007, the idea that we could intentionally change the brain was widely circulated through bestselling books such as Norman Doidge's *The Brain That Changes Itself* and Sharon Begley's *Train Your Mind, Change Your Brain*. Since then there has been an explosion of public interest in neuroscience and a growing stockpile of books and resources attempting to fill that demand.

Having read through many of them, I found that they often contain a fascinating yet overwhelming amount of detailed information about the brain, mostly conveyed in an academic manner that is not obviously actionable. After also digging through some of the original research papers referenced in these books, I found that, in general, researchers tend to present their studies in a way that only other academic researchers can fully understand. Very few researchers (and journalists who write about their studies) actually make the effort to demonstrate and explain in practical and concrete terms how their discoveries apply to everyday life experiences.

Therefore, my challenge as a brain geek and mind-hacker was to find a way to make sense of all this information and create a practical approach that I could put into use for myself and also share more broadly, especially with audiences without in-depth scientific training and knowledge, such as students in inner-city high schools. Eventually, I chose to focus on three emotional states of the brain as a simple framework to connect my state of mind to three different patterns of brain activation (Brain 1.0, Brain 2.0, and Brain 3.0).

My framework on the three emotional states of the brain can be seen as an update on the widely popularized but largely inaccurate triune brain model, which was originally developed by a neuroscientist named Paul MacLean in the 1950s. MacLean had hypothesized that the human brain evolved over time in three main layers, from inside out, such that the structures in each layer enabled new adaptive functions and behaviors. He named the three layers the reptilian complex, the paleomammalian complex, and the neomammalian complex, believing these complexes were responsible for the instinctual behavior of reptiles, early mammals, and higher mammals (like primates and humans), respectively. However, subsequent comparative neuroanatomical studies showed that MacLean's presumptions about the designation and functioning of brain structures in reptiles, early mammals, and higher mammals didn't hold up. Nevertheless, the essential idea that the brain evolved sequentially from inside out is still considered plausible. Today, scientists believe that as the brain grew in size and complexity, it also reorganized such that the functioning and connectivity of older structures evolved to support the functioning of the newer structures.

Unlike MacLean's model, the Calm Clarity framework was never intended to be a comprehensive theory about the evolution of the brain from more primitive to advanced species. The best way to think of the Brain 1.0, 2.0, and 3.0 framework is as a simple tool to help people understand and apply very complex findings from neuroscience to life in an intuitive and practical manner.

I got clear validation I had hit the nail on the head when I began pilot testing the Calm Clarity Program in inner-city high schools. It was amazing to watch just how quickly the students adopted the Brain 1.0, 2.0, and 3.0 terminology in the first session and started using these concepts to make sense of their emotions in everyday life. After I showed the students how easy it was to activate Brain 3.0 using a ten- to twelve-minute mind-hacking meditation exercise I'd created for the purpose, people shared their experiences. The general consensus was that they had experienced such a dramatic shift that they naturally wanted to spend more time in Brain 3.0 and less time in Brain 1.0 and Brain 2.0.

Brain 1.0, Brain 2.0, and Brain 3.0 Are Brainwide Patterns of Activation Rather Than Specific Brain Structures

I want to take a moment to clarify here that there are no boundaries in the brain that mark where Brain 1.0, Brain 2.0, and Brain 3.0 begin and end. Actually, the neural structures that give rise to Brain 1.0, Brain 2.0, and Brain 3.0 overlap in that these brain states represent patterns of brainwide activation, which engage many neural networks in different combinations. In general, from any given moment to the next, neural networks across the brain are firing in many different patterns to enable different mental processes to occur. It's like how a large arena can be configured for different purposes, such as a football game, a soccer game, a concert, a party, or an expo, and how one Lego set can be used to build a nearly infinite number of creations, such as dinosaurs, robots, buildings, and vehicles ranging from cars to spaceships.

What I mean by "Brain 1.0" is an archetypal brain activation pattern in which the neural circuitry triggering the body's defensive mechanisms are strongly aroused. This is commonly called a "fight-or-flight response," but according to Joseph LeDoux, one of the leading researchers on the threat response, the actual defensive reactions automatically triggered by danger are to "freeze first; flee if you can; fight if you must."[11] Therefore, I prefer to call it the freeze-flight-fight cascade to be more precise. The triggering of Brain 1.0 tends to reduce blood flow to the prefrontal cortex (the key structure that enables executive functioning), and is often indicated by a highly activated amygdala; however, this doesn't mean the amygdala is the center of fear in the brain. In MRI studies, the amygdala seems to activate whenever people process emotions in pictures of faces and encounter uncertainty and novel situations, so the amygdala's key function may be to make split-second assessments about whether we are safe with friends or need to be on guard. Regardless of our emotional state, the amygdala continuously operates in the background to read emotions in other people and keep us safe.[12] It may also play a role in encoding memory by using emotions to tag experiences that are relevant to our survival and well-being so we can more easily recall them.[13] When Brain 1.0 is highly activated, we may still continue to engage some areas of the prefrontal cortex and reward circuits to do things like wallow in self-pity or eat an entire tub of ice cream.*

Similarly, what I mean by "Brain 2.0" is an archetypal brainwide activation pattern in which the reward circuitry (located in the basal ganglia and hypothalamus) is strongly activated, but this doesn't mean that the reward circuits are the only structures that contribute to the experience of Brain 2.0. When we are in Brain 2.0, some areas of the prefrontal cortex are deactivated or impaired (such as the neural circuits that enable us to control impulses, think critically, and carefully analyze a situation), but other areas are strongly engaged (such as the motor areas). Regardless of our emotional state, the basal ganglia continuously operate in the background, enabling our autopilot to carry out any routinized motor movements we do while multitasking, such as typing notes while listening to a lecture, and holding objects like a book as we read and utensils and glasses while we engage in dinner party conversation.[14]

Finally, what I mean by "Brain 3.0" is an archetypal brainwide activation pattern in which the prefrontal cortex (the area of the brain inside the forehead that underlies our capacity for executive functioning) is not impaired or deactivated.† Again, this doesn't mean the prefrontal cortex is the only structure that contributes to the experience of Brain 3.0. In fact, when we are in Brain 3.0, we are still actively engaging key structures of Brain 2.0 (our reward circuits) and Brain 1.0 (our amygdala) to help us navigate our daily life, in particular our social interactions. Brain 1.0 and Brain 2.0 can hijack us only when Brain 3.0 is underdeveloped or impaired. When Brain 3.0 is strong, we can wisely harness Brain 1.0 and Brain 2.0 to effectively and efficiently navigate the world. (This will be explained in more detail in part 2.)

Over time, the most frequent patterns of brain activation you experience shape your brain structure, in other words, your connectome. For

---

* To wallow in self-pity, you would employ prefrontal neural networks involved in language and autobiographical memory. To enjoy a tub of ice cream you would employ prefrontal networks and your basal ganglia to coordinate movement and decide which flavor of ice cream to savor.

† For the sake of simplicity, since each hemisphere of the brain has a prefrontal cortex, it is common to use the singular term "prefrontal cortex" to refer to both the left prefrontal cortex and the right prefrontal cortex rather than the plural form "prefrontal cortices." This is also true for many brain structures that appear in each hemisphere: other parts of the cerebral cortex, the amygdala, the basal ganglia, the insula, the thalamus, etc.

example, a person who is predominantly in Brain 1.0 has a connectome in which the neural pathways of the freeze-flight-fight mechanisms function like superhighways and many of the neural pathways of Brain 3.0 that connect into and regulate Brain 1.0 do not get fully developed or become impaired. Similarly, a person who is predominantly in Brain 2.0 has a connectome in which the neural pathways of the reward system function like superhighways and many of the neural pathways of Brain 3.0 that connect into and regulate Brain 2.0 do not get fully developed or become impaired. To be predominantly in Brain 3.0, rather than Brain 1.0 and Brain 2.0, requires activating and strengthening the neural pathways that connect the prefrontal cortex into Brain 1.0 and Brain 2.0 so frequently that these circuits become superhighways. Thus, making a long-term shift into Brain 3.0 represents a significant change in a person's connectome.

Beyond the Brain

Once in a while, a participant in the Calm Clarity Program asks me, "Is there a Brain 4.0?" If I take that question literally, I would explain that in terms of neurobiology, the concept of Brain 3.0 represents the most newly evolved structures of the brain; therefore, if we tried to map Brain 4.0, it would lead us back to the structures of Brain 3.0. However, I intuitively understand this is the real question they are asking: what lies beyond Brain 3.0? And for that question I have a different answer, one that I largely kept to myself until I started to write this book.

The truth is that the most life-changing discovery from my journey to understand how to change and optimize my brain was that human consciousness cannot be entirely explained by the firing of neurons. While understanding the brain does shed light on universal human patterns of behaviors, tendencies, and instincts, it does not solve the far greater mysteries of consciousness. What happened to me as I transformed by shifting more and more into Brain 3.0 is that I started to have unexplainable experiences that forced me to acknowledge that it is very possible that the essence of who and what "I" really am can never be fully explained by neuroscience nor limited to the physical matter making up my brain.

I have to confess that as someone who began this journey as a scientifically oriented skeptic with no religious faith or spiritual beliefs, I was utterly baffled by how it was even possible that my quest for self-actualization through science could lead me to rediscover and reconnect with a "soul" or "Higher Self" that, for the time being, lies beyond the realm of present-day science and technology to investigate. For a long time, my mind struggled to come to terms with this ironic unfolding. I doubled down even more on the science and tried to find ways to explain away or dismiss these experiences as my imagination or as hallucinations, but it was all in vain. The metamorphosis that had already begun couldn't be stopped or reversed. When the light and wisdom of the Higher Self wants to shine through, there comes a point where the ego cannot hold it back or dim it any longer.

I hadn't anticipated that, as my efforts to build up Brain 3.0 yielded fruit and as Brain 1.0 and Brain 2.0 (the primal parts of my brain) quieted down, I would start to intermittently experience a state that could be described as joyful, rapturous bliss and spiritual elevation. While I like to call it "Calm Clarity," I believe the ancient sages in India called it "yoga," a Sanskrit word that means "divine union." They also called it "sat-chit-ananda," which means "eternal truth, consciousness, unconditional joy." It is a state in which my Higher Self communicates freely with me, revealing wisdom far beyond my mind's ability to imagine or conjure up on its own. It is a state in which my soul gives me clarity of vision and guidance that elevates and energizes me with a sense of peace, purpose, and grace surpassing anything I have ever felt before. It is also a state that could be described as "calm intensity" because in challenging situations and crises, that calmness and clarity help ground and center me. In addition, every moment of life becomes profound and significant. Every present moment becomes a gift.

About This Book

This book is inspired by the Calm Clarity Program that I designed to help people understand and shift into Brain 3.0 by directly experiencing how they feel, think, and act differently when they are in Brain 3.0. Originally, the reason I chose the format of an in-person group workshop to share what I'd learned was because the easiest way to activate Brain 3.0 is through positive social interactions and experiential, interactive learning exercises. Being in a warm, nurturing, safe, and supportive group is one of the most powerful means to activate the vagus nerve and elevate oxytocin, key mechanisms that help bring people into Brain 3.0 (which will be explained in chapter 7). I also knew that people needed to directly experience a shift into Brain 3.0 to really understand the difference between Brain 1.0, Brain 2.0, and Brain 3.0 and how they are able to be their best selves only when they are in Brain 3.0.

Using a book to foster a direct experience and embodied understanding is much more challenging because it is not possible to replicate the interactivity of an in-person group workshop using only the written word. Therefore, rather than try to capture the workshop in a book, I have decided to embrace the strengths of the book format to share ideas. Freed from the constraints of having to stay on a fixed time schedule, this book gives me the ability to use storytelling to invite readers to experience the journey to Brain 3.0 through my eyes.

What this means is that while this book has a great deal of overlap with the Calm Clarity workshop program, it is also capable of standing alone. Those who have participated in Calm Clarity training will find this book a helpful refresher that both deepens their understanding of what they learned in the workshop program and introduces more insights on how to continue to foster and develop Brain 3.0 in themselves and in the people with whom they interact.

As the number of people who experience the Calm Clarity Program has grown, I have learned that what stands out to participants are my passion for making brain science practical and concrete and my bringing abstract concepts to life using personal stories and experiential exercises. These strengths have been critical to inspiring participants to embrace neuroscience and meditation, proactively care for and cultivate their brains, and continue to practice the exercises and apply what they learned after the training. I have therefore intentionally adopted this practical, narrative, and experiential approach in writing this book.

The book consists of two main parts, with each containing an introduction that summarizes what is to follow. In part 1, I bring the Calm Clarity concepts to life by interweaving the underlying science with my story and explaining how various insights enabled me to address personal challenges that I've faced throughout my life. The six chapters in this part walk through different phases of my personal journey shifting from Brain 1.0 through Brain 2.0 into Brain

3.0, up until the present day, in a way that helps illustrate how all human beings are predisposed to experience these three emotional states in the course of growing up and navigating the world around us.* We will dive further into how to apply these insights in part 2.

Part 2 presents a "mind-hacker's guide" that shares the insights, tools, and practices that enable me to be more whole and integrated in Brain 3.0. In chapter 7, I share a high-level overview that explains what it means to shift into Brain 3.0 and helps you better understand and appreciate the essential functions of Brain 1.0, Brain 2.0, and Brain 3.0. In chapter 8, the focus is on helping you shift. The different sections are designed to help you understand how your mind works and how Brain 1.0, Brain 2.0, and Brain 3.0 are triggered and activated in your daily life. These sections also provide a series of practices and exercises from the Calm Clarity Program that enable you to activate and strengthen important Brain 3.0 neural networks.

Finally, I close by sharing more about my continuing lifelong journey to deepen my own shift and move toward full integration. Please understand that initially, as a science-oriented and secular person, I felt very hesitant and un-comfortable talking about spiritual topics in public, but I have decided to take the risk of opening up about my personal realizations in order to help more people find their own path to experiencing a higher state of consciousness. Therefore, in the closing chapter, I tie together the scientific and spiritual in-sights that inspired me to create the Calm Clarity Program by explaining a diagram I made to piece together the "cosmic" bigger-picture perspective as I understand it. I end this chapter by sharing the manifesto I use to guide myself to continue to grow and foster Brain 3.0.

In the appendix, I provide a few simple tools to help you incorporate the Calm Clarity insights, meditations, and exercises into your daily life. The

---

* It may be helpful to explain ahead of time without spoiling the plot that my life journey has involved many countries, from the land of my birth, Vietnam, to the refugee camp in Indonesia where I spent my infancy, to inner-city Philadelphia where I grew up, to Boston, where I went to college and started my career, and back to Philadelphia, where I attended graduate school. Then my journey shifted to Asia, where I lived for almost seven years, from 2006 to 2012. During that time, I worked in Beijing, China; Ho Chi Minh City, Vietnam; and Singapore. In the summer of 2011, between leaving Vietnam and moving to Singapore, I was gifted with an eye-opening tour of India that exposed me to India's mystical spiritual traditions and that later inspired me to go on the year-long soul-searching journey in 2012 that brought me to New Zealand, Burma, India, and then back to Singapore. At the end of that journey, I returned to Philadelphia, where I began building Calm Clarity as a social enterprise.

appendix includes a reference guide summarizing the Calm Clarity meditations and mindfulness practices, a template to create a practice plan for fostering Brain 3.0, a template to create a daily routine to activate Brain 3.0, and a recommended reading list for anyone who wants to learn more.

I would like for this book to serve as a resource that contains solid and substantive information on the scientific findings and spiritual wisdom that inspired the Calm Clarity Program. My intention is to share explanations at a level of detail that is useful both for people who are not familiar with the subject matter and for people who have already been exposed to it and want to gain a deeper understanding to apply these scientific and spiritual insights in their lives. Therefore, feel free to spend more time on the passages that most interest you and skim through or skip the passages that seem heavy or dense at the present moment. You can always come back to these passages later when you want to take a deeper dive.

This book contains a significant amount of information and there is no need to try to remember everything. Instead, please focus on what you find most relevant and how you can apply any insights you gain to your everyday life. Each person has his or her own unique way of learning and assimilating knowledge, so please empower yourself to make use of this book as a resource in the manner that makes the most sense to you.

Part I

My Journey Through Brain 1.0 and Brain 2.0 to Brain 3.0

What a man can be, he must be. This need we may call self-actualization. . . . It refers to the desire for self-fulfillment, namely, to the tendency for him to become actualized in what he is potentially. This tendency might be phrased as the desire to become more and more what one is, to become everything that one is capable of becoming.

—ABRAHAM MASLOW, psychologist

Self-actualization is possible only as a side-effect of self-transcendence.

—VIKTOR FRANKL, psychiatrist and Holocaust survivor

One does not become enlightened by imagining figures of light, but by making the darkness conscious.

—CARL JUNG

EXPERIENCE HAS TAUGHT me that the most effective way to get people engaged in learning about the brain is to ground the insights in personal life experience. Therefore, I will explain the key principles in the Calm Clarity Program as I share how I navigated through the obstacle course that life threw at me. My personal journey will demonstrate that what I know about Brain 1.0 comes from having experienced tragedy, what I know about Brain 2.0 comes from many years of misguided attempts to fill an insatiable emptiness inside me, and what I know about Brain 3.0 comes from experiencing an unexpected and reluctant spiritual metamorphosis.

As I tell my story, it will show that I am living proof that a person can be a refugee from a war-torn country, grow up in poverty, be surrounded by vio-lence, and suffer PTSD, and still beat the odds. It will also show that it is possible for people who grow up in traumatic circumstances to heal their brain, change their trajectory, and make a positive impact on the world. My story demonstrates that as human beings, we must not underestimate the resilience of the human spirit—in ourselves and in others.

Lastly, this is a story about how one person's quest to use insights from science to achieve self-actualization inadvertently turned into a journey of discovering the "true nature" of the "self." And since this journey continues to unfold, this book can only tell the first part of my story.

I.

A Traumatic Start

Pain that is not transformed is transmitted.   —Richard Rohr

Although the world is full of suffering, it is also full of the overcoming of it.   —Helen Keller

The wound is the place where the Light enters you.   —Rumi

Let's face it. A lot of terrible things happen every single day to bring people into Brain 1.0. No one is immune from tragedy. We all live in a world that traumatizes us and makes us feel like victims of broken systems, vulnerable to exploitation by con artists, criminals, and crooked politicians. Our environments are full of temptations for drugs, alcohol, food, retail therapy, and other pleasures and thrills as a form of easy escape. At the same time, the world is also full of wonder, majesty, and inspiration that naturally bring people into Brain 3.0. In every corner of the Earth, people are working hard to improve life for their families and to make a difference in their communities.

The fact that the misery and the awesomeness of our planet are not evenly spread out has always been very hard for me to come to terms with. The more I learned about how unfair the distribution of resources can be and how unequal the access to opportunity can be, the more my sense of outrage grew. Seeing signs that the system is not only rigged against the poor and powerless but also set up to exploit them has always triggered my Inner Godzilla.

Being born on Earth is like a lottery. A small lucky proportion are born into loving families blessed with inner resources (like education, emotional intelligence, and resilience) and a degree of financial stability, where parents are generally able to shelter children from the harsh realities of this world for as long as possible, give them a safety net in case they fall, and then coach and equip them to prosper in their careers and to raise a family of their own. Loving, caring, "privileged" families like these naturally develop and pass on Brain 3.0 from generation to generation. But a lot of people aren't lucky. The vast majority of people, like me, are born into families that cannot shield them from horrors. Too many parents are so traumatized that they get trapped in Brain 1.0 and can't help transmitting their pain and suffering to their children just by role modeling what for them are normal behaviors and expectations.

As I dive deeper into Brain 1.0, please keep in mind that what I mean by Brain 1.0 is a pattern of neuronal network firing rather than a specific part of the brain. In the Brain 1.0 pattern, the amygdala, a part of the brain that is involved in reading emotions and looking for signs of danger and threat, is highly activated and puts the entire body into a state of "red alert" that scientists refer to as hypervigilance, a prolonged state of anxiously looking out for danger and threats and not being able to relax (a.k.a. freeze-fight-flight mode). Whenever the body is in this very strong state of stress, there is reduced blood flow to the frontal lobes that help us carry out higher-order mental processes; thus we have less "processing capacity." This is why when we are afraid and anxious, it can be nearly impossible to think clearly, take in and process information, and make sound decisions.

According to the Sanctuary Institute, a pioneer in the area of trauma-informed care, "trauma is defined as an experience in which a person's internal resources are not adequate to cope with external stressors." In the Sanctuary Model developed by Sandra Bloom, the experience of trauma can fall along a wide continuum that includes discrete events as well as ongoing, cumulative, and less tangible experiences such as poverty, racism, discrimination, and neglect. The Sanctuary Model also reframes many of the behavioral symptoms related to trauma as the misapplication of maladaptive survival skills developed to cope with adverse experiences.[1]

In psychiatry, for a person to receive a diagnosis of PTSD, the trauma needs to be related to a direct personal experience of an event that involves actual or threatened death, serious injury, or sexual violation.[2] However, I have found that many people carry painful emotional scars from circumstances that may not fit this hurdle of being life threatening. When we are young children, we are completely dependent on adults for our well-being. Therefore, not having attentive parents, caregivers, and teachers; not having a sense of belonging at home, at school, or in the community; not getting sufficient emotional nurturing; or generally not having the sense of safety and security needed for solid emotional development can be overwhelming and prompt us to turn to maladaptive coping strategies that can later get in the way of our well-being and ability to form healthy and nurturing relationships as adults. Therefore people sometimes use the terms "little-t trauma" or "micro trauma" to refer to distressing experiences that are not life threatening but still inflict psychological pain and suffering.

When we experience trauma or long-term exposure to very stressful conditions, our brains, minds, and lives get reorganized as if the trauma were still going on, such that the trauma contaminates every new encounter and event. The amygdala stays hyperactivated as a "default" state, and the functioning of the frontal lobes becomes disrupted and impaired. Since the left frontal lobe is involved in the experience of positive emotion, when Brain 1.0 disrupts its functioning long-term, we can easily get trapped in a chronic negative emotional state. Furthermore, any bodily sensations that remind people of the trauma easily become overwhelming, so they often adapt by subconsciously dissociating as a way to block, suppress, and numb the painful sensations and memories.

In *The Body Keeps the Score*, a groundbreaking book providing a scientific perspective on healing trauma, psychiatrist and researcher Bessel van der Kolk explains that "traumatized people chronically feel unsafe inside their bodies: The past is alive in the form of gnawing interior discomfort. Their bodies are constantly bombarded by visceral warning signs, and, in an attempt to control these processes, they often become expert at ignoring their gut feelings and in numbing awareness of what is played out inside. They learn to hide from their selves." He further explains, "People who cannot comfortably notice what is going on inside become vulnerable to respond to any sensory shift either by shutting down or by going into a panic—they develop a fear of fear itself."[3]

What's even more sobering is that the negative impact of trauma is not confined to the person who experienced the trauma—it is transmitted across generations. Research has found that experiencing trauma can cause changes in gene expression that can be passed on to offspring. These changes result in an increased risk factor for developing post-traumatic stress disorder.[4] I can testify from firsthand experience that children born to parents who are traumatized have a high likelihood of being exposed to trauma and of seeing traumatic experiences as normal. Unfortunately, the impact of trauma in childhood—what scientists refer to as "adverse childhood experiences" (ACEs)—is even worse than in adults because it negatively affects the development of a child's brain and body in ways that make him or her more vulnerable to chronic illnesses like asthma and diabetes, as well as depression, anxiety, and addiction as they grow up. Unless the cycle is broken, traumatized people often get locked into Brain 1.0 as a way of life, and then pass on this pattern of brain development and activation to the next generation.

While many people have heard of trauma, what many don't realize is just how widespread trauma is, and how close to home it can be. According to the Centers for Disease Control and Prevention (CDC), a study of a large sample of 17,337 people insured by Kaiser Permanente in the late 1990s asked participants whether as a child they had experienced any of of the following ACE indicators, such as "experiencing physical, sexual, or emotional abuse; experiencing physical or emotional neglect; witnessing domestic violence in the home; living with someone who abused substances, was mentally ill, or who was imprisoned/sentenced to serve time; and experiencing parental separation or divorce." The findings revealed that approximately 64 percent, or two out of three people, had suffered at least one adverse childhood experience and that 12.5 percent, or one out of eight people, reported four or more ACEs.[5]

Further, the data unequivocally showed that participants with higher ACE scores had higher rates of obesity, chronic disease, mental health problems, and addiction disorders. Experiencing at least four ACEs became recognized as the threshold for severe trauma because the risk of developing serious health issues and engaging in risky behaviors increases dramatically at that point. According to ACEStooHigh.com, for the group of people who experienced four or more ACEs (and yes, I fall into this category), the risk of developing chronic pulmonary lung disease increases by 390 percent, hepatitis by 240 percent, and depression by 460 percent, and the risk of committing suicide increases by 1,220 percent compared to people with an ACE score of zero.[6] Furthermore, the site states, "people with an ACE score of 4 are twice as likely to be smokers and seven times more likely to be alcoholic" and "people with an ACE score of 6 or higher are at risk of their life span being shortened by 20 years." What researchers found most remarkable was that the participants in the study were relatively affluent and educated white-collar professionals who had health insurance. About 75 percent of the participants had either attended some college or earned a college or graduate degree and about 75 percent of the participants were Caucasian. No one expected that the ACE rates would be so high in this demographic group.[7]

I couldn't help wondering as I looked at the data: what would the ACE scores look like in the area where I grew up? Thankfully, I wasn't the only one who wanted to know. After I returned to Philadelphia, I learned that a group of institutions had already created a task force to conduct the Philadelphia Urban ACE Study.[8] The findings were published at the end of 2013, just as I was getting ready to conduct the first pilot test for the Calm Clarity Program with public high school students in Philadelphia. In this study, the researchers added five new indicators to include traumatic experiences that are common among an urban population but were not included in the original CDC study: experiencing racism, witnessing violence, living in an unsafe neighborhood, living in foster care, and experiencing bullying. The findings showed that 83 percent, or eight out of ten people, reported at least one ACE using the updated indicators (using only the original ACE indicators, the number would be approximately 70 percent, a little bit higher than the Kaiser Permanente study) and that 37 percent, or nearly four out of ten people, had four or more ACEs using the updated indicators (using the original indicators, that number would be approximately 22 percent, almost double the rate in the Kaiser Permanente study).

The report also included a map showing the percentage of the population with four or more ACEs by zip code, which confirmed what I suspected all along: the area of Philadelphia where I grew up and currently live is in the top bracket, with 45 percent or more of the population having a score of four or more ACEs. These revelations from the ACE studies made me ask a few questions: For people in communities experiencing these rates of trauma, wouldn't they need to have such a strong Brain 1.0 to sense and navigate the many dangers in the surrounding environment that Brain 1.0 could become their default state? Knowing these numbers are so high, is it even possible to help people in these situations shift into Brain 3.0? If yes, then how? Then it occurred to me that the primary place where I could find answers to these questions was in my own life story. How did I come out of Brain 1.0?

For me, growing up in an environment that put people in Brain 1.0 was like fighting a steep uphill battle on my own, both in the external world and inside my head. The biggest danger was always the loss of hope—which nearly died whenever my inner demons got the upper hand. Yet by somehow managing to emerge from this gauntlet of adversity in one piece, I've come to embrace my experiences as serving a purpose: resistance training for the soul. For me, overcoming adversity opened a path to enlightenment, which I now think of simply as the ability to shine light into darkness. Early on, there were many moments in which I let darkness disempower me, but I eventually found a way to transform these experiences into wisdom, compassion, and understanding. What happened as I conquered my inner demons was that my emotional and mental immune system got so much stronger and I developed the ability to acknowledge and do something about the ills of our world without getting infected by them. As terrible as trauma is, my life shows that it can be a vehicle for transformation.

Born into Generations of Displacement

For me, being a refugee means having a family tree with a lot of holes and gaps that can't be filled in. It's generally very hard for children to make sense of war, but it was much harder for me to do so growing up because I knew that if the Vietnam War had not happened, my family probably wouldn't have ended up as refugees in the United States, in a country where we had far more opportunities but no roots. Over the years, as I tried to trace my roots, I discovered that I couldn't find any that went very deep. What I learned from piecing together the stories my parents and relatives had shared and placing them into the greater backdrop of world history was that several generations of my family were displaced by a series of wars over the past century. My ancestors didn't get to settle down for long before they had to migrate again.

Assembling a clearer picture of my family's history has enabled me to connect mental health issues, such as anxiety, various forms of addiction, and abusive controlling behavior, which have run in my family for at least four generations, to undiagnosed and untreated trauma. Learning how war, violence, and displacement contributed to a family culture where being in Brain 1.0 was normal has enabled me to be more compassionate and forgiving toward my parents, other relatives, and myself for the amount of trauma and suffering we have yet to heal. I hope that sharing about my family's past will encourage more people to examine and compassionately attend to the deep psychological wounds that their own family may be nursing.

To reconstruct a timeline of the historical events affecting my family, I had to do a lot of digging around on my own—through history books and my personal travels—to fill in the gaps from stories my parents had shared. My best understanding is that from the late nineteenth century through the early twentieth century, upheavals such as rebellions, civil war, and famine drove several waves of my ancestors to relocate from northern China into southern China, and then further south into Indochina to areas that are now part of present-day Vietnam and Cambodia. Because of the repeated displacement, my parents can't name ancestors beyond those who migrated to Vietnam and Cambodia.

On my mother's side, her ancestors settled in the Mekong Delta area in southern Vietnam in the late 1800s. They assimilated into the countryside and sent their children to Vietnamese schools. As a result, my mother's side of the family speaks Vietnamese more fluently than Chinese. On my father's side, his mother moved to Vietnam with her parents when she was a young teenager, after the Japanese invasion of China. Her family settled in Cholon, a bustling Chinese community on the outskirts of Saigon, where she didn't have to learn how to speak Vietnamese fluently. My paternal grandfather was from an ethnic Chinese family who had settled in Cambodia. He moved to Saigon for business and there married my grandmother. I never met my paternal grandfather because he got sick and died very young, when my father was only two years old. After that, my father, his mother, and his older sister would occasionally visit his paternal grandmother and other relatives in Cambodia. Then, when it was no longer safe to travel freely, they lost contact.

I know that my parents and grandparents spent a good portion of their lives in active war zones because beginning with World War II, Vietnam became mired in war after war for almost five decades. In 1940, French colonial rule over Vietnam was disrupted when Japan invaded Indochina and occupied the region until the end of World War II. After Japan withdrew, there was a brief intermission of peace and hope for independence, which turned to disappointment when the Allied powers returned Vietnam to French rule despite the protests of the Vietnamese delegation to the treaty talks. That forced the leaders of the Vietnamese independence movement to turn to the Soviet Union and China for support. The revolutionaries became Communists and fought relentlessly until France finally withdrew in 1954. After that, the Geneva conference created an unstable peace by partitioning the country into two parts: North Vietnam, which was controlled by Communist forces, and South Vietnam, officially named the Republic of Vietnam, which was backed by the United States. Inevitably, Vietnam then plunged into a new phase of civil war that would continue for almost twenty years until 1975, when North Vietnam gained control of the whole country.

My father was about six years old when the French withdrew from Vietnam. As my father grew up, the fighting between North Vietnam and South Vietnam continued to escalate with no end in sight. Since he was the only son of a widow, his mother and older sister did everything they could to keep him out of the war. As they were very poor, what they mainly did was keep him out of sight. My father spent most of his adolescence in hiding so the draft officers wouldn't have him on their radar. To finally get my father exempted from the draft, they saved enough money to buy him a Taiwanese passport, but that meant that he had to completely surrender his identity and destroy his original documentation. From then on, my father never got to use his original name and actual birthdate. I imagine all the hiding and secrecy must have taken a mental and psychological toll on his development and sense of safety, because my father has a degree of paranoia and cynicism that is off the charts, even compared with other refugees from Vietnam.

My mother was a toddler when fighting between North Vietnam and South Vietnam began. As a young girl, she lived through the bombing of her village in the Mekong Delta not once, but twice. Along the road her family had to take to evacuate on foot, the Northern and Southern armies, positioned on opposite sides of the road, shot at each other while the villagers scrambled for cover. Though her family lost their home and possessions, everyone, including her parents and nine siblings, miraculously escaped without harm. Starting over with nothing, her parents had a hard time making ends meet. One day, my mother and her oldest sister (who I call Big Aunt) came home from working in the market to discover that my grandmother had given away two of their youngest sisters. Fortunately, Big Aunt had the courage to challenge her mother's decision. She went searching through the town until she found her sisters and brought them back. After a second bombing destroyed what was left of the village, my mother's family eventually relocated to Saigon.

When the Communists conquered South Vietnam, they renamed Saigon, its capital city, after their leader, Ho Chi Minh. For people living there, such as my family, the end of the war did not bring peace or prosperity. The government started rationing food supplies, such as rice, milk, and infant formula—forcing people to stand in line for hours to get their allotment. Then they issued a unified currency, the Vietnam đồng (VND), which rapidly depreciated. Continuous cycles of hyperinflation and panic made the currency all but worthless, so people transacted in gold instead. Over time, as the Communist Party consolidated its administrative control, the government rolled out brutal policies and divided the spoils of war. People who had supported South Vietnam or were deemed too bourgeois were hauled off to reeducation camps, and their houses and property were confiscated and given to Communist Party officials. No one knew who would be next. In this atmosphere of fear and paranoia, people began to flee by boat despite the dangers of getting caught or dying at sea. The refugees, who became known as the "boat people," felt that the chance to have a better life was worth risking their lives and the lives of their children.

Meanwhile, neighboring Communist countries, Cambodia and China, turned on Vietnam. In 1975, Pol Pot's Khmer Rouge regime came to power in Cambodia and began a genocidal campaign of social reengineering. The Khmer Rouge began to systematically kill everyone who represented the "old society," such as intellectuals, businessmen, professionals, anyone with ties to foreign governments, any residents who were not ethnically Cambodian (such as people of Vietnamese, Chinese, or Thai ancestry, among other minorities), and anyone who practiced religion (including Buddhism, Christianity, and Islam). During their four-year reign of terror, it is estimated that the total death toll came to between 2 million and 3 million people out of a total population of 8 million. Many of these people were killed by being buried alive in mass graves, which came to be known as the "killing fields." Tragically, my paternal grandfather's branch of the family tree in Cambodia has become completely lost to us. We can only guess that our relatives in Cambodia most likely perished in this horrific manner.

In 1976, the Khmer Rouge began to cross over into Vietnam, massacring entire villages along both sides of the border. This prompted Vietnam to declare war on Cambodia and support Cambodian leaders who had taken refuge in Vietnam, such as Heng Samrin and Hun Sen, to overthrow Pol Pot. Because

Cambodia and China were allies, the relationship between Vietnam and China became strained. To put an end to the Khmer Rouge attacks, Vietnamese forces invaded Cambodia at the end of 1978 and defeated Pol Pot in January 1979 and set up a new government, the People's Republic of Kampuchea. However, there was no peace in Cambodia as Khmer Rouge forces continued to fight, so Vietnamese troops could not fully withdraw from Cambodia. Even though Vietnam stopped Pol Pot's genocidal campaigns, the international community punished Vietnam by declaring sanctions and economically isolating Vietnam. Rather than recognize the new government, the United States and the United Nations recognized and supported Pol Pot's regime as the legitimate government of Cambodia until 1992, in spite of his crimes against humanity. Decades later, I still don't know how to make sense of this horrifying instance of the misguided belief that "the enemy of my enemy is my friend."

China also retaliated against Vietnam's invasion of Cambodia by putting up a show of force in mid-February 1979. Chinese forces crossed into northern Vietnam and captured several towns. Having made their point, the Chinese army withdrew in early March, but the invasion renewed a heightened level of tension between the two countries, which has lasted to the present day. In the immediate aftermath, ethnic Chinese living in Vietnam got clear messages that they were no longer welcome. Many saw the writing on the wall and finalized plans to flee in case hostilities escalated between China and Vietnam. (Fortunately, that didn't actually happen.)

When I asked my mom about how she met my father and why she married him, she explained that soon after the Vietnam War ended, my parents were introduced by a mutual family friend playing the role of matchmaker. During that time, my father stood out as a suitor simply because he was one of the few single men who still had all their limbs intact. Part of the reason she agreed to marry him was because he promised that he would take her out of Vietnam. After they married, they waited until after my older brother was born, in 1976, to make their first attempt. Unfortunately, the captain of the boat they planned to use was arrested by the authorities, and the boat was confiscated. Around the time my mother became pregnant with me, the conflict with Cambodia was escalating and tensions with China were beginning to mount. Soon after I was born, Vietnam invaded Cambodia and then conflict broke out in the northern border with China. This prompted my parents to plan another escape attempt. In late spring 1979, my family (my parents, paternal grandmother, older brother, myself, and several aunts, uncles, and cousins) traveled to the seaside village of Bac Lieu in the south on the pretext of visiting relatives and therefore only packed for what looked like a weekend trip to avoid suspicion. Then, in the middle of the night, we secretly boarded an overloaded fishing boat with hundreds of other refugees.

Luckily, this time the escape was a success, but we still had a long and harrowing journey to get to a refugee camp. After making it safely into the open seas, the boat drifted for about a week and ran out of food. People began to starve. My mom said I was so weak I barely moved or cried. Then our boat came across a Malaysian navy vessel and everyone rejoiced. The navy crew offered to tow the boat to a refugee camp and chivalrously invited the women and children on board to be more comfortable. The refugees had no idea they were about to be betrayed by their rescuers. The women who boarded the ship were raped. The crew robbed the rest of the refugees on the boat at gunpoint and confiscated the captain's compass. They pointed us in the direction of the nearest refugee camp and then abandoned us. It's no wonder that my parents have been suspicious toward the altruism of strangers ever since. I often wonder if I subconsciously absorbed their fear because meeting strangers has always triggered an unexplainable sense of terror in me.

That night, the boat got caught up in a tropical storm that, over and over, nearly caused it to capsize. The storm was so fierce that people worried the boat might break apart. My mom told me that people desperately cried and prayed, thinking that death could come at any moment. Families held each other and exchanged what they thought might be their last words. Luckily the storm subsided, and at dawn, we caught sight of land. The boat finally made it to an island in Indonesia, and the Indonesian authorities then brought us to the Kuku refugee camp.

Everyone at the refugee camp was grateful to be alive and to have gotten this far. Nevertheless, life in the camp was far from easy. For starters, there was no running water, and the hygiene was terrible. The area was covered with flies and mosquitoes. When we first arrived, anyone who was so thirsty that they drank water directly from the streams without boiling it first got very sick. Refugees were assigned makeshift huts with thatched roofs and no walls. People had to gather wood to build their own beds, or else sleep on the dirt floor. Malnutrition was rampant. Tropical diseases and food poisoning regularly ravaged the camps. Each day, refugees died. The elderly and young children were especially vulnerable.

As a baby, I didn't get enough nutrients to have much energy or strength, or even to grow a full head of hair. I caught several serious diseases, developed seizures, and came close to death several times. The many ordeals I experienced during this time had a noticeable impact on my development. I mostly stared into space and hardly ever smiled. I was months behind schedule when it came to being able to sit, crawl, walk, and talk (I wouldn't be able to talk until I was six or seven years old). At the time, my parents had no idea if I would turn out normal. The most important thing was that I kept pulling through.

After spending about a year in these conditions, my family finally received notice that America would take us in. We were then transferred to the Galang refugee camp, where refugees were processed for resettlement. The infrastructure at Galang was much more developed than at Kuku but still not what one would call comfortable. When it was time to finally leave the refugee camps, my parents were so grateful that everyone in our family had survived. Many refugees weren't so lucky.

As an infant I was too young to form conscious memories of this time period, but I think these experiences still got encoded in a nonconscious manner into my brain and body. When I was a child, whenever my parents talked about this period of time, what they had endured, and the ugliness of how they'd been exploited and betrayed when they were in such desperate circumstances, it would immediately bring me into Brain 1.0. I would feel overwhelmed by emotional pain, anger, fear, and distrust.

Learning as a child that my parents had undergone such a dangerous and harrowing journey in order to give me and my brother a better future gave me an unbearable sense of guilt that could only be eased by my making a promise that someday, somehow, I would make sure that the sacrifices they'd made were not in vain—that it would lead to something meaningful and worthy enough to redeem my family's suffering and struggles. In hindsight, having this aspiration gave me such a strong sense of purpose that it enabled me to also develop Brain 3.0.

An Unsheltered Childhood

When my family arrived in the United States, we were resettled in a violent, gang-ridden part of Philadelphia. Many of the long-term residents openly resented the influx of Southeast Asian refugees. Racial tension often turned violent. Asian immigrants were regularly robbed, beaten, and sometimes killed. Every month or two, there would be a new tragedy among the refugee community. Some of our relatives and friends ended up in the hospital with knife or gunshot wounds. It didn't take long for my parents to realize we had merely traded one war zone for another. Surrounded by good reasons to be in Brain 1.0, we learned that the American dream was really a nightmare.

My father had a few close calls, which taught him that he had to be more street-smart. He worked as a waiter in Chinatown and didn't get off work until late at night. One night, he was chased by robbers and barely outran them to get home safely. The next day, he saw that the pursuers were waiting for him at the subway stop, so he got back on the subway and crashed at a friend's house. Then for a period of time, I rarely saw him at home because he made arrangements to stay somewhere else after work. We started seeing him again when he eventually saved enough money to buy a used car so he could commute from home to work more safely.

My family also learned the hard way that the Philadelphia public school system offered no protection for refugee children. My parents had enrolled my older brother in the neighborhood public elementary school, optimistically hoping he would get a better education than offered in Vietnam. Instead, one day while he was in first grade, he came home with an untreated broken arm. My mom was shocked to discover that not only had his arm been broken during a fight at school, but his broken arm had gone entirely unaddressed by all the teachers and staff. She immediately took him to the hospital to have it treated. The next day, when my mother went to the school to talk to his teacher about the incident, the teacher merely blamed him, a six-year-old child, for what had happened and refused to take any responsibility. It was clear that my brother was not safe at that school, so my parents turned to the Vietnamese refugee community for help and eventually found support from a Vietnamese Catholic priest who helped enroll my brother at a nearby parochial school through a financial aid program.

Given how stressful the transition was, my parents were naturally preoccupied with figuring out how to make ends meet and keep the family safe. They didn't have the time or resources to worry about why I still couldn't talk. When it was time for me to start school, my mom brought me to kindergarten without explaining what it was. When I saw that she was going to leave me there alone, I panicked. I started crying hysterically and grabbed her feet. After the teaching staff pried me off my mother, I felt so abandoned that I crawled under a table like a feral cat and cried for hours. My kindergarten teacher was so horrified when she realized that I couldn't talk, she called home to recommend my parents send me to a special education program. She suggested I needed to learn sign language because she thought I was mute. However, once she saw that my parents didn't speak English and had trouble understanding her, she realized there were other possible explanations for my situation.

Fortunately, my school decided to first try giving me speech therapy and English as a second language classes. I must have realized the urgency of the situation and took the classes seriously. Those early years of school were a struggle. I had to read every one of my assignments two to three times to understand it. Yet once I learned to talk, read, and write, I was so hungry to learn everything I could about the world, I read every book my school gave me from cover to cover. We didn't have any books at home, so I continued to read my schoolbooks over and over to make sure I understood them. All this effort must have kicked off many changes in my brain that allowed me to catch up in terms of learning and development. By the end of third grade, I became the top student in my class.

Around that time, my family took over a take-out restaurant in a dangerous area in Philadelphia and we lived upstairs. I figured out that it must have been a rough neighborhood because my father used bulletproof glass to protect the take-out window. It was normal for children in Vietnam to work, so no one in my family thought it was unusual that my parents put me to work when I was about eight years old. I had to take orders, do food prep, and, where possible, cook. Since my parents couldn't speak English fluently, I had to help them read the mail and fill out paperwork. It was also part of my job to call the police to report when gang fights broke out, when customers got robbed, and when dealers used our take-out restaurant as a place to sell drugs.

It was hard for me to develop a sense of trust and safety in the world because I regularly watched customers try to cheat and bully my parents. Oftentimes, when these people didn't get what they wanted, they angrily cursed out my parents and threatened to beat them or kill them. If they started punching, kicking, and causing real damage to the building, I would then call the police. Every now and then, guns were fired. I once found a stray bullet lodged in the wall of the restaurant, and another time, I found one lodged in the floor of the second story where we lived. When I was in middle school, a customer was shot in the head while walking out of the restaurant. I will never forget how much blood was left after the paramedics carried him away. It took hours to clean up the mess the next morning. It became normal for me to have recurring nightmares about my family being gunned down or about one of us being hit by a stray bullet as we slept, and to wake up in tears. In my family, we never talked about these experiences and how terrified we felt, so I stoically held all this inside.

Unfortunately, the space inside our home was also unsafe. We lived with my paternal grandmother, who suffered from paranoia and anxiety. Whenever she got agitated, she would scream at us or tell us terrible lies about our mother, and sometimes beat us for no reason. It wasn't until I got a bit older that I learned that my grandmother had been diagnosed with dementia and psychosis. Over the course of my childhood, I witnessed her slow deterioration into a vegetable state. In the earlier years, when she could freely move about, she was often very dangerous. Every now and then, she would chase us with a knife or other weapon, so the only thing we could do was run into a room and lock the door until she got tired, went away, and forgot why she was upset with us. If we told my parents what happened, my father would blame us for not taking care of her, leaving us feeling confused and guilty. Then, after my grandmother broke her leg from kicking in her bedroom door, she became immobilized. This made her safer to be around, but it was depressing to see her in agony. She was rarely ever at peace. Even in her sleep she would moan and scream in grief and terror.

My parents explained that they couldn't put my grandmother in a nursing home because she did not speak English, so she could not communicate with the staff or follow their instructions. The few times she was hospitalized, the staff would end up confining her to her bed or a chair, which would make her become even more hysterical. To my parents, putting my grandmother in a nursing home meant leaving her to die. They didn't have the heart to abandon her like that so they cared for her at home while raising four children and running a take-out restaurant.

As a result, my family couldn't have a normal social life. Everyone had to make sacrifices. After school, I had to quickly finish my homework and then work in the restaurant so my parents could take shifts feeding her, bathing her, and changing her diapers. When both my parents and I worked the restaurant, my younger siblings had to keep an eye on my grandmother and alert us whenever there was an emergency. The strain on the family was often overwhelming. It was clear that my parents needed more help and support. Despite everything I did, it was never enough. They would constantly complain about how good I had it and how much they had to sacrifice for us. I often felt terribly guilty just for being alive and adding to their burden.

As bad as things seemed, I had regular reminders that other people had it even worse. Some of the kids who came to the take-out restaurant had parents who were junkies who didn't take care of them or even feed them. My parents regularly told us horror stories of fathers and mothers who were alcoholics and gambling addicts who stole from their own families. At least my parents were good, caring, responsible people who meant well. They fed us, clothed us, and made sure we had the basics. For the most part, they did the best they could with what little resources they had to provide a life for us that was far better than the life that they had had in Vietnam.

The problem was they were under so much pressure that once in a while, they cracked. All it took sometimes was a straw to break the camel's back. When my mother or father lost their temper, my siblings and I never knew how much they would take out their anger on us by screaming, cursing, and beating us. These episodes were terrifying because it was clear they were not in control. It was like a monster took over their body. On some occasions, they even threw things, like cans and boxes, at us as we ran for cover. Years later, as I studied the brain, I realized my parents were experiencing an amygdala hijack, when the Inner Godzilla completely takes over and shuts down Brain 3.0—something that I would later come to understand from firsthand experience. These incidents taught me that good people who have experienced extreme trauma, but have never had the opportunity to process and heal those scars, can lose their senses and behave in devastating and destructive ways, even to people they love.

Finding a Path in Brain 3.0

While there were clearly many reasons why I spent a lot of my childhood in Brain 1.0, it wasn't always so bad. Similar to how the weather in Seattle or London is generally rainy and foggy, but there are still many moments when the sun breaks through the clouds and there are occasional beautiful blue-sky days—there were many moments when Brain 3.0 cleared away the fog of Brain 1.0. Despite how stressed out my parents were most of the time and how often they were in Brain 1.0, they were also the primary people who modeled Brain 3.0 for me. Every day, I learned from their sacrifices and work ethic how to discipline myself, take care of others, and work in service to something much greater than myself. Seeing them face severe hardship, yet calmly and quietly push forward every single day, always gave me the strength to keep going.

It was my father who first demonstrated to me how to harness the power of Brain 3.0 to solve a crisis. It happened when I was about four or five years old, before I learned how to talk. During a visit to my aunt's house, I followed my older brother and cousins to the 7-Eleven next door. The problem was that I was holding in my hand an unopened Kit Kat bar that my aunt had just given to me. When I tried to leave the store, the clerk grabbed me and accused me of stealing the candy bar. I couldn't defend myself, so my brother tried to explain that I had it before coming into the store, but the clerk wouldn't believe us. One of my cousins went to tell my father. Meanwhile, I was so afraid of going to jail for a crime I didn't commit, I began wailing in panic. When my father arrived, he calmly suggested the clerk compare the UPC numbers on my Kit Kat bar with the bars in his store and if the numbers didn't match, it meant my bar was from a different box. The clerk did the comparison and saw to his amazement that my father was right. He apologized and let me go. From that moment onward, I became fascinated by how my father had handled the situation so brilliantly. How did he know about UPC codes? That incident inspired in me a strong curiosity to learn how the world works so I could think my way out of problems.

I loved learning so much I eagerly embraced school as an opportunity to come into Brain 3.0. I somehow managed to compartmentalize between school and home, so that I could get into Brain 3.0 at school no matter how crazy things got at home. While I was at school, I never talked about my life at home because I didn't want anyone to call child protective services to break up my family and put me and my siblings into foster care. No one had any clue about the chaos I had to deal with at home and how little support I got from my family.

As I got older, I was disappointed and frustrated countless times by my parents' traditional Asian mindset toward girls. They told me over and over that a girl is simply supposed to live at home, help her parents, and stay out of trouble until she gets married. It outraged me when my mother explained that in traditional Chinese culture, girls marry into the husband's family, so investing in a daughter's education or future is seen as a waste. Therefore, families put their resources into nurturing and educating their sons, while neglecting their daughters. This explains why generations of women in my family did not have much of an education and didn't know how to value one.

Even though we were now living in America, my parents brought this baggage with them. There was clearly a double standard in how they treated my brother and me. They pushed him to excel but were indifferent toward my academic achievements. They kept sending clear messages to me that they didn't see any point in my working so hard and didn't want me to be too ambitious, because if I was too accomplished and intelligent, no one would marry me. My parents' discouragement and devaluation of me for being a girl only angered me. I became determined to prove that I could succeed on my own and that I wouldn't need a husband.

I got the opportunity to prove my "worth" when I got into Central High, the top magnet school in Philadelphia. The smartest kids in the city went there, so it was a very intimidating transition and no one expected that I would be more than an average student there. I continued to love learning and soaked in as much knowledge as possible. Then, when the class rankings first came out in my sophomore year, I learned to my surprise that I was ranked at number two among a class of about eight hundred students. The news amazed everyone in my family. Yet, even then, my parents continued to nag my brother to excel but tell me to set my expectations lower. They also told me they would not let me leave Philadelphia for college because girls could not be safe away from their parents. Nevertheless, when I found out about a prestigious all-expenses-paid summer development program for high-potential students of color, which I thought was too awesome to miss, I brashly applied anyway in the hopes that if I got in, I could convince my parents to let me go. It only caused me heartache because I did get accepted, but my parents refused to budge, forcing me, bitterly and resentfully, to turn it down.

At the end of my junior year, my high school had a college book award assembly and gave me the Harvard Book Award.* After the ceremony, my guidance counselor, Barry Goldstein, explained that he had nominated me for the Harvard Book Award because he thought I would be a great fit for Harvard College and recommended I apply, which stunned me. I lived in a neighborhood where kids were lucky if they got to attend community college, so going to a place like Harvard seemed very far outside the realm of what was possible. I was amazed he thought I had a shot at getting in. At home, when I told my parents what had happened, they had no idea what Harvard was, so I had to explain that it was the top college in the United States and possibly the world. Still, they were mainly appalled at the idea of it being a seven-hour drive away, in Boston, Massachusetts. I explained that it was a long shot and convinced them to let me try my luck with the application since I could get application fee waivers and it wouldn't cost them anything.

In hindsight, given that I had to figure out the college application process on my own, it's nothing short of a miracle that I got into Harvard and also got offered a Mayor's Scholarship to the University of Pennsylvania. When the acceptance letter from Harvard arrived, everyone was in shock. Sadly, the atmosphere in my house was more like a funeral than a celebration. My family wanted me to stay in Philadelphia, as enrolling in Harvard meant I would not be able to help and support them. I felt bad, but Harvard was a once-in-a-lifetime opportunity. How could I turn it down? Besides, I really wanted to take a break from my family and explore the wider world outside of Philadelphia.

I decided to call Harvard to explain why my leaving home would cause financial hardship for my family. I then asked if they could consider making the financial aid package more generous, because otherwise I would have to stay in Philadelphia and go to Penn. Harvard responded by giving me a few thousand dollars more in grant funding. Once my parents saw how determined I was, they reluctantly agreed to let me go on the condition that I

---

* This is now called the Harvard Prize Book.

promise to become a doctor. I didn't particularly want to be a doctor, but they had me in a corner, so I made the promise with my fingers crossed. From then on, I didn't want my parents to have any power over me, so I started planning to be as financially independent as possible by getting a job on campus.

Imploding at Harvard

When I started college in September 1996, I was unaware of the consistently high prevalence of mental health conditions among Harvard students because mental health was not yet widely discussed, and the National College Health Assessment, which tracks statistics on mental health, only started in 2000. The earliest assessment data I could find for Harvard comes from a 2003 *Harvard Crimson* article, which stated that 47.4 percent of Harvard students "reported feeling depressed at least once during the past academic year, and about a third say they've felt overwhelmed 11 or more times—the highest category choice."[9] Unfortunately, mental health is still a major concern at Harvard today. According to a 2015 *Harvard Crimson* article, "around 40 percent of Harvard College students have used [the university's] mental health services [at least once] over their four years and between 20 and 25 percent of undergraduates visit the mental health services each year."[10]

Sadly, mental health issues are not isolated to Harvard. According to the fall 2016 National College Health Assessment, 61.9 percent of college students in the United States reported feeling "overwhelming anxiety," 39.1 percent reported feeling "so depressed that it was difficult to function," and 11.2 percent "seriously considered suicide" in the previous twelve months.[11] Given how widespread anxiety and depression are among college students, I want to shed light on how being in a stressful, high-pressure Brain 2.0 college environment can trigger and/or exacerbate mental health conditions by sharing more about my own experience.

When I first arrived at Harvard, it was like walking on air. I had to keep pinching myself to make sure it was not just a dream. Everything was so new and exciting. I had never been immersed in such an intellectual and affluent environment before. I could tell I didn't fit in, but I couldn't really put my finger on why. In the beginning, my obliviousness was a good thing because it allowed me to flourish academically. My first semester, I took courses to fulfill the standard pre–medical school requirements, which were probably among the most challenging classes offered to freshmen, and still made the honors list.

Then, in February of my freshman year, my grandmother passed away. Her death hit me like an earthquake inside. As I recall, that was when the uncontrollable PTSD symptoms and breakdowns started. I felt like a horrible person for not being with my family, for feeling relieved that she was finally released from such a terrible life, and for being glad that my family could finally move on. I had no idea how to deal with my grief and my guilt. My normal coping mechanism of ignoring and suppressing my feelings only caused me to unravel and dissociate at unexpected times. I cried through the night. Even during the day, I had uncontrollable crying spells that lasted hours. I withdrew from social events, isolated myself, and beat myself up about everything. I got locked into Brain 1.0 and couldn't get out of it. In that state of mind, I began to fixate on how I didn't fit in at Harvard. I struggled to build connections that went beyond the superficial. I became more and more socially isolated.

One of the most devastating things that happened to me during this period was learning that the mother of my best friend at Harvard didn't want her to be friends with me. Tina (not her real name) was a brilliant Taiwanese American student who had attended a prestigious private school on the West Coast. Her mother was from a very prominent family in Taiwan and set very high expectations for her. After one of her mother's visits to see her at Harvard, Tina shared that she was very upset about what her mom had said about me. Apparently, her mother accused me of being a Vietnamese person pretending to be Chinese. I was completely offended because that made no sense. I asked: If I were Vietnamese, why wouldn't I be proud of being Vietnamese? What would be the point of pretending to be Chinese? Then she explained that her mom thought Chinese people were superior to Vietnamese people, so a Vietnamese person would fake being Chinese to be higher class. I explained that it doesn't work that way; if I were really Vietnamese, I would resent the Chinese and completely own being Vietnamese. I explained that my family spoke the Teo Chiu dialect and had lived in Vietnam for many generations, so we'd assimilated a lot of Vietnamese culture and probably spoke with a funny accent that was different from that of Chinese people from other parts of the world—but that was nothing to be ashamed about. Then Tina explained that it didn't matter because her mother had a ranking of the different Chinese groups and she placed the Teo Chiu at the bottom.

Then I realized that the point was that her mother didn't approve of her spending time with me because I was beneath her in class. Tina explained that she didn't agree with her mother, that she also was upset with her mother for what she'd said, and that she'd told her mother she would continue to be friends with me. It was such a strange and confusing conversation to process. It was the first time that I realized people could actually look down on me, even though I was a Harvard student, for coming from a low-income background. Tina was my best friend, so I felt both deeply wounded by what she'd shared and also deeply sorry for Tina because her mother sounded like a monster. I knew from other stories she shared that her mother suffered from bipolar disorder and, throughout her childhood, her mother's episodes had caused her a lot of pain.

Nevertheless, because of what Tina said, I became much more guarded about sharing my background with other Harvard students. Given how I grew up feeling like an unwanted refugee in a violent neighborhood, this incident triggered many earlier traumatizing experiences of fear, insecurity, and discrimination and led me to associate feelings of distrust and danger with people at Harvard. In Brain 1.0, I instinctively created thicker armor to protect myself and I didn't know how to take off this armor even if I wanted to. This made it practically impossible for me to build a solid social support network in college.

By that point, financial stress—or rather, distress—had already become a never-ending source of social anxiety for me. In college, all my spending money came from what I earned through my work-study jobs. Since I had such a tight budget, I couldn't stop obsessing over maximizing every dollar. If I let even the smallest amount go to waste, I would ruminate and beat myself up over and over. One time, when I discovered I had lost a few dollars while doing laundry, I compulsively searched every corner of my room and cried for hours because I couldn't find the money. Whenever classmates invited me to go out, I regularly made excuses not to go because I just didn't have the money to spend. When people eventually stopped inviting me, I was more relieved than sad.

Whenever I found myself with a group of classmates and the conversation shifted to topics that revealed dramatic class differences, I would go quiet. When people talked about things like their favorite family vacations, prestigious summer camps they had gone to, exciting trips they were planning for spring break, how they'd started investing in the stock market, how their trust funds were set up, or attending expensive concerts, I had nothing to add. How could I tell them that I had never, ever had a family vacation because my parents worked practically every single day of the year—even holidays—to make ends meet? How could I tell them that I had never been to summer camp because my parents needed me to work? How could I tell them I couldn't go out with them because I couldn't afford it? I couldn't bear the thought of them looking down on me or using this information to put me down.

It didn't help that Harvard had an "every man for himself" type of Brain 2.0 culture. People were generally too busy to take notice if anyone was in distress. It was normal for people to fill every minute with so much activity that they didn't have time to breathe. Although not everyone was this way, the pressure to keep up the façade of having your act together and being superhumanly productive affected everyone at Harvard. It was not a place where a young person could feel safe enough to really open up and be vulnerable. Many of my peers were so perfectionistic and hypercompetitive, I felt stressed out just by being in the same room with them.

I think spring semester of freshman year is generally when the realization that it is much more difficult to be the best at Harvard hits all the students hard. Some people reacted by finding a healthier basis for their self-esteem, some people reacted by becoming even more neurotic about distinguishing themselves in an area where they could stand out, and some people reacted by getting depressed. As far as I could tell, most people experienced some combination of these three reactions and the anxiety and depression in the student body was palpable and contagious. Thus, Harvard became the perfect storm of a place for me to self-destruct and wallow in self-pity. I felt like no one there could possibly understand or care about my pain and what I was going through. I tried to tell myself I didn't need friends and could make it on my own, but it wasn't true. I was in bad shape.

On top of that, the institutionalization of Brain 2.0 in Harvard's culture made it increasingly rare to naturally activate Brain 3.0. I couldn't stand how people turned every interaction into an opportunity to prove their intellectual superiority. I saw it among students, teaching assistants, professors, and administrators. The use of jargon, run-on sentences that spanned an entire page, references to obscure papers and concepts, all seemed designed to perpetuate a privileged inner circle and keep knowledge out of the grasp of ordinary people. I felt suffocated in pretension. My love of learning, which had pulled me through grade school and high school, withered in the elitist Brain 2.0 atmosphere of Ivy

League academia. In my head, all I wanted to say was "fuck it." Harvard's mission to educate citizen-leaders started to seem like self-promoting Kool-Aid that I did not want to drink. In my depression, I came to see Harvard as a finishing school for very sheltered and helicoptered young people to indulge in the notion that they were entitled to a lifelong superiority complex because they went to Harvard. I didn't want to assimilate into this ivory tower but I also needed to find a way to stick it out until I got my college degree.

Being lost in such a Brain 2.0 environment plunged me further into Brain 1.0 because I felt like I was falling more and more behind in a rigged race where everyone was leaving me in the dust. Every instance I saw of how my peers were groomed and coached by their parents to succeed only made me more conscious of how disadvantaged I was in comparison. Everywhere I looked, I found evidence that the odds were stacked against me and that I was set up to fail. My fixation on how unfair the world was made me angry, bitter, and cynical.

Trips home didn't help. I didn't know how to explain to my family and friends that going to Harvard was anything but a golden ticket. How could I tell them I was miserable? What's worse, my family accused me of becoming uppity and being ashamed of them. When I saw how insecure they were, I realized they couldn't understand what I was going through or support me through it. Further, what they said got to me. I felt like I didn't belong anywhere, like I had lost my tribe and my identity. I didn't know who I was anymore. I had not anticipated that going to Harvard would require paying such a huge price. Without a social support system or a mentor to guide me through this transition, I felt more and more lost and got angrier and angrier at the world. I sank even deeper into Brain 1.0.

By the fall of my sophomore year, I was no longer able to do problem sets or write papers without dissociating and falling apart. I had the foresight to see that if I continued to take science classes, I would be at risk of failing out. I knew I had to find a major I was still capable of doing the work for. I was okay with dropping off the pre-med track, since my heart was never in it, but I was determined to graduate on time. After exploring several different departments, including History of Science and Anthropology, I eventually chose to become an art major, which at Harvard goes by the fancy title of Visual and Environmental Studies (since it's so long, everyone referred to it by the acronym VES). I chose art because the act of creating somehow brought me back into Brain 3.0. It didn't matter what medium—drawing, sculpture, painting, photography, or theater. While I was making art, I got immersed in the moment and lost track of time. It was the only reliable way to break out of Brain 1.0. Plus, being an art major gave me the perfect cover for being weird and different, as well as a surer path to graduation.

What I didn't know at the time was that the hippocampus, a structure in both the left and right hemispheres of the brain that is involved in the processing, storage, and retrieval of spatial and episodic information,[12]* and is therefore critical to learning, is particularly sensitive to stress and trauma.[13] Recent discoveries that the hippocampi are key hot spots for neurogenesis throughout our adult lives led scientists to hypothesize that the generation of new neurons here may be linked to the ability to learn and encode new memories.[14] Unfortunately, research on animals has shown that chronic stress and trauma seem to disrupt the functioning of the hippocampi and the mechanisms for neurogenesis. In research on humans, brain scans of people suffering from PTSD show that they have hippocampi that are smaller in volume than a control group composed of healthy people who had not experienced trauma.[15] In hindsight, these changes in the hippocampi could explain why having PTSD made it difficult for me to take in new information and retrieve information from memory (and why I have never been particularly good at spatial navigation).

In the end, changing my major only provided a temporary respite from my symptoms. Eventually, I started to have panic attacks about how I would make a living or pay off my student loans as an artist. Then one night I had a horrible panic attack that made me feel like I was about to explode. For hours, nothing I tried could give me any relief. So finally I took myself to the ER, where they gave me Ativan, and, to my surprise, it worked.

Sitting alone in the clinic in the middle of the night, I asked myself: What if the root of my problems wasn't so much that the world was horrible or that Harvard was toxic—what if the problem was me? Why could I not appreciate anything anymore? Maybe the way I perceived the world was distorted. It seemed that I had "lenses" that enhanced the negative and filtered out the positive. That night, I finally came to the conclusion that if the problem was inside my head, I couldn't pull out of this on my own. I needed to get professional help.

---

* Episodic memory is the ability to remember the details of an episode or event in one's life that took place at a particular time and place and to remember lessons that one has learned.

2.

Becoming a Mind-Hacker

We see the world, not as it is, but as we are—or, as we are
conditioned to see it.
— Stephen Covey

Any man could, if he were so inclined, be the sculptor of his own
brain.
— Santiago Ramón y Cajal, father
of modern neuroscience, Nobel laureate in medicine

The great thing in all education is to make our nervous system
our ally instead of our enemy.
— William James, father of American psychology

As I shared in the preface, when the psychiatrist explained that my
symptoms were the result of early childhood traumas causing my brain to
malfunction, my entire perspective on life changed. Learning that my inner
misery was tied to malfunctioning brain circuits put me on the path of self-
empowerment and self-healing through science. It was the beginning of a
lifelong personal mission to understand as much as I could about the brain
and experiment with different methods to bring my brain to full functioning.
This search led me to eventually understand that Brain 1.0 and Brain 2.0 trap
me in self-limiting patterns and that I had to shift into Brain 3.0 to experience
genuine self-actualization.

In college, the impact of neurochemistry on the way I felt, thought, and
acted was made real to me when I started taking antidepressant medication.
The psychiatrist had warned me that we might have to try different antidepressants and cycle through different doses to find one that worked for me.
Plus, it would take time for my body to acclimate to the medicine and dosage,
so for each change it could take several weeks before we could observe the
effects. He suggested we start with Zoloft at a moderate dosage level. Strangely,

during the first two weeks, I developed an anorexic compulsion to starve my-self to death. I lost my appetite and could hardly bring myself to eat during mealtimes. This really stood out to me because up until then, I had never had any eating disorders and had never dieted or obsessed about my weight. After my psychiatrist adjusted the dose, the anorexic episode went away. A couple of months later, I felt noticeably less negative and anxious. It was clear the drugs made a difference, but they also had bad side effects like weight gain, dizziness, fatigue, trouble sleeping, and stomachaches. I still didn't have much interest in life and my normal vitality and zest didn't come back. I began to wonder if there was another solution.

Getting treatment was an anthropological immersion into the American mental health system. The more I saw, the more I became concerned. I disliked how patients were treated as cases and how my appointments were so trans-actional. I often sensed that my psychologist and therapist were just checking boxes and did not seem emotionally invested in what would become of me after I left Harvard. I was alarmed that they seemed fine with the idea of my being medicated for the rest of my life. By that point, I realized the medication only helped me feel less awful—which is a good thing, but it's not the same as feeling good (being less in Brain 1.0 is not the same thing as being in Brain 3.0). It frustrated me that the treatment model put all the emphasis on medi-cation but little to no emphasis on building skills to more effectively manage the hardships that caused the stress and anxiety in my life. In therapy sessions, talking about the traumas I'd experienced forced me to relive them but did nothing to help me cope with my family's continued financial struggles and dysfunctional patterns. These observations led me to conclude that if I wanted to make a full recovery, I would have to take ownership of my healing journey. It was time to build upon the progress I had made under medical attention and to make use of the valuable things I had learned from my health care professionals to take my healing to the next level.

Falling in Love with Brain Science

Thankfully, diving into brain science rekindled my love of learning, which helped me start to feel like my real self again. What I learned made clear that I needed to take better care of my brain physiologically. While I had not yet fleshed out the Brain 1.0, 2.0, and 3.0 framework that I use now, I had already begun to intuitively sense that at that time in my life, the neural circuits I would later call Brain 3.0 had become a scarce resource, the neural circuits I would later call Brain 1.0 had become my dominant mode, and letting my Inner Godzilla run free was making my life worse. What I learned about the brain helped me see that I had to take it easy when Brain 3.0 was exhausted, and make efforts to regenerate those circuits, to prevent my Inner Godzilla from taking over.

After learning that the brain needs rest as well as oxygen and high-quality fuel, both of which have to be pumped in via blood by the circulatory system, I made lifestyle changes to provide more of these to my brain. I gave myself a relatively early bedtime of eleven p.m. and stopped pulling all-nighters, unless absolutely necessary. I began eating healthier (I cut out a lot of junk food) and focused on eating more tryptophan-rich foods like turkey to help my body produce more natural serotonin, an important neural transmitter that Zoloft helped increase amounts of in the brain. I started exercising regularly for the first time in my life. Exercise, in particular, brought much more oxygen-rich blood to my brain, which gave me more clarity.

In my digging around for brain research, I stumbled upon a field known as affective neuroscience, which changed my life by giving me big clues as to what was going on inside my brain. "Affect" is the word psychologists use to refer to the experience of emotion, and affective neuroscience is the study of the neural mechanisms for human emotion. As Richard Davidson explains in *The Emotional Life of Your Brain*, this field got started in the 1970s, around the time that he, as a graduate student at Harvard, set up the first EEG study of brain activity underlying human emotion.* The inspiration to do so came to him after reading about a research paper on patients with brain damage, which had found that damage in the left frontal region of the brain induced pathological crying and that damage in the right frontal region induced pathological laughter.[1]

This led Davidson to hypothesize that the prefrontal cortex (abbreviated as PFC), which again is the part of the brain inside the forehead, is involved in emotions (each hemisphere of the brain has a PFC, and they are referred to individually as the left PFC and the right PFC). He wondered whether positive

---

* EEG stands for electroencephalogram, an instrument that detects and records electrical activity in the brain using electrodes (sensors in the form of small, flat metal discs) attached to a person's scalp.

emotion could be linked to the left PFC and negative emotion could be linked to the right PFC. He also guessed that depression could be explained by a malfunction in the left PFC. Since fMRI had not yet been invented, he did not then have access to a noninvasive technology to observe brain activity. So Davidson worked with his colleagues to develop a method using electroencephalogram (EEG) sensors on the scalp to measure electrical signals from the brain as a proxy for neural activity.[2]

Davidson's EEG studies revealed that, contrary to the prevailing view of the day, what was considered to be the rational brain, the prefrontal cortex, actually played a key role in emotions. Further, the EEG readings confirmed that the prefrontal cortex of each hemisphere had opposing roles: the left PFC activated during positive affect and the right PFC activated during negative affect.[3] Decades later, when more advanced brain imaging technology such as fMRI became available, brain scans showed more precisely that when people experience emotions, specific regions of the left or right prefrontal cortex are activated depending on the emotion.[4]

As Davidson continued to research emotions in the brain, he found that individuals vary in the level of baseline activity in the PFC of each hemisphere and confirmed that there is a connection between depression and very low activity in the left PFC. The data also allowed him to link people's variances in the relative activity between the left PFC and the right PFC to personality traits—what Davidson refers to as "emotional styles." His studies with infants confirmed that differences in emotional styles are already detectable soon after birth. People with higher baseline activity in the left PFC are naturally optimistic and resilient, meaning they recover more quickly from stimuli that trigger stress. People with higher baseline activity in the right PFC tend to be more cautious and reactive and take longer to recover from stress-provoking stimuli (this clearly described me).

Here is how scientists think it happens: The amygdala, an almond-shaped structure that sits under the cerebral cortex inside the temporal lobe of each hemisphere, is involved in processing information gathered by our senses and matching it to previous experience to make an instantaneous determination whether we are in danger. Depending on what it picks up, it turns on an instantaneous emotional reaction. Bessel van der Kolk calls the amygdala the "brain's smoke detector" because its central function is to identify whether incoming input is relevant to our survival. The amygdala is very sensitive to

signs of danger, so if it picks up anything threatening, it triggers strong negative emotions such as fear, anger, or disgust. This may also activate the right
PFC to detect and anticipate the threats and dangers, so we can take appropriate actions to protect ourselves.[5]

In the brain, the network for resilience involves a pathway that carries
inhibitory signals from the left PFC to calm the amygdala. As Davidson explains, "activity in the left prefrontal cortex shortens the period of amygdala
activation, allowing the brain to bounce back from an upsetting experience."[6]
But if the level of activity in the left PFC is not strong enough to calm the
amygdala, or the connection between the left PFC and amygdala is impaired,
then the amygdala stays on high alert and the right PFC continues to scan for
reasons to justify the need for continued vigilance. As I shared earlier, a highly
activated amygdala is a key part of the Brain 1.0 pattern. If a person tends to
be in Brain 1.0 over long periods of time, his or her amygdala may become
larger because it is continuously overactivated.

When I learned about these findings, I realized that my symptoms could
be tied to the specific areas of the brain identified in the research. I guessed
that my left PFC could be severely underactivated, that my amygdala and my
right PFC could be overactivated, and that the connection between my left
PFC and my amygdala could be malfunctioning. I also wondered if this pattern had already developed when I was a baby refugee, which might have made
me more vulnerable to developing depression later on.

I decided to work on building up the neural circuits in the left PFC for
positive emotions, so I could more easily calm my amygdala and get myself
unstuck from negative spirals. By far the most effective mind-hacking idea I
came up with was to laugh more. I started spending time every day watching
comedy shows or movies that made me laugh hysterically. It worked. Laugh-
ing actually helped power up my left PFC to calm my amygdala. Instead of
beating myself up whenever I experienced a setback, I created a habit of finding something funny in the situation or my own behavior to laugh at. Conversely, I also decided to reduce my exposure to things that trigger and activate
Brain 1.0 in order to weaken those neural circuits. So I started cutting down
on horror movies, violent thrillers, and tear-jerking dramas.

I have since learned that as long as there is an equal or higher amount of
baseline activity in the left PFC, having high levels of baseline activity in the

right PFC is not necessarily a bad thing. This is because high-baseline right PFC activity is correlated with the ability to make realistic assessments, detect problems, see negative aspects of a situation, imagine what can go wrong and make backup plans, and more accurately calculate the risks of failure; whereas high-baseline left PFC activity is correlated with the ability to be optimistic, identify opportunities, appreciate the positive aspects of a situation, imagine what can go right, and take risks when the rewards are worth it. To have the best chances of fulfilling our aspirations, we need to have strong capacity in both the left and right PFC. The left PFC enables us to move forward and pursue opportunities for gratification, growth, and self-enrichment. The right PFC serves as a bullshit detector and helps us to stop ourselves from taking foolhardy risks where the downside outweighs the upside. It gives us self-control to resist and control urges and impulses that would lead to "bad" outcomes. It is our right PFC that allows us to delay immediate gratification and resist temptation.

It turns out that high activation of the right PFC is associated with an effective anxiety management strategy called defensive pessimism. Defensive pessimists channel their anxiety to imagine possible worst-case scenarios and the many ways that things can go wrong. Then they prepare by taking action to prevent these problems and laying out contingency plans in case things still go wrong. A large proportion of very successful people are defensive pessimists.[7] In contrast, optimists tend to have high failure rates because they either ignore or underestimate and underprepare for the obstacles, risks, and hurdles they need to clear in order to achieve a goal.[8]

As you might guess, I am speaking from firsthand experience because I spent many years of my life anxiously imagining the worst things that could possibly happen every time I would try something. Even now, when I spend more of my life in Brain 3.0, I continue to find defensive pessimism to be a very beneficial strategy. Whenever I make important plans, I give myself time to list what can go wrong and what I can do about it. Today, I don't get bogged down in negativity when doing this exercise because I have enough activity in my left PFC to have a balanced view. By intentionally activating my left PFC, I can also appreciate what can go right and visualize how carrying out the plan ties into my long-term aspirations and goals. This gives me the fuel to stick it out even if I run into setbacks. I often intentionally switch between priming the left PFC and the right PFC so I can see opportunities for growth and also anticipate problems and prepare for them. Having both left and right PFC functioning in balance enables me to move forward with my eyes open.

I often find it helpful to use a car analogy to describe the importance of these two parts of the brain: I think of the left PFC as the gas pedal and the right PFC as the brake pedal. If the gas tank is empty or the gas pedal doesn't work, then the car won't move. If the brake pedal doesn't work, then I can't drive safely. It's very likely I would hit other vehicles, pedestrians, or buildings, and that I (along with all the people I might injure) would end up in the hospital or dead. In order to arrive at my intended destination safely in one piece, I have to alternate between pressing the gas pedal and the brake pedal. Both need to be fully functional. The thing is, the way my brain got wired, I naturally have a lot more braking power than gas power. That means I need to pay more attention to refueling my gas tank regularly and maintaining the gas pedal in good working condition. (To see this metaphor illustrated, you can jump ahead to figure 5 in chapter 7 [page 206].)

Tuning In for Biofeedback

By far, one of the worst symptoms of my depression was a condition called anhedonia, which is the inability to feel any pleasure or positive emotions. I experienced it as an extensive and pervading feeling of numbness. A strange consequence of that emotional numbing was that it also impaired my capacity to make simple choices like what to eat, what to wear, or even what pen to use. That made everything I had to do so much slower. It was like moving around while bogged down in quicksand.

A big light came on when I discovered research that explained why: emotions guide our brain's decision-making processes around liking and preferences, and without guiding emotions, we can get lost in endless, paralyzing analysis. I also realized that my experiencing anhedonia was likely the result of a lifetime of not learning how to cope with negative emotions in a healthy manner. As a child, when terrible things happened, I would become overwhelmed by pain, frustration, anxiety, guilt, and/or anger. I often felt like I had no other option to deal with these feelings than to turn off all my emotions so I could think my way to a solution. Because I purposefully disconnected my head from my body over and over again, I eventually became completely out of tune with my body. By pushing these emotions away, I became even more afraid of them.

During this time of recovery, I learned to see my physical and emotional state as a fascinating biofeedback and neurofeedback system. By observing my patterns, I realized that negative emotions were messages from my body that needed to be registered. I learned the hard way that running from or suppressing negative emotions didn't work because they would then well up unpredictably and cause me to dissociate or lash out at people. Instead, it was more effective to let myself feel, acknowledge, and process these negative emotions, in a way that didn't involve stewing in them.

As I allowed myself to sit with these emotions, I came to see them as messages signaling that there was an important disconnect between my head and my body that needed my conscious attention. I learned to recognize my feelings and give myself time to process them. This meant going inward to understand and acknowledge the underlying issue, listening to the message, and then making choices and taking action in line with the message. Taking the time to do this helped me become more harmonized inside. I think it must have also strengthened the neural pathways connecting Brain 3.0 to Brain 1.0, so it became easier to calm myself whenever Brain 1.0 got triggered.

At Harvard, my body was so exhausted from the constant bombardment of stress that anything that triggered a sense of being overwhelmed could cause me to dissociate and get hijacked by Brain 1.0. I learned to listen to my body, in particular to my energy levels. I began to notice what gave me energy and what drained me. Whatever energized me and lifted me up, I would continue to do more of. Whatever drained me and pulled me down, I had to let go of and stop. And when I had no energy, I learned that beating myself up didn't help. Instead, I had to be compassionate toward myself and accept that I was in a state in which I could not think clearly. If I had to make a decision or do homework and my brain got foggy, I would take a nap or do something relaxing until I had regenerated enough Brain 3.0 to continue.

As I observed the patterns of my feelings, thoughts, and behaviors, and their impact on my state of mind, I realized that a lot of unnecessary pain and suffering came from wallowing in self-pity, anger, or resentment. Whenever I let myself marinate in negative thoughts and emotions, I would feel drained and even more isolated and hopeless from further activating Brain 1.0. Even though I felt entitled to these negative emotions, I saw they were weights that were better off shed. I set an intention that whenever I caught myself ruminating on negative events and worries, I would let go of what was eating me up before my energy levels were drained.

In retrospect, the best way I can describe what I learned to do is "to tune in to my inner compass." Becoming more centered, in turn, gave me the inner strength I needed to resist the herd effect at Harvard. I stopped benchmarking myself with my peers and, as a result, stopped feeling trapped in a losing race. By deliberately taking myself out of Brain 2.0, I also spent less time in Brain 1.0. Whenever I started beating myself up, I would imagine an inner sage, a Greek philosopher like Socrates, asking rhetorical questions to help me to think differently about the situation. In addition, I took deliberate steps to exercise and strengthen my neural circuits for positive emotions by regularly setting aside time to laugh by watching comedies or reading jokes and humorous stories. This revitalized the part of my brain that can appreciate the irony and humor in everyday life. Soon I was able to laugh at myself, my flaws, and my mistakes, and to stop taking everything so seriously.

Through these proactive efforts to hack my mind, the voice in my head evolved from that of an inner demon to that of an inner coach, scientist, and, at times, stand-up comedian. As my left PFC got stronger, it became easier to not always fixate on the negative, but to also see and feel the positive. I became much more compassionate toward myself, my brain, and my body. By tuning in more and more to my inner compass, I reintegrated my head, heart, and gut. As I became more integrated, the dissociation happened less and less frequently and the fog of anhedonia lifted. In this manner, I slowly built up the neural circuits in Brain 3.0 that could pull me out of Brain 1.0, thus making my emotional immune system much stronger. (At that time, I had not yet built up the neural circuits in Brain 3.0 that could pull me out of Brain 2.0—this would come later on.)

Recognizing a Much Larger Problem

Graduating from Harvard was such a big relief, like finishing a long, arduous, painful marathon in last place. I will always remember Harvard as a place where the gap between the privileged and the disadvantaged felt so wide, it felt impossible to close. No matter how fast I tried to run, I could never catch up, and I felt a constant sense of panting for air. So I have come to associate Harvard with a feeling of suffocation, of not getting enough oxygen into my body. It always felt like I could breathe more air outside Harvard than in it. Therefore, it was so liberating to finally leave that place behind and return to the "real world," even if for me that meant inner-city Philadelphia.

After I came home, I began to talk cautiously about my experiences with friends and relatives. It was eye-opening to learn that I was not the only person who'd struggled in college. A lot of people shared that they'd often felt unhappy and hadn't felt like they belonged. Financial stress didn't help. After using what little money we had from working to pay for school fees and buy books, we couldn't afford to go out and socialize, which meant we didn't make many lasting friendships. It's no wonder so many people got depressed. Quite a few of my friends needed to take time off. Several of my cousins dropped out of college and never went back.

These conversations helped me realize that the difficulties I experienced are common among first-generation college students. Unfortunately, the shame of poverty and the stigma of mental illness drove many of us to hide our struggles. We compensated by putting up a front. However, when the challenges became overwhelming, the risk of falling between the cracks was heightened by our notion that reaching out for help meant we were weak. The self-defense mechanisms that had enabled us to get through high school without much support from our families and communities worked against us in college. The walls we'd built to protect ourselves and hide our emotions also prevented us from learning and practicing the interpersonal skills needed to form healthy relationships with classmates, mentors, and advisers and from seeking help and support when we were in over our heads. In college, we discovered that we had become prisoners inside our own walls, and we had no idea how to free ourselves from captivity. We didn't know how to outgrow these patterns that held us back.

At that time, my experiments with mind-hacking were still at a very early stage, and my thinking had not yet crystallized into a form that was readily sharable. I didn't presume that what seemed to work for me could be helpful for anyone else. My own emotional state was still too raw and fragile. I was still concerned about the possibility of a relapse. I did not feel confident I could handle a big increase in my stress levels.

I found it frustrating that these issues were so taboo that no one wanted to openly talk about them, despite the fact that the dropout rate for low-income college students is stunningly high, at about 50 percent or more. I had a gut feeling that trauma had to play some role in these dramatic statistics. It was obvious from the levels of violence, crime, and drug addiction in Philadelphia that trauma had to be very widespread. The fact that so many young children were relentlessly exposed to trauma and were often innocent bystanders to violence could help explain why three to four out of ten kids in Philadelphia didn't finish high school, let alone apply to college.

At that time, it was the year 2000 and the initial ACE (adverse childhood experiences) study had only recently been completed. The impact of long-term exposure to trauma on the development of children was not yet widely understood, not even among psychiatrists and therapists. Up until that time, most research on treating PTSD had focused on adult survivors of acute trauma (trauma from a one-time event or a short period of exposure), who were otherwise healthy, emotionally stable individuals with a functional social support network. After the ACE study, researchers started paying more attention to people who grew up in violent situations like war zones or very abusive households where a parent or guardian was often the perpetrator of violence and there was no way to escape or protect oneself. Researchers realized that the conventional treatments for PTSD were usually ineffective in these cases. Oddly, many victims of abuse had a tendency to reenact trauma or attach even more to their perpetrators to gain a sense of emotional stability.

Seeing that there were very clear differences between the impact of acute trauma and the impact of chronic trauma experienced by children growing up in dangerous conditions, researchers such as Bessel van der Kolk began to advocate that a new trauma diagnosis be created, under a name like "developmental trauma disorder" (DTD), "complex post-traumatic stress disorder" (C-PTSD), or "complex trauma," to make clear that people healing from acute trauma and people healing from chronic childhood trauma need different treatments. However, their efforts failed to convince the American Psychiatric Association to include a new diagnosis category in the latest edition of the Diagnostic and Statistical Manual of Mental Disorders (DSM-5).[9] Even today, almost two decades after the first ACE study, there is still very little being done to address the widespread nature of developmental trauma in our society, especially in the many inner-city communities where over 45 percent of the population has experienced four or more ACEs. For the most part, society continues to blame, stigmatize, and punish people suffering from trauma rather than help them find effective forms of healing.

Getting Back on My Feet

Back to the year 2000, as a fresh college graduate, having returned to inner-city Philadelphia and seeing the impact of poverty and violence everywhere, I felt strongly compelled to put myself in a position to be able to do something to make a difference. However, with limited resources, connections, skills, and experience, I felt useless and clueless. Before I could be an effective agent for change, I needed to focus on building skills and gaining financial stability. To begin with, I needed to secure a job that would let me break the cycle of stress and financial distress that ran in my family.

I only had a six-month grace period for my college loans, which meant that I needed to start paying them off at the end of the year. During my senior year of college, I passed on recruiting because I wasn't in a good frame of mind, and adding more stress might have made things worse. Now that I'd come home and felt better, I needed to start thinking about a career, but having trained as an art major, I didn't know where to start. I felt like I was probably in the very small minority of Harvard students to graduate without a job, a seat at a graduate school, or even a plan. Despite Harvard being the top-ranked institution in the world, Harvard's Office of Career Services was not very helpful or user-friendly. At that time, if I wanted a list of alumni in a particular field to reach out to, I had to put in an order and pay a rather hefty fee. I wondered whether perhaps the Office of Career Services was historically weak because Harvard students traditionally came from such affluent and accomplished families with such strong referral networks that they didn't have to rely on the university for support.

Looking back, I still think of it as a miracle that I graduated from Harvard jobless and yet somehow ended up building a successful business career. At the time of graduation, I had no clue how to think about building a career mainly because I had had zero exposure to a corporate work environment. When I moved back to Philadelphia, to earn income, I called a temp agency and started taking assignments for ten dollars an hour. I did things like manning the reception desk at a law office, filing paperwork at a financial planning office, and doing data entry for the programming department at Comcast. Many of the people I worked for were pleasantly surprised to get a Harvard graduate to fill their assignment.

Thankfully, I had Harvard classmates who had figured out how to launch a career. When I reached out to my friends for advice, I learned that many of them had gone into management consulting. Listening to them talk about how much they enjoyed their work because it gave them exposure to the business world and the opportunity to understand and solve the problems that affect major corporations, I realized that it sounded exactly like the type of training I needed. But I was sure I wasn't qualified. Luckily, these kind and generous friends convinced me that I had what it took to be a good management consultant, talked me through the recruiting process, and made me believe it was possible for me to get through all the hurdles.

One of my dear friends, Kibi, worked at the Monitor Group, a premier consulting firm founded by a group of professors at Harvard Business School. I liked how Kibi described Monitor as "a home for idealists who wanted to change the world," so Monitor became my first choice. With coaching and support from Kibi and other friends, I submitted applications to the major consulting firms. Several firms, including Monitor, invited me to do first-round interviews, which involved solving a series of brain-boggling case questions in a short amount of time. I somehow got an offer from the Monitor Group and started working in the head office in Cambridge, Massachusetts, in January 2001.

Interestingly, after graduating from Harvard, I found it much easier to get to know former classmates who also worked at Monitor or lived in the Boston area. We shared a preference for being simple human beings rather than pretending to be superhuman. I had often thought of myself as bringing down the average at Harvard, but I would eventually learn from my friends that that wasn't how they saw me. The truth was, many of them were preoccupied with their own inner demons. As time went by, I came to learn from heart-to-heart conversations with friends from affluent families that there is a dark side to privilege: the pressure to live up to your parents' expectations for you and to earn the esteem of your social circle through achievement. The pain of feeling they weren't unconditionally loved for who they were also wreaked havoc on many people's lives. Their upbringing and social conditioning somehow locked them into Brain 2.0.

From these conversations, I realized that my having no expectations to fulfill, and no pressures to follow in anyone's footsteps, had been a precious gift. The fact that nothing was handed to me on a silver platter because of whose child I am means that my achievements are mine. Today, this gives me a sense of earned independence and autonomy that I've noticed few people seem to truly enjoy. It also serves as a wellspring of self-confidence, authenticity, and groundedness. But to get to this point of self-acceptance, I had to first run through the long grueling maze of Brain 2.0.

3.

An Unhappy Pursuit
of Happiness

Beware the barrenness of a busy life.   —Socrates

Instead of dedicating your life to actualize a concept of what
you should be like, ACTUALIZE YOURSELF. The process of
maturing does not mean to become a captive of
conceptualization. It is to come to the realization of what lies in
our innermost selves.   —Bruce Lee[1]

What you get by achieving your goals is not as important as
what you become by achieving your goals.   —Zig Ziglar

WHEN I STARTED working at Monitor, my inner demons were no
longer a danger to my life, but they were far from tamed. As an art major, I felt
like a fish out of water at a prestigious management consulting firm. I had to
learn a completely new vocabulary and teach myself how to succeed in an
unfamiliar professional culture. I couldn't help feeling like an impostor, paranoid that my coworkers would discover my incompetence at any moment.

I was surrounded by brilliant peers, many of whom had studied economics
or business and seemed groomed by their upbringing to make partner. Almost
everyone I met came from a well-to-do family, and many of them had parents
who held prestigious and influential positions in the world. In contrast, the
adults in my extended family worked as janitors, waiters, cooks, clerks, hair-
dressers, or nail artists while the more entrepreneurial ones ran mom-and-pop
shops. So my Inner Critic (which we'll discuss in chapter 8) insisted over and
over that I had to work extra hard so people would have no clue that I had come
from a background where people mainly come into this type of workplace to
clean the office or deliver food.

Just like at Harvard, at Monitor it was rare to meet people who had uneducated parents and grew up in poverty, who could not rely on their parents for financial support or career guidance, or who had to help their parents pay bills and cover health care costs. I continued to be very self-conscious about coming from a very disadvantaged background. After feeling so left behind at Harvard, I felt compelled to prove to myself that from now on, I could not only stay in the race, I could be a star performer.

This obsession fueled me to become a workaholic who tied my sense of self-worth to my performance. It led me to completely lose sight of why I had come into the business world in the first place: to build skills to make a difference. As I became fixated on proving myself and winning the rewards and recognition that validated my competence, I got swept away into Brain 2.0 and then more or less stayed in that mode for the next twelve years. I continued to stay on top of research about the brain and to use that knowledge in a Brain 2.0 manner to push my performance to a higher level. During that time, I stockpiled a long list of achievements and accomplishments, yet never felt at ease with myself because I knew that people around me had a much larger stockpile.

In my early thirties, it started to sink in that no matter how many amazing things were listed on my resume and no matter how many awesome deeds I checked off my bucket list, I still wasn't very happy. By that point, I had gotten an MBA from Wharton, arguably the world's top business school, and become a senior investment director at one of the largest asset management firms in Southeast Asia, which had over a billion dollars in assets under management. I was making more in one year than my parents could make in one decade. Although I had become an example of the American immigrant success story, I felt such an unexplainable emptiness inside me. I really didn't like the person I had become, and I didn't like the people I had to work with. I kept reminding myself that thousands of MBAs were drooling over my job and the career trajectory ahead of me, but all my attempts to hack my mind to love what I did failed to change how I felt.

For a long time, whenever a reward was dangled in front of me—like a bonus, a promotion, a new deal to negotiate, or the recognition of being the first female guest of honor in a yacht race from Hong Kong to Vietnam sponsored by my company (yes, that happened)—I couldn't help chasing it. I got such a high from being good at what I did, from kicking ass and being regarded as a VIP, that even if I didn't love my job, I still felt compelled to be a star at it. Yet by the end of 2010, even those thrills began to wear off. Despite how much I tried to dig into my competitive instincts, I lost my fire. It was like my gas tank remained empty no matter how many times I tried to refuel it. I think what happened was that after I'd proved to myself that I could keep up with my peers in the career race, the race itself began to feel meaningless. No matter how many rewards I won, I seemed to be getting nowhere toward real happiness and fulfillment. It was like some part of me figured out that what I thought was a race was really a hamster wheel, and this part wouldn't let the rest of me go back to pretending I didn't know I was really staying in place.

Once again, I turned to neuroscience to understand what was going on. What I learned (which I will explain in a moment) confirmed that the path I was on was pointing toward an emotional dead end. I concluded that the only way out of this feeling of futility was to find meaning and purpose. It was finally time for me to pursue my initial aspirations after college to make a difference for young people growing up in circumstances similar to what I'd faced. But I still didn't know how I could do that. I would ultimately decide to leave my finance career behind so I could give myself time to meditate on how to fulfill those aspirations. Yet, before I could do so, I had to grow tired of chasing every carrot dangled in front of me just because it was there.

How "Carrots" and "Sticks" Work in the Brain

Scientists have found that all creatures, including the simplest single-celled organisms such as the amoeba, share two primal instincts: approach food and other necessities for survival, and avoid danger. That led evolutionary biologists to theorize that positive and negative affect evolved to reinforce these instincts among organisms capable of sentience (meaning they can feel emotions and perceive sensations). Animals feel positive emotions and sensory pleasure as a "reward" for activities that promote their own survival and the survival of their species, such as eating, drinking, and having sex, so they will go out and do more of those activities. Animals feel negative emotions and pain when there is a threat to their survival and when they are wounded or vulnerable, so they take defensive measures to protect themselves and heal.

Human beings are no exception to this instinctive pattern. The fact that emotions serve an evolutionary purpose means that positive emotions are not necessarily good and that negative emotions are not necessarily bad in and of themselves. Positive affect is a signal for us to approach what we anticipate will be positive or pleasant experiences, and negative affect is a signal for us to avoid what we anticipate will be negative or unpleasant experiences. We have to keep in mind, however, that these signals are not always accurate or reliable. We can also influence a person's behavior by priming (intentionally activating) either their "approach" system or their "avoidance" system before exposing them to a task. In one of my favorite books on behavioral economics, *Thinking, Fast and Slow*, Daniel Kahneman describes a study on how priming affects a person's decision to accept a small but risky gamble. If people are primed to think about it as a slim chance of a financial gain, they are more likely to accept the gamble; in contrast, if people are primed to think about it as high risk of financial loss, they are more likely to decline.[2]

Depending on our life circumstances, if we grow up in families that have more resources for growth and positive experiences, we'll exercise "approach" circuits more. If we are faced with scarcity and have learned many negative lessons from painful losses, we'll exercise "avoidance" circuits more. Based on inborn traits and the impact of life experience on the brain, people tend to develop a "preference" for responding to approach signals or avoidance signals, just like being right-handed or left-handed. This is why some people respond more to carrots and others respond more to sticks. People with approach as a dominant motivation tend to take more action when presented with opportunities to grow and gain resources. They respond more quickly and with more energy to motivational messages that emphasize learning new things and experiencing new situations, winning prizes, earning more income, getting promoted, etc. People with avoidance as a dominant motivation system tend to take more action to prevent losses and painful experiences. They respond more quickly and with more energy to prevent bad outcomes like not letting their health worsen, not losing their income, not letting their quality of life worsen, etc. They are more driven by fear of loss and discomfort than by desire for gain and pleasure, and thus do more to maintain status than to improve status.[3]

Brain imaging studies show that people with a dominant approach motivation have higher activation in the left prefrontal cortex, while people with a dominant avoidance motivation have higher activation in the right prefrontal cortex.[4] That means that while two people could be doing the same thing, their brain activation could show that they are actually doing it for different motivations. For example, take two students applying to an elite college: one may see it as a life-changing opportunity to learn and grow, while the other feels an obligation to apply because his or her parents and grandparents went there, and the family would be upset if the student doesn't carry on the family legacy.[5]

You might have guessed by now that the avoidance system is connected to Brain 1.0 and the approach system is connected to Brain 2.0. When Brain 1.0 gets activated, it reduces activity in the left prefrontal cortex, emptying the gas tank, so only the braking system functions. When Brain 2.0 gets activated, it reduces activity in the right prefrontal cortex, taking the braking system offline so our impulses and risk-taking urges go unchecked. This is how the activation of Brain 1.0 and Brain 2.0 can impair different hemispheric functions of Brain 3.0.

My theory is that people generally shift back and forth between approach and avoidance. Stage of life can also dictate which system is more dominant. For instance, during adolescence, Brain 2.0 develops rapidly, which manifests as a strong tendency for teenagers to seek out and engage in risky behaviors. (I will come back to this topic shortly.) It is possible that people who are raising families tend to have a stronger braking system because they are very concerned about taking risks that could negatively affect their children. But their hypervigilance can lead to an overactivation of Brain 1.0, which puts them in such a state of constant worry and dread that they lose their ability to enjoy life.

We've already talked about Brain 1.0 in detail, so let's turn our attention to Brain 2.0. This pattern of brain activation is essential to the survival of the human species because it prompts us to seek food and water, find a mate, and expend lots of energy to better our situations. If Brain 2.0 did not give us a feeling of anticipation and pleasure, we might not bother to do hard work to ensure our survival and growth—why would we take any risks that put our lives in danger if we didn't feel internally rewarded for doing so? In cases where the underlying structures of Brain 2.0 are severely damaged, people lose their sense of drive and motivation. They don't bother getting out of bed. They don't even search for food when they are hungry. They only eat when food is placed in front of them.[6] Therefore, Brain 2.0 is a fundamental part of being human and getting stuff done.

However, getting stuck in Brain 2.0 mode can also cause a lot of problems. When Brain 2.0 is overactivated, we can get locked into tunnel vision about getting one specific thing to the point where nothing else seems to matter. Without that one thing, we feel like life is not worth living. We become so anxious and obsessed about getting it that we compulsively do whatever it takes to achieve our goal, even if it hurts the people we care about and causes large-scale destruction. If we do get it, we may experience a euphoric high, but the craving is only temporarily satisfied. Once the high wears off, we find ourselves in need of another fix. This is essentially what happens when we develop addictions that destabilize our brain chemistry. Yet our capacity for addiction is not confined only to substances like sugar, alcohol, or cocaine. We can also become addicted to socially constructed rewards, like praise, status, achievement, and wealth.

I have now learned from life experience that people who seem like winners in a Brain 2.0 paradigm aren't necessarily happy, secure, or healthy. The truth is that letting Brain 2.0 run wild can result in lives of quiet desperation. I've met many successful people who have high blood pressure and insomnia and feel like they always have to overcompensate to prove themselves. Many are disconnected inside and thus don't feel a sense of greater meaning or purpose in their lives. They just focus on getting through their overcrammed schedules and long checklists of tasks, and then when they're finished, celebrate by in-dulging to excess in some form of hedonistic gratification. Too often, this is what the motto "work hard, play hard" leads to if we don't make conscious choices about what we work toward. This world is full of eye-opening stories of incredibly successful people who still struggle inside and become dependent on antidepressants, alcohol, drugs, and other forms of self-medication to fill an empty void inside that no amount of achievement, wealth, or status can fill.

For example, Michael Phelps experienced this crisis shortly after becoming the most decorated Olympic athlete in history at the London Olympics in 2012. The world naively assumed that this level of achievement would lead to great happiness and fulfillment, but in the fall of 2014, Phelps was arrested for drunk driving. His loved ones intervened as it became clear that he was self-destructive and didn't want to live anymore. This unraveling demonstrated that Phelps, as a swimming prodigy who had turned himself into a gold medal–winning machine, had not been prepared for how success, fame, and an extravagant lifestyle overactivate Brain 2.0 and destabilize the dopamine system. After Phelps retired from swimming following the London Olympics, he increasingly turned to gambling and drinking to keep Brain 2.0 stimulated.

In fact, when he was arrested for DUI, he had just left the Horseshoe Casino in Baltimore.

When he entered a rehab facility called the Meadows, he began to unpack how he had come to this point in his life. In an interview, he shared his insights: "For a long time, I saw myself as the athlete that I was, but not as a human being." There, at the Meadows, he finally connected to his own humanity. He shared, "I would be in sessions with complete strangers, who know exactly who I am, but they don't respect me for things I've done, but instead for who I am as a human being. I found myself feeling happier and happier. And in my group, we formed a family. We all wanted to see each other succeed. It was a new experience for me. It was tough. But it was great."[7] Essentially, what Phelps is describing is how the group support and his own soul-searching reflections at the Meadows enabled him to shift into Brain 3.0, and thus become more integrated and whole. In rewiring his brain to feel true happiness, he tapped into a renewed sense of purpose and joy and reconnected to his love of swimming. Then in the 2016 Rio Olympics, Phelps demonstrated that his shift into Brain 3.0 enabled him to become an even greater athlete and human being.

To a much less extreme degree than Michael Phelps, the anticlimax and emptiness I experienced after achieving what I defined as success also led me to self-medication and self-destructive patterns. When I first got into mind-hacking in college, my focus was mainly on finding ways to come out of Brain 1.0, but I had not yet learned how to bring myself out of Brain 2.0. In many ways, shifting out of Brain 2.0 is much more challenging than shifting out of Brain 1.0. No one likes being in Brain 1.0 since it usually feels miserable, so people are naturally motivated to come out of it. In contrast, we often feel high when Brain 2.0 gets activated, so we want to stay high for as long as possible. Actually, we want the high to last forever. Unfortunately, a high eventually ends with a crash at some point. After seeing this pattern play out so many times, I started to wonder: why do things that feel so good often turn out to be so bad for me? To unpack this, I had to dig deeper into the brain.

Dopamine and the Reward System

Reptiles, birds, and mammals all have analogous brain structures that are involved in approach and avoidance: the basal ganglia and the amygdala, respectively. The amygdala and the basal ganglia are located near each other in the brain, in an area called the limbic system (the word "limbic" means border or transition, which is how scientists decided to name the structures under the cerebral cortex and above the brain stem).

The "basal ganglia" is the name given to a large area of the limbic system at the base of the forebrain that is composed of several clusters of densely packed neurons called nuclei. These clusters are involved in movement, learning and habit formation, reward seeking, and addiction and compulsive disorders. It's important to note that several of these clusters produce the neurotransmitter dopamine. Since dopamine is best known for its role in reward seeking and addiction, researchers often refer to the dopamine system (which includes all the areas involved in making dopamine and the neural circuits that use dopamine as a neurotransmitter) interchangeably as the "reward system."[8]

What is less well-known is how important dopamine is to everyday healthy functioning because it plays a crucial role in movement and coordination, drive and motivation, attention, learning, and working memory, as well as reward anticipation, reward-seeking behavior, and addiction. Without sufficient levels of dopamine in the brain, it is much harder to learn, feel motivation, and move and take action toward any goals. When there is a malfunction in the dopamine system, people become more vulnerable to addiction, compulsive disorders, anxiety, or manic episodes. In addition, movement disorders such as Parkinson's disease are linked to the degeneration of the dopamine-producing cells in the basal ganglia.[9] Tourette syndrome, on the other hand, is linked to excessive dopamine.[10]

Scientists discovered the reward system in the 1950s, while using an electrode to probe the brain of a rat. When they stuck the electrode in a part of the brain whose functioning they didn't yet understand, they noticed a very peculiar reaction by the rat. It was as if the rat blissed out. Then it kept returning to the exact place where it had gotten stimulated, anticipating more of the same kind of stimulation. The researchers had not seen a rat behave that way before, so they designed an experiment where the rat could press a lever to activate the electrode. One of the rats in the experiment pressed the lever about 7,500 times over a twelve-hour period, refusing food, water, and even sex to repeatedly press the lever.[11] Intrigued, the scientists tried another experiment in which they electrocuted the floor underneath the lever to see if that would deter the rat. It didn't. The rat compulsively pressed the lever until its feet burned.[12] Scientists later realized that the area they were stimulating was very sensitive to the neurotransmitter dopamine, which was how they came to refer to all the parts of the brain that produce this type of reward-seeking response when stimulated as the dopamine system or the reward system.

Initially, the scientists mistakenly attributed the rat's behavior to pleasure and bliss. Later, when similar experiments were repeated with human beings, the subjects exhibited the same compulsive behavior, repeatedly stimulating the electrode while neglecting personal hygiene and social commitments. Researchers learned from interviewing these subjects that they were experiencing intense desire and craving rather than happiness. One of them explained that the feeling was like she was about to get what she wanted but it never actually came—it was like an urgent, overpowering feeling of anticipation.[13] Researchers have since confirmed that dopamine does not necessarily make people feel good—it often makes people feel agitated, tense, restless, and anxious so that they feel compelled to take action to alleviate their agitation. Brain imaging studies have found that the reward system is highly activated in people suffering from addiction and obsessive-compulsive disorders.[14]

Scientists studying childhood brain development have found that while the reward system is present in babies, its development accelerates during adolescence. According to Laurence Steinberg, one of the world's leading experts on adolescence, puberty triggers a dramatic increase in the concentration of dopamine receptors, which renders the reward system more easily activated. In fact, the nucleus accumbens (a structure in the basal ganglia that is one of the most intensely activated areas of the brain in the experience of pleasure) actually gets bigger in size as we grow from childhood into adolescence, and then shrinks as we grow into adulthood. This makes it much easier for a teenager to experience a rush, thrill, or high from life's simple pleasures. It also means teenagers have a very hard time controlling their urges and impulses because it feels extremely good to give in to them.[15] As a result of these brain changes, it is physiologically normal for teenagers to spend a lot of time in the state of Brain 2.0.

The development of Brain 2.0 during adolescence also "unleashes" many primal mammalian social instincts, and whenever we follow these instincts, we are rewarded with a squirt of dopamine. These internal rewards drive us not only to want to be part of a tribe, but to crave being a valued member of the tribe. According to David Rock, author of *Your Brain at Work*, having more status and influence corresponds to having higher levels of "feel-good" neurotransmitters, such as dopamine and serotonin, in your brain, while lower status corresponds to lower levels of dopamine and serotonin.[16]

Brain 2.0 enables us to navigate our social environment and makes us instinctually aware of implicit networks of influence and power. Because of Brain 2.0, we can't help but track the construction and deconstruction of in groups and out groups, better known as cliques. This heightened sensitivity to social validation, popularity, and status can cause teens to become obsessed with appearances as a means of social signaling, thus becoming fixated on having the right clothes, the right shoes, the right hair, the right bag, the right car, etc. Teens are particularly vulnerable to social contagion because seeing whatever their peers have can automatically activate their dopamine circuits so that they want to have it too. Because of these changes in their brains, teenagers can't help being superaware of social cues signaling who or what is cool, and who or what is not cool. And fortunately, once the window of adolescence passes, our sensitivity to the regard of our peers also decreases.

As one can imagine, there are downsides to having one's self-worth and well-being dependent on external validation, especially from other adolescents, who as a group are going through major bouts of hormone-fueled mood swings that they don't really understand or control. An overactivated reward system creates a tendency toward tunnel vision, which is often experienced as all-or-nothing thinking. When teens get what they want, they experience a huge surge of dopamine and other endorphins. It's a natural high that makes young people feel like they are invincible. Conversely, when what they are fixated on doesn't turn out the way they want—for example, if they learn that their crush does not like them back, if they don't get invited to a party, or if they don't make a team—they tend to feel like their life is over and collapse into Brain 1.0. As they mature, people learn to get out of tunnel vision and see the bigger picture, but for many perfectionistic high achievers, all-or-nothing thinking can become ingrained as a lifelong habit that makes them miserable, even when most of their life is going well.

Interestingly, fMRI studies reveal that approach and avoidance are more like two sides of the same coin. Physiologically, the amygdala and basal ganglia are interconnected through a number of pathways, which allow Brain 2.0 to trigger Brain 1.0 and vice versa. When people are fixated on getting a specific outcome that involves factors they don't control and is thus possibly beyond their reach, they get stressed out. Their amygdala gets highly activated, putting them into a freeze-flight-fight state. Conversely, when people get fixated on avoiding an outcome they perceive as terrible, the amygdala puts them into freeze-flight-fight mode, and neural circuits in Brain 2.0 also activate to coordinate motor function to take immediate action. If they succeed at preventing that negative outcome, they also get a surge of dopamine. In either approach or avoidance mode, when the outcome does not turn out the way people want, they often find relief in food, alcohol, or other drugs to gratify their reward circuits with a chemically induced dopamine fix.

Our tendency to swing between Brain 2.0 and Brain 1.0 creates huge emotional roller-coaster rides until we develop enough Brain 3.0 to break the pattern. Here is where it is helpful to keep in mind that the areas of the brain that constitute Brain 3.0 are the slowest to develop—the frontal lobes only fully mature in our midtwenties. This sequenced development of the brain means that by the time the structures of Brain 3.0 mature, many of us have already hardwired into our autopilot very strong neural circuits that correspond to habitual patterns driven by our Inner Godzilla and Inner Teen Wolf. Therefore, we need to invest conscious effort to become aware of these habits and patterns. By doing that, we build and strengthen neural pathways from Brain 3.0 structures into Brain 1.0 and Brain 2.0 structures, enabling us to calm and deactivate these neural circuits.

Although the maturation of the frontal lobes gives adults more capacity to "quiet" Brain 2.0, it doesn't mean people make use of that capacity. When things feel good, we get much more pleasure from just going along with whatever our dopamine circuits reward us to do. It takes a long time before we realize through experience that getting a dopamine fix in response to a craving is not as fulfilling as the deep, lasting happiness we can get in Brain 3.0. This is why using and strengthening the neural pathways that allow Brain 3.0 to tame Brain 2.0 is a lifelong endeavor. No matter how strong Brain 3.0 is, it is always a challenge to resist a tempting dopamine surge offered by Brain 2.0.

The built-in challenge is that when we are hijacked by the Inner Teen Wolf in Brain 2.0, there is reduced blood flow to the frontal lobes, where self-control and executive functioning reside. In particular, when the inhibitory function of the right PFC gets deactivated, we become much more vulnerable to getting swept away by impulses and addictions that undermine our long-term health and well-being. In our consumer culture, companies have developed sophisticated techniques to strongly activate our reward circuits to get people to quickly buy their products and services.

Therefore, to protect ourselves from being unwittingly manipulated, to break unhealthy habits, and to reduce addictive tendencies, it is essential to learn how to shift out of Brain 2.0 into Brain 3.0. The challenge is that it requires breaking out of how we've been conditioned to behave since we were babies.

Conditioned to "Succeed"

Ivan Pavlov, a Russian scientist who received the Nobel Prize for medicine in 1904, is best known for discovering conditioned learning by accident, while he was studying the physiology of digestion in dogs (for which he won the prize). To study digestion, he developed a way to measure and track the levels of saliva that dogs secreted. One day in the lab, he noticed that the dogs were salivating whenever the person who brought them food entered the room. That inspired him to set up an experiment to see if the dogs could be conditioned to salivate using a different neutral stimulus. He decided to ring a bell whenever a dog was being fed. Then later, when he rang the bell when no food was being served, the dog still salivated. This showed that the dog had unconsciously learned to associate the bell with food. With this finding, Pavlov opened up a new field of research on what has since become known as classical conditioning theory.[17] It didn't take long for scientists to prove that conditional learning also applies to human beings and that it explains how the use of rewards and punishment as reinforcement can have a long-term impact on behavior.

I believe Pavlovian conditioning explains a common phenomenon called code-switching, which is when people who speak multiple languages or dialects switch how they speak when they are around different groups. I unconsciously internalized code-switching to navigate between the very different worlds I was a part of. I speak with a perfectly fluent American accent with my colleagues and friends, but when I speak with my parents, aunts, and uncles, a heavy Chinglish accent comes out of my mouth that I don't seem to consciously control. The weird thing is that I can't make myself do the Chinglish accent when I am with my friends or colleagues. It only comes out when I'm around my older relatives.

After I learned about code-switching, I started to realize that it didn't just apply to how I talked. I had actually developed different personas when I was with my older relatives, with close friends, and with colleagues. I was a completely different person in an executive boardroom than I was walking around my neighborhood in Philadelphia. I could be so different in how I talked, thought, and behaved, that whenever people from my different social worlds collided, I got confused about what to do or say—I would be embarrassedly frozen with anxiety and discomfort until the social worlds uncollided.

It's taken a lot of reflection to understand that these paralyzed responses are caused by a code-switching breakdown. I believe what happened was that in Brain 2.0, my Inner Teen Wolf would intuitively figure out what behaviors and qualities got positive social reinforcement and what got negative social reinforcement, and would then create an appropriate "persona" for me to adopt in that specific social sphere to maximize getting the most social carrots and minimize getting hit with social sticks. Whatever qualities my parents cherished, I naturally shined with them at home. Whatever qualities the culture at school rewarded and admired, that was what I shined with my classmates and teachers. My subconscious mind seemed to figure out how to adapt to each group and hide the part of me that didn't fit the mold. Thus, whenever my various social worlds collided in a way that one world's carrots was another world's sticks, my Inner Teen Wolf could not act.

This was a scary thing to discover about myself because it meant that I was mindlessly letting whoever I was with dictate how I thought, how I talked, and what I did—it meant that I didn't have a stable sense of self because I was a social chameleon. After I became aware of this pattern in college, for a long time, I couldn't make up my mind as to whether it was a bad thing. Code-switching got me a lot of carrots and enabled me to run in very diverse social cliques on campus. Wasn't having a lot of carrots and many diverse groups of friends a good thing? It would take another decade for me to learn the hard way that this extreme degree of code-switching would lead me to become an unanchored, miserably successful workaholic.

After I started working, I instinctively built a new persona that exemplified what Monitor considered ideal. It just happened on its own—I don't recall ever consciously deciding to do it. All my colleagues had to do was reward a behavior, and next thing I knew, I was doing it. Since project teams spent so much time together, many teams gelled into temporary pseudo-families and socialized together. Often, team members talked about things that were going on (or not going on) in their personal lives and shared about their families and friends. These were the conversations where I was especially quiet and did not contribute, but interestingly, no one seemed to notice. People seemed to project onto me an upbringing that was similar to theirs.

After some time, I started to observe that most of my colleagues and I must have shared a habit of overdelivering to compensate for our fears and insecurities. That was the only way to explain why almost every one of us stayed at work until close to midnight every day and had no healthy boundaries between our work and personal lives. We repeatedly canceled weekend plans with friends and family in order to respond to last-minute requests from managers to run extra analyses. We were typically on standby twenty-four hours a day, seven days a week. Anyone who wasn't willing to do this was not a team player. If you let the team down, it would be reflected in your performance review and it would affect your bonus. There was no way my Inner Teen Wolf could abandon the wolf pack.

Working as a management consultant was brutal. The typical number of hours my teammates and I had to put in every week ranged from 80 to 120. Whenever I wasn't sleeping, I was basically in work mode. In order to establish credibility and to be regarded as a valuable contributor on my teams, I always put in extra time at the beginning of every project to get smart on the client's industry. My job involved running very complicated, detailed analyses and researching thousands of tiny inputs to build into complex financial models. As we approached the dates when we had to deliver presentations to the client's board of directors, we often had to pull many all-nighters to synthesize the analyses from the various work streams and polish the final PowerPoint presentation. It was tedious and grueling. Yet, no matter how much personal sacrifice was demanded, I couldn't let the firm down. I had to keep performing. I was addicted to the carrots. Most of all, I was addicted to being a person who could win carrots.

After a bumpy first year in which I slowly found my footing, my managers started to tell me they considered me a strong performer. Even clients were raving about my work and giving me hints that they would gladly offer me a job to join their teams. Seeing that I was not at any risk of getting fired allowed me to relax just a little bit. I learned that a solid reputation at the firm created a positive spiral. Managers started to request you specifically for their best projects. People offered more mentorship so you would choose their projects in the future. Given that my entire life revolved around work, I was relieved that all my efforts and sacrifices were not entirely in vain. My Inner Teen Wolf was very happy with this validation and continued to weave a persona of "star management consultant" that I compulsively played.

I also noticed that my Inner Teen Wolf had a way of unconsciously absorbing the desires, values, and habits of whoever I spent time with. It was like being infected by their brain. For instance, I don't normally consider myself a competitive person; yet when I interacted with hypercompetitive people, I would feel a strong urge to kick ass. I don't consider myself a materialistic person; yet when I saw the lifestyle that my colleagues had, I wanted to live that way too. When I saw that almost all my female friends and colleagues had a Coach handbag, I also had to get one. I don't consider myself a tech geek; yet when I saw that all my colleagues and friends had fancy cell phones, I had to get one too. It wasn't long before what was normal for my friends and colleagues became what was normal for me. I found myself feeling entitled to whatever they felt entitled to. I got used to eating at expensive restaurants, taking car service, drinking fine wine, eating foie gras, etc. Thanks to my Inner Teen Wolf, I assimilated seamlessly into Monitor's culture.

Since it was expected that everyone in management consulting get an MBA, in a few years, I found myself applying for business school. When I got accepted by the Wharton School, the management team at Monitor was so pleased that they gave me a return offer that would cover my tuition to entice me to come back to the firm after graduation. It was incredible to reflect and remember that in January 2001, I had entered Monitor as a clueless art major from a low-income family, desperately worried about making a mistake and getting fired, and in June 2004, I was leaving as a star performer in whom they had decided to make a long-term investment.

While at Wharton, I figured out that everyone in my class was salivating over jobs in private equity and venture capital, so I caught the bug too, even though I didn't yet know anything about what these career tracks involved. I took a class on venture capital and went on a student-organized trek to visit venture capital firms in Silicon Valley. On the trek, it seemed nearly impossible for a woman of color without any connections to break into this field except by first building a start-up that grew into a successful multi-billion-dollar giant. I decided I would be better off being more sensible about a career with better odds. So I looked at my bucket list and saw that I really wanted the opportunity to work abroad in Asia and reconnect with my roots. Finishing business school seemed to be the perfect window of opportunity to try to make that happen because I was still single. Else, if I waited, I could end up married with kids and never get to fulfill this dream. So I told the people at Monitor I would only come back if they could transfer me to Asia. They agreed and relocated me to their Beijing office in September 2006.

I loved being in Asia, but Monitor Asia was nothing like the firm I knew and loved in the United States. I realized there was very little oversight from the head office in Cambridge because the global leadership team seemed to have no idea that what took place in their Asia division diverged greatly from what would have been considered acceptable in the North American offices. After I started speaking up about my concerns in an effort to improve the situation, I saw to my disappointment that no one in a leadership position in Asia or North America was interested in taking action. So in spring 2007, when a recruiter contacted me about a role as the director of private equity investments for the largest asset management company in Vietnam, I moved on without looking back.

At that time, Vietnam was one of the hottest emerging markets. Coincidentally, Wharton decided to hold its upcoming Global Alumni Forum in Ho Chi Minh City in 2008 and my company had already committed to be one of the lead sponsors. Soon after I arrived, I found myself asked not only to help organize but also to emcee the event. It was incredible that I, a person who had no idea what private equity was before going to Wharton, stumbled into a very rare and coveted role to run private equity investments for the largest asset management firm in the country where I was born. My classmates and all the alumni who came to the forum were clearly impressed. Although I was only in my late twenties, I was not only the most senior woman among the investment staff at my company, but I was also one of the most senior women in the finance industry in Vietnam because the economy was so young. This type of distinction could not have been possible in larger and more developed markets like the United States or even in China. Just by virtue of being in a frontier market, I was a trailblazer.

As I was in my late twenties, my brain's frontal lobes had fully matured and I should have had much more capacity to be in Brain 3.0. However, I really wasn't putting it to good use. In a way, it was like having arm muscles but no upper body strength because I never exercised them. I had become conditioned by and habituated to living in a microcosm of elite overachievers—a world I could never have imagined as a child. I was having such a good ride in Brain 2.0, why bother changing?

Tiring of the Inner Teen Wolf

Little did I know that the four years I lived in Vietnam would be so unsettling, bewildering, and agitating that it would force me to become disgusted with living in Inner Teen Wolf mode. I would become so fed up with that way of life that I would completely turn away from that trajectory and instead set off on a life-changing adventure in uncharted terrain to learn how to develop the "muscles" of Brain 3.0 and search for a deeper and more meaningful happiness.

There's no better place than a developing country to see human nature in its rawest form. When there is limited rule of law and no social expectations to behave ethically, there is nothing to hold back the Inner Teen Wolf or the Inner Godzilla. In this type of environment, everywhere you look, you get a close-up view of the dark side of society: people cut out of the economy, having to make a living in the black market or by selling their bodies. I could not come to terms with how I saw human beings exploit each other's suffering and desperation. I could not come to terms with learning how human beings use their power and authority to bully others and extort kickbacks and other perks. I could not come to terms with living in an environment where integrity didn't matter.

Over the four years I lived in Vietnam, it took such a hefty toll on my own mental and emotional well-being that I could not have faith that many of the people with whom I had to interact for business purposes would act ethically. During this period of time, I met so many people who openly cheated on their wives or husbands that it eroded my faith in the sanctity of marriage. It grossed me out when married men hit on me.

At social gatherings, it wore me down that people, by sharing the latest rumors in the market, would constantly remind me that my firm had a terrible reputation. In the beginning, I often responded by explaining that it could just be malicious gossip, because there was no evidence of any wrongdoing, but my closest friends kept warning me, "where there is smoke, there is fire." The problem was, these issues weren't isolated to my firm. All the investment

firms in Vietnam had similar challenges. As someone who took pride in having integrity and values, it was impossible to have peace of mind about building a long-term career in Vietnam. This forced me to realize that it would not
be possible for me to ever feel fulfilled if I remained in Vietnam.

Despite the trappings of success and my growing pile of carrots, I had to
acknowledge that my life had started to feel empty and purposeless. My attempts to use techniques from positive psychology to focus on all the positive
aspects of my job and savor positive experiences couldn't change this feeling.*
After I paid off all my student loans, these feelings became stronger. All my
tools to try to stay engaged with my job and find motivation to come to work
became less and less effective over time. I began to recklessly slip all the way
into Brain 2.0 to escape. I started to party, splurge on expensive items, drink,
and get high more and more to pass the time. But all those efforts to distract
myself and self-medicate were fruitless. Nothing seemed to fill up the sense
of emptiness that was growing inside of me. Then one day I looked in the
mirror and didn't recognize myself. What I was seeing was not who I wanted
to be. I realized I could never find real happiness living like this in Brain 2.0.
It was a wake-up call that I would keep losing touch with myself if I stayed.

From there, I went into a period of introspection that revealed to me that
the main reason my heart had brought me first to China and then to Vietnam
was to form a visceral connection with my roots. Growing up in the United
States, I felt too unanchored from my family's heritage. I needed to live in Asia
to trace my family's history and to put our experiences into the larger context
of geopolitical history. I needed to walk on the ground that my parents and
ancestors walked and breathe in the air they breathed. Once I had completed
this emotional healing process, my heart started telling me it was time to move
on. By the end of 2010, I finally stopped resisting the message and accepted
that it was time to start planning to leave Vietnam.

But my problem was that I didn't know what to move on to. Since college,
I had always had this vague idea that one day I wanted to start a nonprofit to
help students from disadvantaged backgrounds, but I also had this notion that

---

* According to the Positive Psychology Center at the University of Pennsylvania, "Positive Psychology is the scientific study of the strengths that enable individuals and communities to thrive. The field
is founded on the belief that people want to lead meaningful and fulfilling lives, to cultivate what is
best within themselves, and to enhance their experiences of love, work, and play." To learn more, see
https://ppc.sas.upenn.edu.

I would have to get rich first before I could effectively do anything philanthropic. I wasn't yet rich. I was earning good income, but since the financial crisis, firms across the industry, including mine, had cut back significantly on bonuses and compensation. It seemed like it could take a long, long time before I could build up enough wealth to do what I wanted.

Instead, I started thinking about a smaller goal. I started saving up as much money as I could for what I called a "fun-employment fund." I figured I might need to buy the time and freedom to discover what my life purpose was. I hoped that one or two years' worth of salary would be enough to get some answers. I didn't really have an idea of whether or when I would ever put this fund to use, or how I would use it; I only sensed that it was something I wanted to have as soon as possible, just in case. To be honest, I wondered whether something as vague and intangible as a life purpose could ever really materialize. It sounded like a mythic, idealistic, and romantic notion—like a daydream. I thought of myself as too much of a realist to chase daydreams. I didn't anticipate that the universe was about to show me otherwise.

4.

A Skeptic Stumbles on "Enlightenment"

Whence come I and whither go I? That is the great unfathomable question, the same for every one of us. Science has no answer to it.
—ERWIN SCHRÖDINGER, Nobel laureate in physics

We are in the position of a little child, entering a huge library whose walls are covered to the ceiling with books in many different tongues. The child knows that someone must have written those books. It does not know who or how. . . . The child notes a definite plan in the arrangement of the books, a mysterious order, which it does not comprehend, but only dimly suspects. That, it seems to me, is the attitude of the human mind, even the greatest and most cultured, toward God. We see a universe marvelously arranged, obeying certain laws, but we understand the laws only dimly. Our limited minds cannot grasp the mysterious force that sways the constellations.
—ALBERT EINSTEIN[1]

Our soul is in its nature entirely independent of body, and in consequence . . . it is not liable to die with it. And then, inasmuch as we observe no other causes capable of destroying it, we are naturally inclined to judge that it is immortal.
—RENÉ DESCARTES[2]

SOMETIMES I WONDER whether it would be more accurate to say that I found Calm Clarity or that Calm Clarity found me. This is because, in hindsight, it was as if a series of unusual dominoes fell into place to prompt me to leave my investment career to go on a soul-searching quest through India. In telling this story, I need to explain that throughout my education and the conventional phase of my career, I was the last person I could have ever imagined going on a spiritual journey.

To help you appreciate the degree of change that was involved, I have to point out that until my early thirties, I was neither religious nor spiritual—it could probably be argued that my views leaned toward the antireligious end of the spectrum. My personal philosophy could be best described as scientific materialism (or scientism) because I considered science to be the only reliable, credible source of knowledge. Since I was a child, I have always been naturally oriented toward science and empirical evidence, and have always preferred reading nonfiction over fiction.* In fact, my preference for science and for rational analysis was so strong that I was genuinely baffled by how people could believe in things like horoscopes, fortune-telling, and superstitions that have no scientific merit.

Further, growing up in poverty with scarce resources conditioned me to be pragmatic about where I focused time and energy. My attention naturally veered toward the more immediate aspects of life—the bills I had to pay, the tasks that needed to be done, and the deadlines that had to be met. The less tangible something seemed, the less clear the payoff was for the effort involved, and the less attention I gave it. Thus, I paid very little attention to religious or spiritual matters (and anything that seemed fluffy).

This lack of attention was not a result of lack of exposure. I was baptized Catholic and attended parochial school from kindergarten until eighth grade. I was also exposed to Buddhism and the legends around the Buddha through my extended family and the Southeast Asian refugee community of which we were a part. As far as I could see, both religions involved going through the motions of performing rituals that didn't make any obvious sense. It was also frustrating that whenever I asked questions of my parents, grandparents, aunts, and uncles, I learned that they did these things because that was what they were taught to do by their parents, but they didn't know why either. I eventually concluded that these rituals were done just to honor tradition. As a child, I never felt any clear sense of spirituality and didn't much care for tradition, so as I grew older (and graduated from my Catholic elementary school), I opted out of going to church and to temple except for special occasions like weddings and funerals.

---

* For people who are familiar with the Myers-Briggs Type Indicator, my type is INTJ (Introversion Intuition Thinking Judging), a profile that is particularly rational and data driven.

When I was younger, I was so restless and fidgety that meditation had no appeal to me whatsoever. I could never sit cross-legged for more than a few minutes before my legs fell asleep, and the pins-and-needles feeling that arose would make me even more restless and miserable. The idea of doing nothing but sitting still, cross-legged on the floor, to become enlightened like the Buddha seemed ridiculous to me, especially because I found the concept of enlightenment too elusive. As a refugee who started life in poverty and felt the stinging stigma of being on welfare, I had no interest in becoming like the Buddha, a prince who renounced his kingdom, wife, and child and spent the rest of his life begging for food. My biggest priority was to be financially successful so I could enable my family to have a better quality of life.

Looking back, I think I turned against religion when the sex abuse scandal within the Catholic Church became front-page news. I was so disgusted and outraged by how church leaders covered up and protected pedophilic priests that I got fed up with organized religion in general. As a headstrong young person, I swung to an extreme position, embracing a technocratic worldview that regarded religion as an antiquated and anachronistic holdover from an earlier, prescientific era. Basically, I threw the baby out with the bathwater.

To clarify, I think the phrase "agnostic indifference" best captures my previous position toward religion rather than the word "atheism." Being an atheist involves actively denying the existence of God. I couldn't take that stand as I recognized that it was impossible to either prove or disprove God's existence. However, whether or not God existed, I saw no sign that God took an active role in my life or in human affairs. God seemed indifferent. Therefore, why does it matter if there is a God? What is the point of investing all that energy connecting with a God who isn't involved in our lives? I concluded that human beings were on our own to survive on our planet.

Besides, as someone who witnessed and experienced so much trauma and violence and lost an entire branch of my family to the large-scale genocide that took place in Cambodia, I found it much easier to assume that we lived on a God-forsaken planet than to try to rationalize how a "God" who is benevolent, loving, and just could let so many fucked-up things happen to innocent people on this Earth. How could an omnipotent, omniscient, loving "Creator" make this world so full of suffering, pain, tragedy, and conflict, then watch the human beings he or she made in his or her image fail to rise above the darkness and brokenness of our society, and then condemn masses of people to eternal damnation for not being able to heal our brokenness or to light a path out of the darkness into which we were born? This paradigm for God made no sense at all and I revolted against it.

Little did I know that at this turning point in my life, this worldview was about to be shattered. Ironically, embarking on a journey to understand how meditation practices can change the brain changed my life forever because it led me to discover and acknowledge that whatever consciousness is, it cannot be reduced to activity in the brain. All my assumptions about living in a random universe where there is no overall plan or purpose and no influence from an active "Creator" were about to be overturned.

The series of transformative mystical experiences I was about to undergo forced me to admit that there is much more to the mystery of life, consciousness, and the universe than science can account for. Yet at the same time, these experiences also made me appreciate even further the important role that science can play in shedding light on how *consciousness* (meaning the *soul*) intertwines with the brain and the rest of the body to create sentience—the subjective inner experience of being alive and having a body. These epiphanies would lead me to realize that science and spirituality are two branches of one tree and that we need both to understand why we are here and how to make good use of the relatively brief time we have on this Earth.

Ushered into India

As I shared earlier, my tough childhood conditioned me to keep my nose to the grindstone and to reflexively dismiss religion and spirituality as being irrelevant to modern life. So what made me change my worldview? It started in the summer of 2011 when I received a new job offer and a very generous gift of a free tour of India. From thereon, unpredictable, remarkably synchronistic happenings seemed to shout: you cannot dismiss what is unfolding as random coincidence, you are connected to something much bigger than yourself.

Soon after I made up my mind to leave Vietnam, three things happened simultaneously:

1. My boss made a cost-cutting decision to eliminate my position and gave me a decent severance package.

2. I got a job offer from a billionaire who ran his own social impact investment platform in Singapore.

3. I received an invitation from Mr. Kumar, the chairman of one of the portfolio companies I had invested in, to visit India as his guest.

It was amazing how right at the moment one door closed, a new door opened. The social impact role seemed like a dream job come true, as it involved using my private equity skills to make investments in high-potential social enterprises across developing countries in Asia. It was an "exploding" offer (one that must be accepted within a couple days or else expires), so I didn't have time to do any background research on the rather secretive firm and the even more secretive billionaire in charge of it, or to reach out to former employees. Although I knew in my heart that my long-term future was back in the United States so I could be nearer to my aging parents, I also knew that job openings to work in social impact investments in developing countries were very rare. I was handed an extraordinary opportunity to learn about social enterprises in emerging markets, and the experiences and insights I would gain could put me in a much better position to later become a social entrepreneur. But I would be going in blind. I had very little information on the very private billionaire and the team I would be working with. In the end, I decided the risks were worth it, so I accepted the offer.

As part of the transition process, I had to notify the portfolio companies I worked with and resign from all the board seats I held on behalf of the firm in Vietnam. That was when Mr. Kumar, the head of one of these portfolio companies, surprised me by giving me an amazing thank-you and farewell present. Mr. Kumar was the chairman and majority owner of an industrial agriculture company in Southeast Asia, into which I had made a significant investment soon after I arrived in Vietnam. After we closed the deal, I sat in on board meetings. Over the four years Mr. Kumar and I worked together, we built up a strong working relationship and a great respect for each other. When he learned that I had never been to India, Mr. Kumar repeatedly offered to take me there. However, I was concerned about creating a conflict of interest, so I kept declining.

When Mr. Kumar heard I was leaving, he wanted to finally make good on his promise to bring me to India. He offered to personally arrange an all-expenses-paid tour of India for me as his guest and made clear that he would take it as an insult if I didn't accept, as I had no more reason to be concerned about conflicts of interest. I was so touched by the generosity and sincerity of his invitation that I gratefully said yes.

As our business relationship shifted into a friendship, we began to talk more about personal matters. I was taken aback to learn that this successful businessman was also a devout Hindu who credited his success to his very strong spiritual orientation. I had naively assumed that being spiritual and being good at business were diametrically opposed. In hindsight, one of the most precious lessons I got from Mr. Kumar was the realization that this simplistic notion is not true. Spiritual people can be great at building and growing businesses.

Mr. Kumar personally curated the itinerary based on what he wanted to show me about the India he loves and proudly calls home.* He made sure I would visit world-famous historical landmarks, such as the Taj Mahal and the Amer Fort outside Jaipur, and also included lesser-known sites important to locals. The ten-day whirlwind tour he planned took me through New Delhi, Agra, Jaipur, Hyderabad, Bangalore, Mysore, and Shravanabelagola.† I had never heard of the last two places, so if Mr. Kumar hadn't put them on my itinerary, I might have never learned about them—which would have been unfortunate because it was the lasting impact of what I saw at these two sites and the questions that arose from visiting them that would later bring me back to India on a self-directed quest to find answers.

In Mysore, the major tourist attraction was the palace museum. There, I learned that before the modern-day Republic of India became established as a country in 1950, the ruling family of Mysore had earned a reputation throughout India as exemplars of enlightened leadership. In particular, I became fascinated by the last king who served a life term, Krishna Raja Wadiyar IV (1884–1940). According to the palace museum literature and guide, his reign is often described as the Golden Age of Mysore because it was a period of great industrial development and social progress. Krishna Raja Wadiyar IV was highly regarded by his contemporaries, including Mahatma Gandhi, as a

---

* Mr. Kumar's schedule didn't allow him to accompany me, so instead, he assigned his personal assistant to escort me throughout the tour and checked in with me on the phone and also in person where our itineraries crossed.

† Bangalore, Mysore, and Shravanabelagola are located in the state of Karnataka in the southwestern region of India.

saintly philosopher-king who made great strides in alleviating poverty and improving public health, strengthening the education system, and developing the economy. Furthermore, he was such an adept businessman that when he died in 1940, he was one of the wealthiest people in the world, worth more than $400 million (a fortune that would be equivalent to almost $7 billion in 2017).

By this point, I had already visited several palace museums in major cities in India, and they had started to blur together. Nonetheless, the palace of Mysore stood out because it was the only one I saw decorated with esoteric illustrations of yoga positions and teachings. Up until then, I had thought of yoga as a trendy exercise class offered at a gym or yoga studio that involved a lot of stretching and flexibility and wearing tight-fitting clothes made of Lycra. However, what I saw there at the palace made me see yoga differently. These drawings hinted that the Wadiyar legacy of enlightened leadership may have been tied to a deep yoga practice. I could palpably sense that someone who lived in this palace once revered these drawings. There was something sacred about them.

When I did more research afterward, I learned that Krishna Raja Wadiyar IV's grandfather Krishna Raja Wadiyar III (1794–1868) was a devoted yogi who had created an ambitious illustrated treatise, called the Sritattvanidhi ("The Illustrious Treasure of Realities"). Containing instructions and depictions for 122 yoga poses (called asanas), it is considered the first major compendium of asanas in modern history. My guess is that the yoga illustrations I saw at the palace are most likely from or related to the Sritattvanidhi.

Krishna Raja Wadiyar IV must have inherited his grandfather's passion for yoga because he asked Tirumalai Krishnamacharya (1888–1989), who is now widely regarded as the father of modern yoga, to teach the royal family. The king became his primary patron and even set up a yoga school on the palace grounds for Krishnamacharya to run. During that time, Krishnamacharya must have found and studied the Sritattvanidhi, which probably inspired him to become one of the first yogis to make the practice of asanas a core element of a yoga practice (other forms of yoga such as Kriya Yoga don't really emphasize asanas). In 1934, when he published his first book, Yoga Makaranda ("Nectar of Yoga"), he listed the Sritattvanidhi as a source. By far, the largest chapter in the book is on asanas. It provides detailed instructions for forty-two asanas along with photographs of Krishnamacharya and his students demonstrating the poses. The king also funded Krishnamacharya and his students to travel throughout India to give yoga asana demonstrations, which led to the revival of yoga asana across the country. The lineage created by Krishnamacharya would come to be known as Ashtanga yoga, and his top students, such as K. Pattabhi Jois (1915–2009) and B. N. S. Iyengar (1927–present), would go on to become world-famous ambassadors of modern yoga.*

Learning about the Wadiyar dynasty's impressive legacy made me curious to ask: What does "enlightened leadership" mean? What role did yoga play in the lives of these kings, and how did it affect their leadership? Did practicing yoga help them make wise and skillful decisions? Being exposed to their passion for yoga instilled in me a genuine desire to learn more about traditional yoga as it is practiced in India rather than how it is consumed in the United States. However, on this whirlwind trip, there was no time for me to check out any yoga schools in Mysore. (This would happen later, after I moved to Singapore.†)

The next morning, Mr. Kumar arranged for me to visit Shravanabelagola, a small town north of Mysore and west of Bangalore. Mr. Kumar explained on the phone that Shravanabelagola was a special place for local Indians and that he felt I would enjoy seeing it. It is a sacred Jain pilgrimage site, which is home to the Gommateshvara Bahubali, the largest monolithic statue in the world. It sits at the top of Vindhyagiri Hill, and seeing it requires climbing about seven hundred steps barefoot, as wearing shoes or sandals is prohibited on the sacred grounds. I also learned from an online search that in a 2007 poll conducted by the *Times of India* on the "Seven Wonders of India" the Gommateshvara Bahubali topped the list with the most votes from readers.[3]

It was a beautiful clear day and, as someone who loves challenges, I eagerly climbed the steps just to prove that I was fit enough to do so. When I got to

---

* Both attracted large international followings and put Mysore on the map as a major site on the yoga tourist circuit.

† By some odd coincidence, when I later moved into an apartment in Singapore, I discovered there was a yoga studio next door. When I went to check it out, I was stunned to learn it was actually an Ashtanga yoga school run by an American student of the late Pattabhi Jois. Months later, when I decided to undergo yoga teacher training, I followed the recommendation of a friend in Vietnam to contact a specific teacher in Singapore. He turned out to be an Indian student of B. N. S. Iyengar and trained me in that Ashtanga yoga lineage. What intrigues me most about Ashtanga yoga is the fact that many of its famous practitioners have lived (and practiced) well into their nineties, and some, like Krishnamacharya, became centenarians.

the top of Vindhyagiri Hill, I was both stunned and amused to find the nearly sixty-foot statue was of a completely naked Bahubali in the standing meditation posture in which he achieved enlightenment. Given that Indian society still frowns on women wearing shorts or baring shoulders, I couldn't help feeling embarrassed to gaze up at this statue in a crowded public space because it was impossible to look at the statue without seeing his genitalia. The statue depicts snakes crawling on Bahubali's feet and vinelike plants growing around his legs and arms. As someone who is afraid of snakes, I was freaked out by this, but the guide explained that this was meant to illustrate that Bahubali was so still in his meditation that the animals and plants treated him like he was a tree.

From what the guide explained, the story of Bahubali goes like this: His father was Rishabha, a great king who is regarded as the founder of Jainism, an ancient religion that possibly precedes the founding of Buddhism by thousands of years. When Rishabha decided to pursue enlightenment, he abdicated the throne and divided his kingdom among his one hundred sons. After he left, his eldest son, Bharata, demanded that all his brothers turn over their kingdoms to him. Bahubali was the only one who stood up to his older brother and actually defeated him. Nevertheless, after the victory, Bahubali lost interest in worldly affairs and decided to follow in his father's footsteps to seek enlightenment. So he handed over his kingdom to Bharata anyway. Eventually, Bharata would also renounce his kingdom to pursue enlightenment.

Looking out from the top of Vindhyagiri Hill, the guide then pointed me to the top of a nearby hill, where another cluster of ancient temples stood. He explained that they date back over two thousand years and could possibly be the oldest surviving active temples in India. They can be traced to the time of Chandragupta (340–298 BCE), who in 322 BCE established the great Mauryan empire, which was one of the first empires to unite most of present-day India into one state. The guide explained that Chandragupta converted to Jainism and, at the end of his life, he also abdicated his throne in favor of his son Bindusara (the father of Ashoka, who is widely hailed as the greatest king in India's history). As a renunciate, Chandragupta then traveled with his teacher to Shravanabelagola, where he spent the remainder of his life in meditation.

Learning this about Chandragupta made me wonder: how many rulers in India's history abdicated their kingdom to find enlightenment? This seemed to be a recurring theme. Assuming what the guide shared with me about Rishaba, Bahubali, Bharata, and Chandragupta was true, that meant that the Buddha was neither the first nor the last ruler in India's long history to renounce his title and all his possessions to pursue a spiritual quest. Furthermore, the depiction of the enlightenment of Bahubali in the Jain tradition was so strikingly different from the Buddhist iconography with which I was more familiar, that it prompted me to ponder more about what "enlightenment" really is—and was it something I should care about? On the road back to Bangalore, I wondered: Is enlightenment a physiological state that can be objectively evaluated? Could enlightenment be validated by brain scans? Are all forms of enlightenment the same, or are there different types?

This line of thinking led me to ask even more questions that I had never thought about before: In what ways did the Buddha's teachings differ from those of Jainism and Hinduism? How did the story of the Buddha and his teachings eventually become so widespread throughout the world, such that Buddhism became the major cultural religion of China, Tibet, Mongolia, Vietnam, Thailand, Cambodia, Laos, Bhutan, Burma, Sri Lanka, and Japan? Yet in contrast, why did Jainism and Hinduism remain concentrated in the Indian subcontinent? And why and how did Buddhism die out in India, where it was born?

Later that day in Bangalore, I met Mr. Kumar and his wife for dinner. Mr. Kumar was especially pleased to hear that I enjoyed seeing Shravanabelagola because he intentionally wanted to show me hidden gems in India that were not on the typical international tourist itinerary. In time, the conversation touched upon the interrelationships between Hinduism, Jainism, and Buddhism.

Mr. Kumar explained that since Hinduism is the parent of Jainism and Buddhism, they can be considered as one family. He also shared that Jainism is so similar to Hinduism that there isn't a clear line marking where Jainism diverges from Hinduism. Many of the practices and rituals are very similar. In fact, many Jains visit Hindu temples and celebrate Hindu holy days and vice versa. In contrast, the followers of the Buddha intentionally differentiated their practices from Hinduism, ensuring that Buddhism evolved into a stand-alone religious tradition. He also explained that regardless of their religion, Indians revere the Buddha as a great sage, and many Hindus even worship him as an incarnation of the Hindu deity Vishnu.

By explaining this, Mr. Kumar wanted me to understand that Hinduism does not reject or condemn other religions—its view is that all religions reflect different perspectives on a greater truth. He shared that what he finds wonderful about Hindu philosophy is that it finds harmony and unity among plurality. This is why he believes so many different spiritual traditions can peacefully coexist in India.

I was intrigued by Mr. Kumar's description of Hinduism as a dynamic, flexible, and open spiritual tradition. This was very different from the Western conception of religion, which is often presented in all-or-nothing terms. Yet, I had to take what he said with a heavy dose of skepticism because his idealistic view of Hinduism seemed disconnected from the harsh realities of life for people in India who happen to be born into the lowest caste, the Untouchables. Nevertheless, what he shared instilled in me a desire to come back to India on my own in the future to probe deeper into the intertwining evolution of Hinduism, Jainism, and Buddhism.

Visiting so many temples and sacred sites in India as part of this tour had a cumulative effect of making me curious to better understand what all the fuss with enlightenment was about. Why would so many rulers abdicate their kingdoms to pursue this state, whatever it was? And why did they not choose instead to channel their enlightenment to rule their kingdoms with more wisdom and compassion? It seemed to me the world was and still is in desperate need of enlightened leaders rather than hermits or renunciants.

While these were very interesting questions, I had no idea when I would next find the time to explore them. As this trip came to an end, I had to turn my focus to my upcoming move to Singapore and transition into a new job. No one could have predicted just how soon I would end up back in India to continue the spiritual immersion that had only just begun.

Renouncing the Brain 2.0 Paradigm

While I did not know it at the time, I now see that the spirit of Bahubali, Chandragupta, and the Buddha must have somehow gotten under my skin. In the next few months, things were about to unfold that pushed me to get so fed up with life in Brain 2.0 that, like them, I walked away from everything I knew to search for a way to live that was in line with my purpose and values. This genuine renunciation of Brain 2.0 in turn led me to experiences and insights that transformed me and continue now to help me realize much more of my potential.

Like many of my peers, I had planned my career according to a Brain 2.0 belief that I needed to first become extremely successful and rich before I could make a difference. Perhaps this was something I unconsciously absorbed from the examples set by early American tycoons such as John Rockefeller, Andrew Carnegie, Henry Ford, and Milton Hershey, as well as contemporary billionaires such as Bill Gates, Mark Zuckerberg, Michael Dell, and George Soros, who became philanthropists after they built huge fortunes. However painful what was about to unfold in my next job was, it was worth going through to free myself from this way of thinking.

When I started my new position in Singapore, I felt very fortunate to have this unparalleled opportunity to work with a philanthropist to create social impact. It was also thrilling to travel to Indonesia and Bangladesh for the first time to learn about the firm's ventures there. Furthermore, the firm immediately arranged for me to get a one-year business visa to India so I could look at potential investment prospects and support the firm's current ventures there. In the early days, I found it very exciting to have a job that was aligned with making the world a better place.

However, by the end of my first month, I could sense that something was off. The atmosphere at the office was filled with a sense of dread, fear, and anxiety. As I examined the social impact portfolio's track record, I saw a strange pattern that nearly all of the social ventures this firm got involved with imploded within a few years. I learned from the veterans on the team that this was partly because the billionaire was extremely impatient. If he didn't see the results he wanted within a year or two, his impulse was to shut down the entire venture, laying off everyone the venture employed. The fact that only a tiny percentage of the staff in the firm had been there for more than three years indicated that employee turnover was unusually high. Since I was a newbie, my colleagues advised me that the key to lasting at this firm was to keep one's head down and stay out of the billionaire's radar (which they likened to the "Eye of Sauron"). They also warned me that if I ever had to interact with him, to never, ever disagree with him.

I was grateful for their advice, but this was not the way I was wired. It was clear that staying in this type of work environment would be a waste of my time and energy. Nevertheless, in the hopes of creating a breakthrough and of turning a bad situation into a learning opportunity, I started preparing a white paper for the billionaire that presented my point of view on how to build an effective social impact platform. When it was finished, I sent it to him as my resignation. To my surprise, he replied that what I'd shared was brilliant, and he immediately offered me a new role working alongside him in the office of the chairman; he asked me to meet with him the next day.

When I was summoned to his office, I hoped that something positive could come from the encounter. However, as soon as he started speaking, I realized that the dysfunctionality I saw in the firm really did come from the top. He began the conversation by sharing that his primary motivation for getting into social impact was to outdo Bill Gates. As I listened to him, it became apparent that he was not sensitive to the plight and suffering of people living in poverty at the bottom of the pyramid. I was struck by what seemed to be an absence of compassion. He didn't seem to care about the destructive impact of his decisions to shut down ventures in the communities he claimed to serve, and he didn't seem to care about what would happen to the people hired by the ventures.

That was when it hit me that what this organization was doing was not social impact investing—it was more like toxic impact investing. This was Brain 2.0 disguised as Brain 3.0. It was like a case study illustrating the destructive effect Brain 2.0 can have on a social impact investment platform and the surrounding community.

Although it was not a positive interaction, in hindsight I am grateful that I had the opportunity to meet and better understand this self-proclaimed philanthropist and to learn his true motivations. Our conversation made clear that I had to cut my losses and move on. We both quickly came to the mutual conclusion that it would be best for us to part ways. When I walked out of the office for the very last time in December 2011, I was smiling from ear to ear, filled with relief and joy. I was so grateful that I had the freedom and ability to live on my own terms and felt sorry for all the people who continued to work for him in that toxic environment.

My friends who saw me through this difficult period could sense that I was in a state best described as shell shock. Nevertheless, I could see the silver lining. Those few months had cured me of the Brain 2.0 tendency to glamorize success, wealth, and power. This massive disillusionment pushed me to question society's adoration of wealthy people as heroes and to reexamine my own fixation on becoming wealthy. I would never again idolize anyone just because they were worth billions of dollars. Moreover, I learned that being a billionaire could be a liability. Having vast quantities of wealth enabled that person to live in a bubble that was completely divorced from reality, much like the ruler in "The Emperor's New Clothes." He was surrounded by obsequious people whom he paid to obediently flush his money down the toilet on disastrous projects. Most importantly, the experience made me question the correlation between one's wealth and one's potential to make a difference. It was not as clear-cut as I had assumed.

This experience laid the groundwork for me to come to realize that being in Brain 3.0 might be more important than being a billionaire when it comes to creating a positive impact on the world. From reading Einstein's letters and articles, I saw that this realization aligned with a comment Einstein once made on wealth: "I am absolutely convinced that no wealth in the world can help humanity forward, even in the hands of the most devoted worker in this cause. The example of great and pure characters is the only thing that can produce fine ideas and noble deeds."[4]

After the euphoria of parachuting out of that terrible work situation faded, I then had to decide how to move forward. For the first time in my life, my gas tank felt completely empty. I was beyond burned out. All the usual quick fixes that used to work (sleep, exercise, good food, clubbing, shopping, drinking, watching movies) didn't revive me.* To further complicate matters, I had also come to a point in my career where there was no longer a clear line of progression. I could come up with no obvious next step for what I should do to advance in my career.

As I thought about the future, I knew the clock was ticking for me to move back to the United States. My parents were getting older and needed help to manage their health issues. Going back would be a big transition personally and professionally, and the longer I waited, the harder it would be. Several people who had spent many years working abroad warned me that returning expats often find it very challenging to find employers that value their international work experience, no matter how stellar, because positions in the United States are much more specialized.

---

* In hindsight, it's now obvious to me why: besides sleep and exercise, these activities activate Brain 2.0. I had to charge up Brain 3.0 to regenerate and reconnect to my inner spark.

If that was the case, I wondered, should I just embrace this transition as an opportunity to finally become a social entrepreneur? I had been waiting for years to pursue my dream of helping young people overcome adversity. But was I finally ready to do this? I didn't know. I still couldn't see a clear path for how I could move in that direction. I still had no idea what the social enterprise I'd build would actually do. Further, I was terrified of failing and becoming financially bankrupt. In the meantime, wouldn't the safest thing be to get another job and keep earning money? Why jeopardize my employment prospects by being out of work?

Since I had no clarity on what to do on the career side of things, I decided to take a small break to go travel. The only thing that was clear in my mind was that I wanted to see more of the Asia-Pacific region before I moved back to the United States. A good friend of mine was getting married in Melbourne, Australia, in January, so I began 2012 by going to her wedding. There I met up with a couple of girlfriends and together we went onward to visit New Zealand for the first time. We drove around the South Island, climbed the Franz Josef Glacier, and hiked the Routeburn Track, one of the most beautiful alpine trails in the country (and in the world).

The stunning landscapes of New Zealand filled me with awe and rejuvenated my spirits. On one particularly beautiful morning spent hiking the Routeburn Track, as I watched the rising sun slowly evaporate the misty clouds around the mountain peaks, I felt myself completely at peace for the first time in years. It was so striking that I then asked myself: How come I only feel this good and this whole on vacation? Why not redesign my life so that feeling this type of well-being would be normal? What was actually stopping me from giving myself permission to feel more wholeness and fulfillment? No one was holding a gun to my head, forcing me to be a miserable overachiever. I was allowing myself to be held hostage by my Inner Godzilla and Inner Teen Wolf.

In the self-reflection that continued during the hike, I realized that my family and I still carried so much fear from having been desperately afraid, hungry, and poor as refugees. I saw that my mind was constantly racked by irrational fears that no longer had any basis in my current reality—fears of running out of food, of running out of money, of being unwelcome and marginalized, of being displaced again, of having to start over again and again. I then realized I had built up a number of irrational habits to alleviate the anxiety caused by these fears.

For example, I compulsively carried snacks everywhere, ate up excess food rather than have any go to waste, hoarded rather than threw out things I had no use for, saved almost ten times more than I spent, and explored every high-paying job that headhunters reached out to me about, even if I felt no real interest in them—just in case. Before I left for Australia, I had applied for an investment position in Singapore because I thought it would be irresponsible not to. During the hike, I realized that if I were to get the offer, I would feel compelled to take it because turning down a good job would bring up so much anxiety, regret, and guilt. The fact that I was secretly wishing that the company would offer the job to someone else to save me the agony of declining it helped me to see that my heart really wasn't in it.

During the hike, I came to the resolution that it was time to finally shed these fears so they could no longer drive me to make decisions that would repeat self-limiting patterns. After barreling forward as an overachieving workaholic for so many years, I needed to give myself more time to digest how much my life had changed since college. Somehow, in the process of succeeding in my career, I had lost touch with my sense of self. My personal development had been uneven and spotty relative to my career progression. I needed to do a lot more self-reflection and soul-searching to really understand and unpack all the self-limiting beliefs that kept me from doing what I wanted to do most: start a social enterprise.

By the end of the trip in New Zealand, I had decided to extend my break and go to Burma next. It was one of the few countries in Southeast Asia that I had yet to visit and I really wanted to see it before leaving Asia. I also made up my mind to move back to the United States by the end of 2012. So I gave myself permission to possibly take off all of 2012 to explore, reflect, and discover more about my life purpose.

The Quest Begins

When I returned to Singapore in early February, I immediately began planning a trip to Burma. I booked the plane tickets and tour for early March. That gave me about a month to do more research for this trip and also research and plan where else to go and what else to do before returning to the United States at the end of the year.

At the time, Burma was finally opening up, and I eagerly wanted to visit it before it would be irreversibly changed by tourism and economic development. I had heard that in many ways, Burma was like a time capsule, as many parts of the country still didn't have running water or electricity. By going there, I could have a sense of what Southeast Asia looked like in the 1950s—the time period in which my parents were young children. I had also heard that Burma is one of the last countries in Southeast Asia where Buddhism is still a major part of the people's way of life. So I looked forward to hopefully seeing a more "authentic" Buddhism in action in Burma.* Since most of the tourist attractions in Burma are historical Buddhist landmarks, I decided to read up on Buddhism so I could understand more of what I would see and have a richer and more meaningful experience.†

At that point, I didn't even know the name of the historical Buddha off the top of my head, much less any details about his life. Through online research, I finally learned that his name was Siddhartha Gautama (and how to spell it) and that he had lived in the sixth century BCE in a region that is now around the border of India and Nepal. As a child, I had learned that there are two main branches of Buddhism, Theravada and Mahayana, but I had never learned what made them different. I only knew that my family's ancestral tradition was Mahayana, because that is the form prevalent in Vietnam and China. As I started reading up on Buddhism, I learned that the Theravada tradition could be traced back to the life of the historical Buddha. Theravada teachings are based on the Pali Canon, the "official" historical collection of teachings of the Buddha that were transmitted orally for five centuries and then finally written down in the first century BCE. (Pali is the language believed to have been spoken in the region in northeastern India where Siddhartha Gautama lived 2,500 years ago.) The Theravada form spread south and east from India into Sri Lanka, Burma, Thailand, Laos, and Cambodia.

---

* Although China and Vietnam were once Buddhist countries, Communism and industrialization have eroded many of the Buddhist institutions and values. Particularly in China, the Communist government prioritized economic development and building military power. It promoted atheism and condemned all religion as backward. During the Cultural Revolution, many Buddhist temples, along with their libraries, were destroyed, and many monks were put in reeducation camps. Nevertheless, Buddhism managed to survive in China but lost its influence as the vast majority of the Chinese population became secular. In Vietnam, the government has been more tolerant toward Buddhism, mainly because it put it under state control.

† Previously, when I had visited Angkor Wat in Cambodia, I didn't know enough about its history or the religious iconography to fully appreciate the incredible artwork on the monuments.

The Mahayana branch emerged sometime after the Pali Canon was recorded. The earliest Mahayana sutras date to the second century CE and were written in Sanskrit, as that was then the dominant language for scholarship in India. My understanding is that a schism emerged because the Theravadans refuted the authenticity of the newer Mahayana scriptures and chose to strictly follow the Pali Canon. Mahayana Buddhism spread north from India into China, Tibet, and Nepal and then through China to Japan, Korea, and Vietnam. Given the much larger population of these countries, Mahayana Buddhism is practiced by many more people today than Theravada Buddhism.

Learning the history of Buddhism was the easy part. Learning the teachings of Buddhism was much more difficult. The biggest obstacle for me to understand Buddhism within my own ancestral heritage has always been the language barrier. Having grown up in America, I became fluent in English but did not learn Chinese or Vietnamese beyond a preschool level. Even at many Chinese and Vietnamese Buddhist temples in America, they speak only Chinese or Vietnamese. In order to finally understand Buddhism, I had to search for whatever teachings were available in English. I ended up diving into Tibetan Buddhism because it seemed to be the most "user-friendly" to Westerners, with dozens if not hundreds of well-written books in the English language published specifically for an international audience by great teachers like the Dalai Lama, Pema Chödrön, and Mingyur Rinpoche.

Following recommendations from friends, I read *The Art of Happiness* by the Dalai Lama. I was very impressed by how grounded, rational, and logical the Dalai Lama seemed, and by how much he embraced science. His open-mindedness was very refreshing. I was particularly inspired by a *New York Times* op-ed he published on November 12, 2005, titled "Our Faith in Science," in which he stated: "If science proves some belief of Buddhism wrong, then Buddhism will have to change. In my view, science and Buddhism share a search for the truth and for understanding reality. By learning from science about aspects of reality where its understanding may be more advanced, I believe that Buddhism enriches its own worldview."[5]

Intrigued, I then looked more closely at the Mind and Life Dialogues between renowned scientists and the Dalai Lama, which had been convened on an annual basis since 1987 and eventually led to the formation of the Mind and Life Institute to promote research on contemplative practices. Since then, dozens of rigorously designed studies on meditation have been conducted, and hundreds of articles sharing the results have been published in major media outlets and newspapers. I became particularly fascinated by the discoveries that meditation could change the brain in beneficial ways that increased focus and reduced stress and anxiety. As I did more research online about this topic, I also became captivated by articles on the positive impact of teaching meditation in schools located in low-income communities where students deal with high levels of toxic stress and trauma.

Curious to experience these benefits, I started looking for instructions and videos online. Unfortunately, most of what I found didn't resonate with me. I was disappointed that despite all the scientific research being done on meditation, the meditation teachers I came across did not incorporate the science. Worse, some teachers seemed to promote New Age pseudoscience, which I found to be shaky whenever I tried to find the source of their claims.

I also suspected that the way meditation was being conveyed in these videos was not entirely "pure." Perhaps, like Chinese take-out and fortune cookies, what these teachers were presenting was a form of meditation created for Western consumption. Although I actually love Chinese take-out and fortune cookies (after all, I grew up in a take-out restaurant), I knew I couldn't be satisfied with only the fortune cookie version. I needed to also experience meditation in its "authentic" traditional form. As I thought about how to do that, it occurred to me that I still had a one-year visa to India. Why not go to India? Why not go directly to the "source," to where these practices were born?

Two things almost stopped me from doing this: anxiety and guilt. It seemed so unlike me to suddenly hop on a plane to India just to learn how to sit still. A large part of me felt very anxious about taking so much time off from my career. It didn't matter that I had paid off all my student loans and saved up funds specifically for this purpose. I felt guilty about doing something so clearly self-indulgent and financially irresponsible. What if doing something like this would make me less employable? Did I really want to take this risk? What could I possibly gain from this experience anyway? Was this just a silly daydream? Didn't I need to forget this foolishness, grow up, and come back to the real world?

The turning point came when I learned that two of my professional heroes, Steve Jobs (the cofounder of the tech giant Apple who had passed away only a few months earlier) and Daniel Goleman (a bestselling author and thought leader on emotional intelligence who'd helped spearhead the Mind and Life

Dialogues), had traveled to India as young men to learn how to meditate. In fact, it could be argued that their journeys in India were seminal experiences that later enabled them to make a profound impact in their respective domains. What if, like them, the insights I could gain in India could change the course of my life for the better? By reframing this as an opportunity to follow in their footsteps and invest in my personal and professional development, it became easier to justify giving myself permission to go. I had already come to the realization that I needed to carve out time for reflection and soul-searching, so I might as well learn effective techniques to do exactly that.

The next step was figuring out where to actually go. Fortunately, a friend recommended a retreat center called Tushita based in Dharamsala, a city in northern India that is home to the Dalai Lama and the Tibetan government in exile. On Tushita's website, I found out about other retreat centers nearby. In parallel, as I did research on the Buddha's life, I learned that there is a famous pilgrimage itinerary route marking the key milestones of his life. I thought to myself, I might as well go see these places while in India. So thanks to the Internet, in a couple weeks, someone like me who'd started out knowing almost nothing about the Buddha and Buddhism was able to sketch out an amazing itinerary.

Since I hadn't yet decided when or where I would end my trip in India, I simply bought a one-way ticket from Singapore to New Delhi. From there I would take a train to Dharamsala, where I would spend a month doing three retreats, for approximately ten days each, at three different centers: the Tushita Meditation Centre, the Thosamling Nunnery, Institute, and Retreat Center, and the Himachal Vipassana Centre. Afterward, I planned to see the Buddhist pilgrimage sites to better understand the Buddha's life. To travel to these sites, I would return by bus to New Delhi, and from there, fly to one of the two sites that had an airport: Bodhgaya or Varanasi. From what I could tell from my searches, many pilgrimage itineraries involved starting at one of these sites, driving to the other sites in a loop that ended at the other airport city, and then flying out from there. Initially, I planned to fly into Bodhgaya to start the pilgrimage tour there and then end it at Varanasi. I would then fly to Mumbai to visit a friend from business school. As luck would have it, my itinerary would be dramatically revised after meeting someone who would invite me to Kolkata, where my life, my sense of self, and my sense of time and space would be forever transformed.

In parallel, as I prepared for my upcoming trips, I decided to start exposing myself to Buddhism and meditation by checking out centers in Singapore. I primarily spent time with two groups: the Amitabha Buddhist Centre, run by the Foundation for the Preservation of the Mahayana Tradition (FPMT), the same umbrella organization that oversees the Tushita Meditation Centre, and another place called the Centre for Inner Studies.

The people who greeted me at both centers couldn't help being curious about an American visitor randomly showing up at their doorsteps and were happy to give me a tour and answer my questions. When they asked me what I did for a living, I shared that I was in transition and looking to understand my purpose in life. The people at the Centre for Inner Studies recommended an upcoming course because it had helped them gain insight into their life purpose.

I learned that the people there practiced a modern spiritual lineage developed by a Chinese Filipino businessman reverently referred to as Master Choa Kok Sui, or by the initials MCKS. The courses MCKS developed fell into two broad categories: a spiritual path called Arhatic Yoga (which blended together teachings from Buddhism, Christianity, yoga, qigong, and Theosophy) and a form of energy healing called Pranic Healing (which seemed to mix elements of qigong, Reiki, and various faith healing and shamanic practices). The thing that got me was the fact that most of the people at the Centre for Inner Studies were professionals. I had never seen such a gathering of down-to-earth businesspeople openly interested in spiritual topics and energy healing before. And the most bizarre thing was that I felt strangely at home among them.

The course they recommended I take had an intriguing title called "Achieving Oneness with the Higher Soul" that both attracted and repelled me because it sounded really out there. It was clearly outside of my comfort zone, so my first inclination was to say no, but I reconsidered because the people assured me that it had been a powerful experience for them and that it was offered in Singapore only a few times a year because it had to be taught by a senior Indian teacher from Kolkata, who they said was amazing. I had never met an Indian spiritual teacher before and was both very skeptical and very curious about whether it was even possible for a two-day meditation course to give me insight on my life purpose. As I had nothing else scheduled on those dates, I decided to give it a try. I could always leave early if I found it too weird.

On the first day of the course, I learned the instructor's name was Shailesh, and like MCKS, he was a very successful businessman with a strong spiritual orientation. When he wasn't teaching meditation workshops, Shailesh ran a securities trading company. I was also impressed by how large the class was. It seemed there were more than a hundred people present, and many of them were professionals and entrepreneurs—it was strikingly not the stereotypical New Age spiritual crowd. I was partnered with a British serial entrepreneur whose most recent venture specialized in developing technology to combat maritime piracy. He was in the midst of a spiritual awakening and was earnestly exploring many different spiritual teachers and groups to find answers.

From the start, I took most of the content with a grain of salt. Shailesh spent a good amount of time explaining chakras and the anatomy of the human energy body, which I found hard to digest. It seemed like I was the only one in the group who had not previously taken any of the energy healing courses offered at the center, so at the points when Shailesh asked people to scan energy with their hands, I was completely lost. However, my partner patiently walked me through what he was doing and shared fascinating stories about how the Pranic Healing techniques actually worked and how he regularly did healings for his children when they weren't feeling well.

Near the end of the program on the second day, we did the "meditation on the blue pearl" that is meant to build a stronger connection with the "Higher Soul," which Shailesh explained often appears to our mind's eye as a blue pearl. To my surprise, during the meditation, I did see blue-violet patterns that seemed to swirl around a radiant blue star in the center, making it seem like my consciousness was floating through a tunnel toward some sort of ethereal light at the end of it. Then when I opened my eyes, to my amazement, I saw an outline of luminous white light around Shailesh and the other people in the room. I couldn't believe my eyes. This phenomenon lasted for only a few minutes but was enough to completely throw me off and force me to reconsider my earlier doubts on what Shailesh shared about the energy body.

Shailesh could tell that I was very skeptical at the beginning of the course and was noticeably happy to see that my unexpected experiences from the meditations were bringing me around. I was clearly fascinated by Shailesh's embrace of spirituality as a businessman and was curious to learn more about how he integrated these two aspects of his life. After the course ended, I told Shailesh that I was about to travel through India and asked whether it was possible for me to take more courses while I was in India. To my surprise, Shailesh invited me to come to Kolkata to study there as his guest. I normally don't agree to visit people I barely know, but I felt strangely compelled to say yes to this opportunity. After all, I had seen Shailesh radiantly lit up with white light. It was too extraordinary not to explore further. I took a leap of faith and gratefully accepted his invitation.

Because Kolkata is located in eastern India, it made sense to rearrange my tour of the Buddhist pilgrimage sites to begin in Varanasi and end in Bodhgaya, which is the closest site to Kolkata, so I could then take the train from Bodhgaya to Kolkata. I looked at likely dates for arriving in Kolkata and asked Shailesh if the weekend of May 5–6, 2012, would be a good time, but he responded that there would be no classes that weekend because of Vesak. I had never heard of Vesak before, but from an online search, I discovered that Vesak marks the anniversary of the Buddha's enlightenment, which took place in Bodhgaya, and is therefore one of the most important times of the year to be in Bodhgaya. I couldn't help feeling amazed by how changing my trip to go to Kolkata led to my inadvertently ending up in Bodhgaya right around the time of Vesak! This was like ending up in Bethlehem on Christmas Eve by pure chance, without knowing anything about Christmas! I took this as an auspicious sign that accepting Shailesh's invitation to go to Kolkata was the right decision.

Nevertheless, it was hard for me to blindly accept what Shailesh taught in the class without searching for other explanations for what I experienced. I discovered through online research that it is normal to see a light show when we close our eyes. This widely documented phenomenon is known as a phosphene, which is defined as a "sensation of light produced by mechanical or electrical stimulation of the peripheral or central optic pathway of the nervous system."[6] I found a detailed explanation in an article on the Vision Eye Institute's website: "The neurons in our visual system are busily sending signals to the brain via what's known as the thalamus. So, even when we are in total darkness, just resting our eyes or even when we are asleep, there's always something to see. . . . This constant activity can create splashes of color or patterns that can change quite randomly." The article went on to explain: "These phosphenes can be heightened by all kinds of everyday stimuli, such as an intense sneeze, hearty laughter, coughing, blowing your nose or standing up too quickly. Other factors such as low blood pressure or low oxygen intake can increase the visual show even further."[7]

I also learned that seeing an outline of bright light around people is a widely documented phenomenon known as the corona effect. It is commonly reported by people who experience migraines and headaches. The thing is, I don't suffer from migraines or headaches, and the coronas I saw were only "visible" for a brief period of time following a meditation. Regardless, learning about phosphenes and the corona effect didn't necessarily negate what Shailesh shared. The reality is there are no reliable means to prove or disprove that the blue pearl or blue swirls I have come to see regularly when I close my eyes are signs that I am connecting with my "Higher Soul" or that the bright outlines of light I see around people are their energy bodies. Given the lack of certainty and validation, I concluded that there was no reason to take these phenomena too seriously.

Baffled by Buddhism in Burma

In early March, when I traveled through Burma, I found it greatly beneficial and rewarding to have just done research for my upcoming pilgrimage tour to the sites where Siddhartha Gautama lived. For the first time ever, I could look at the artwork on the temple walls and make out what story or scene from his life it was depicting. It was similar to how my Catholic upbringing enabled me to appreciate Renaissance artwork when I visited churches and museums in Italy.

However, in another area, I was greatly disappointed. Having been impressed by the Dalai Lama's portrayal of Buddhism as "reasoned faith," I went to Burma hoping to find more role models of this approach. Instead, I found that my guides and other people I met during my tour seemed to blindly accept and repeat the stories and beliefs handed down to them, no matter how implausible they were. Similar to my skeptical reaction as a child to Catholic stories of miracles, I had a hard time suspending disbelief.

My trip to Burma forced me to start making a conscious effort to distinguish between accounts of Siddhartha Gautama as a historical figure, which have some degree of historical validity, and apocryphal legends surrounding the Buddha, which cannot be historically verified and may have gotten further exaggerated in the retelling by successive generations of Buddhists. Therefore, to reflect this discernment going forward, I will use the name, Siddhartha Gautama, to refer to the historical person (and what historians and archeologists have uncovered about him) and to use the title, the Buddha, to refer to the mythological, hagiographical* figure to whom Buddhists have attributed many superhuman powers and fantastical feats.

For example, during my visit to a remote area of Burma called Mrauk U, which is often called the Angkor Wat of Burma because of the beautiful stone temples in that area, I learned that the locals believed that the Maha Muni image, the most famous statue of the Buddha in Burma, was built by the Buddha himself. According to the local legend, during the Buddha's lifetime, he and a large entourage of five hundred monks visited the ancient Arakan kingdom of Dhanyawadi, which was located in this area. Since I had learned from my research that the historical Siddhartha Gautama spent his whole life in the region that is now northeastern India and Nepal and I had not come across any historical records stating that he traveled to Burma, I asked: How could he have possibly visited this kingdom?

The guide explained that the Buddha and his entire entourage all came by astral travel, meaning they visited in spirit and did not come in their physical bodies—as if this detail makes the story more plausible. According to the legend, the king made a request to the Buddha that he allow them to create a likeness of him that they could honor and pay homage to. The Buddha granted his wish and then used his special powers to create the statue so that it became an exact likeness of him.

Even without taking into consideration the radical concept of astral travel, what I heard about this statue contradicted the pervading archeological and historical understanding of Buddhist artwork. Earlier in my tour, when I was visiting Bagan in the north, my guide explained that the earliest Buddhist artwork in Burma appeared Indian in style and that a distinctly Burmese iconography emerged much later, probably around the eighth century. Thus, I was puzzled that the Maha Muni image did not resemble an Indian depiction of the Buddha, which made me question how old it actually was.

Furthermore, I had also learned from my research that Siddhartha Gautama explicitly instructed his followers not to create or worship images of him.

---

* The term "hagiography" refers to an idealized biography of a holy person, usually developed from hearsay rather than historical research.

Instead he told them that if they needed images to revere, they should honor the Bodhi Tree and the wheel (called "chakra" in India) as symbols of his teachings, which are collectively referred to as the "Dharma." According to historians, the earliest statues of the Buddha can be traced to the first century CE and the Greco-Buddhist kingdoms founded by Greek soldiers who invaded India with Alexander the Great and then chose to settle in the area that is now present-day Pakistan and Afghanistan. These early depictions of the Buddha, many of which were made in the city of Gandhara, have a striking resemblance to ancient statues of Greek gods. After other kingdoms saw these beautiful statues, they hired the Gandharan artists to create statues for them. Thus, the art form spread across India and then into neighboring countries.

Overall, I enjoyed traveling through Burma as a tourist because the beauty of the historical monuments was mesmerizing and I had a great time getting to know the Burmese people and fellow travelers I met. Nevertheless, the trip reinforced my skepticism toward Buddhism as a religion, and I started to worry about whether I would encounter a similar disappointment with Buddhism as "reasoned faith" in India. I had to tell myself that, whatever happened, this soul-searching quest would not be a waste of time. The important thing was that I would gain the firsthand experience to make conclusions for myself, and at the very least, traveling through all these sites in India would make for an incredible adventure.

Mystified in Dharamsala

According to Tushita's website, its name means the "Place of Joy." While I optimistically hoped my experience there would live up to its name, I never expected to experience a state of blissful elevation unlike any experience of joy I had ever felt before. As I will explain, after that experience, it would become impossible to go back to life as I used to know it.

Timing-wise, my retreat at Tushita coincided with the stay of Lama Zopa Rinpoche, the cofounder and spiritual director of the FPMT who is widely revered and respected as a highly realized teacher. The staff explained that he was recovering from a stroke and was in Dharamsala to receive care from Tibetan doctors and healers. It was otherwise very rare to find him at Tushita because he is usually traveling to give teachings and resides mainly in Nepal and in the United States.

Given his delicate condition, the staff instructed all the retreat participants to respect his privacy and explained that Lama Zopa Rinpoche would not be accepting any appointment requests. They said there was a chance that if he felt well, he might decide to give a teaching to our class and that was the most we could wish for. Therefore, from how they presented the situation, it sounded like even though Lama Zopa Rinpoche was on the premises, in all probability, none of us would meet him, let alone have a conversation with him. Yet somehow I did.

The retreat I signed up for was a ten-day Introduction to Buddhism course. I was part of a class of about eighty people from around the world. The group was impressively diverse, with participants from North America, Latin America, Europe, Australia, the Middle East, and Asia. I guessed that it was a fully sold-out class because we filled up all the seats in the "gompa," the Tibetan word for the large main meditation hall where we assembled every day for lectures and meditation. Our teacher was Venerable Namgyel, a Western monk from Australia with a wickedly hysterical sense of humor. He shared that many years ago, he had been a businessman, and after being exposed to Tibetan Buddhism, the impact was so profound that he decided to become a monk.

From the beginning, I had a hard time accepting many of the teachings. I was particularly doubtful of the teaching that there are six realms of rebirth. The notion that humans could be reincarnated as animals completely baffled me. After a couple days, I realized I wouldn't get anything out of being there unless I was willing to suspend disbelief. What finally helped was trying to reframe these teachings as metaphors rather than to take them literally. I also discovered a large stockpile of books inside the gompa, which were written by various Buddhist teachers from different traditions, such as the Dalai Lama and Thich Nhat Hanh, and decided to use the time I had at Tushita to take as deep a dive and get as wide a perspective as possible by reading books by different teachers. It became my habit to ask deep searching questions of our teacher, Venerable Namgyel, and to stay up late reading in the gompa to find my own answers. Often, I was the last person to leave the gompa at night.

One night, I ended up reading until close to midnight without realizing the time had flown by. When I returned to my dorm room, everyone else was already asleep. Soon after I crawled into my bed, I heard chanting coming from outside. I happened to have the bed next to the window, so I looked out and saw it was coming from a small group of four people, all wearing monastic robes, circumambulating the gompa as they chanted.* In the middle of the group, there was a figure that resembled pictures I had seen of Lama Zopa Rinpoche. Thinking this might be the only opportunity I would ever have to see him in real life, I decided to go outside to take a closer look.

At first, I watched them from a respectful distance, standing outside the cafeteria building across from the gompa. Then a Tibetan night guard walked over to me and asked me in a friendly tone if I knew who that was. I responded, "Is that Lama Zopa Rinpoche?" He nodded. At first, I expected that he would be a disciplinarian and tell me to go to bed and leave Lama Zopa Rinpoche alone, but instead, he suggested that I go join them. It didn't even occur to me that that was an option, so I asked in surprise, "Are you sure? Would that really be okay?" He nodded with a big, friendly smile. With his encouragement, the next time the group came around, I made eye contact with one of them, a very warm and friendly nun from Australia, who later introduced herself as Venerable Sarah. When I gestured with my fingers to ask if I could join them, she smiled happily and waved for me to come over. So I enthusiastically ran to them.

Once I caught up to them, everyone acknowledged my arrival with warm smiles as they continued to chant mantras. I followed them at the end of the line, a few steps behind Lama Zopa Rinpoche's attendants. Because I couldn't make out what they were chanting, I decided to simply put my hands together in a prayer position and quietly follow along. I figured that as long as I proactively sent out good vibes, I would more or less get the essence of the ritual right.

As we circumambulated the gompa, I was struck by how joyful, happy, and energized I felt. It was odd because I only had the vaguest idea who Lama Zopa Rinpoche was and had no prior relationship with him or much understanding of Tibetan Buddhism in general. I'd only learned about him about a month earlier when I first visited the FPMT center in Singapore and picked up a few books containing edited transcripts of his teachings. At that center, I had heard the people who considered him their teacher rave about how marvelous and wonderful it was to be in his presence and why that motivated them to make such an effort to see their teacher in person at least once a year. There, in his presence, I began to sense why they felt that way.

---

* Circumambulation is the act of walking around a sacred spiritual object or building in a clockwise direction. This practice is a very common devotional ritual in Buddhism and Hinduism.

When the circumambulation ritual eventually came to an end, Lama Zopa Rinpoche stopped and revealed his curiosity about this random stranger who had joined them. He said hello and asked for my name and where I was from. His English was surprisingly good, but because of the stroke, he spoke slowly and slurred many of his words. At various points, Venerable Sarah, who often serves as his interpreter and secretary, jumped in when I needed help understanding him. After a bit of small talk, he began to talk about the Dharma, which is the word that Buddhists use to refer to the teachings of the Buddha. I recognized what he said to me as being almost identical to one of the teachings he gave that was captured in a book I had read; yet hearing him say it to me in person had a surprisingly more profound impact on me.

I remember only the gist of what he communicated, not his exact words. He began by saying that I was very lucky to find the Dharma because in the West, many people don't know what real happiness is. He then mentioned a few examples of what he considered misguided attempts to find happiness, such as shooting oneself out of a cannon, jumping out of planes, and climbing mountains. Although my guess is that he probably says the same things to everyone, I strangely felt like he was sharing a message specifically for me. It felt like he was calling out my bucket list, which actually included things like skydiving, mountaineering, and other extreme challenges (but nothing about shooting myself out of a cannon). Afterward, as I went to bed, I asked myself: What if he is right about the things on my bucket list? Could my conception of what would make me happy be so terribly misguided? I also wondered: What is this real happiness he alluded to? Would I ever experience it? And why would he say I was lucky to find the Dharma when I was still learning about Buddhism and taking everything with a grain of salt? Besides, what if he says the same thing to everyone, so I shouldn't read too much into it?

The next day, to my surprise, Venerable Sarah came to find me during a break to give me gifts from Lama Zopa Rinpoche: posters he'd blessed for me to put on my altar. I didn't know how to explain that I didn't have an altar (and I still don't), so instead I gratefully said "thank you." As she handed them to me, I recognized that one of the posters was a photo of the famous Buddha statue at Bodhgaya. So I mentioned to her that I would be seeing this statue in person soon because I was planning to visit the Buddhist pilgrimage sites and would be at Bodhgaya in early May. She got very excited upon hearing this. Then she explained that it was very unusual for someone to get the attention of Lama Zopa Rinpoche like this, so she had inferred that he must have seen that I have special karma. She seemed to take my upcoming pilgrimage as some sort of proof about this hunch and shared that she would tell Lama Zopa Rinpoche about my plan to do the pilgrimage.

The following day, Venerable Sarah came to find me again. This time she brought a blessed image of Green Tara from Lama Zopa Rinpoche with the instruction that I was to pray to her for help and guidance (at that time, I was not familiar with Green Tara and honestly had no idea what to do with the image except frame it as a beautiful souvenir of my trip). Venerable Sarah also shared that Lama Zopa Rinpoche wanted to give me instructions for the pilgrimage and that she would have to transcribe these and send them to me by e-mail later. I gladly gave her my e-mail address. Again, she emphasized that Lama Zopa Rinpoche must have seen something special in my karma because this was very uncommon.

I felt very honored and blessed to receive these gifts from Lama Zopa Rinpoche. At the same time, I wondered what karma had to do with it. I was puzzled by karma, specifically why Buddhists tend to attribute anything that happens to karma. I noticed that if something fortunate happened to a person, it was common for Buddhists to say it must be karma, and if something unfortunate happened to a person, they would also say it must be karma. If karma simply means cause and effect, it's like stating the obvious because every single thing that happens is a result of causes and produces effects. In contrast, gravity pulls everyone and everything on this planet to the ground, but we don't find any value in reminding people that gravity is at work when we jump or fall. So why do Buddhists consider it meaningful to rhetorically attribute things that happen to people to karma?

Yet maybe she had a point. Part of the reason I was able to notice Lama Zopa Rinpoche that night was because I happened to get assigned the bed by the window in the only dorm room in the gompa building. All the other dormitories were far enough from the gompa that people staying in them might not have been able to hear the group chanting that night. Oddly enough, during registration, after my six roommates and I were assigned to this room, the Tibetan staff person who showed us to our room picked up only my luggage. Then when he entered the room, instead of dropping my luggage at the entrance or on the first bed, he walked all the way over to the far side of the room and placed my suitcase on the bed by the window. It is amusing to ponder that if it weren't for this Tibetan staff person choosing the bed by the window for me, I might not have connected with Lama Zopa Rinpoche that night. Perhaps this is how karma works—through what might seem like random mundane domino effects that aren't so random after all.

Soon after that night, the staff announced that Lama Zopa Rinpoche was feeling well enough that he wanted to give a teaching to our retreat class. Predictably, the staff commented that our class must have had very good karma because he was not often at Tushita and rarely gave a teaching there. Everyone felt very blessed and excited.

On the day of the teaching, Lama Zopa Rinpoche began by sharing almost the exact same lines he'd said to me that night: we were all very lucky to find the Dharma and experience real happiness because so many people are looking for happiness in the wrong places, like climbing mountains and shooting themselves out of cannons. He then opened the floor to questions. Someone raised a hand and asked him to explain the Buddhist concept of "emptiness," which is the English translation of the Sanskrit word "shunyata" and is one of the most abstract and difficult Buddhist teachings to grasp. As I recall, it may have been the only question that got asked because Lama Zopa Rinpoche's response to this one question filled up the remainder of the session.

Unfortunately, I found that the slurring of words from the effects of the stroke made it nearly impossible for me to follow Lama Zopa Rinpoche's teaching. Instead, I decided to try to tune in to the essence of what he was trying to express. Then something surprising took place inside my mind. It was like a voiceless inner teacher began to explain the concept of shunyata to me. It was conveyed through what can best be described as "thought forms." It was like receiving information packets in my mind in a format that was a precursor to sound, language, or pictures, similar to the way memories arise in the mind. As this happened, I "saw" that all life on Earth and all the structures on Earth, natural or man-made, are made mostly of space. All are made of the same micro components: atoms and subatomic particles, which come together and disintegrate in continuous cycles. There is only one energy field in which all these particles combine, separate, and recombine—a field of infinite potential. I then saw that this field is a living, "intelligent," conscious energy that is evolving these particles into the many diverse forms that are present on our planet: inert structures, plants, animals, and human beings.

As I integrated the insights that came to me with the traditional Buddhist teaching on shunyata, this is how I would explain the understanding that arose in my mind: shunyata means recognizing the world as a projection of consciousness, similar to a hallucination or dream, but different in that there is a collective projection that we as individuals all engage with and contribute to. It involves understanding that how an object appears to us depends on how we perceive it and the meaning we project onto it, that is, the conceptual label we give to it. If we change the conceptual label, we interact with the object in a completely different way. Shunyata also involves recognizing that everything, including our lives, is produced by interdependent arising—innumerable causes and factors coming together to produce all phenomena, like a giant unending domino effect in which all phenomena are interdependent and interconnected—and that all composite forms eventually break apart. Therefore, no physical object exists by itself in an absolute reality.

The wisdom of shunyata involves understanding that everything in our physical universe is interdependently interwoven and therefore impermanent, and that our true essence, or "Buddha Nature," as Buddhists like to describe it, is that of a higher consciousness that transcends conditioning and physical existence. By recognizing and experiencing our Buddha Nature as pure potential, it becomes easier to develop equanimity toward the myriad forms and experiences that the field of infinite potential produces.

As these insights crystallized in my mind, I felt a subtle explosion of energy at my heart and at the top of my head. It felt like my consciousness went through the roof. I was still in my body, but beyond it at the same time, and also overcome with a sense of profound, deep, joyful bliss that I had never felt before. It radiated from my heart and the crown of my head into everyone in the hall. My heart was so full of joy, I felt like it could burst. I also lost track of time. It felt like time froze, but it was possible that the peak of this experience might have lasted only a few seconds. As the explosion faded, I started to feel my consciousness settle back into my body and become smaller and manageable again.

That brief moment of expansion made me see "Due" was not the real "me." By that, I mean that I saw finally that "Due" represents only a tiny fraction of all that I am. The real "me" was this ineffable superconsciousness that I had caught a momentary glimpse of being. I realized to my amazement that "Due" functions like a kind of autopilot created to help the part of my consciousness infused into this human body navigate life on this planet called Earth. It might have seemed like "Due" was in charge, but with the veil temporarily lifted, I realized that the true pilot is this vast indescribable superconsciousness, which animates "Due" and is beyond "Due" at the same time.

Going from normal consciousness to superconsciousness was like going from a five-foot-high view of the world (that's about how tall I am) to looking at my life on Earth from outer space. The impact it had on me was very similar to the "overview effect" that many astronauts describe experiencing when they look back at the Earth from orbit—seeing this sublime planetary big-picture perspective gives them a profound understanding of how the Earth is really one unified global system; the powerful and overwhelming sense of awe and oneness they feel heightens their reverence for the unity, coherence, and fragility of life on our planet. Many have reported feeling a sense of planetary or cosmic consciousness that transcends political divisions.[8] Similarly, this short, fleeting experience of being my "Higher Self" let me see that I was interconnected with everything unfolding on our planet, and I could never go back to the illusion of being just "Due."

Although the peak didn't last very long, the sensation of being blissfully light-headed remained with me for weeks. I couldn't stop smiling. In that state, everything seemed so delightful and amusing. I could spontaneously laugh at any moment for no particular reason. It magically felt like there was a little sun above my head that could never stop shining. In the aftermath of that experience, in spite of the effusive joy, my mind was in a state of confused disorientation and shock. I had no idea how to process this experience or put it into words.

Everything that I thought was true about the universe and life had to be revisited. It laid bare that my lifelong bias in favor of science and against religion had to be reexamined. I also wondered: Where did that inner voice or inner knowing come from? It didn't feel or sound at all like Lama Zopa Rinpoche. Was it possible that being at Tushita and being in Lama Zopa Rinpoche's presence activated something dormant inside me? Did my experience mean that there really is such a thing as an inner teacher and that there really is a Buddha Nature? Furthermore, did living primarily in Brain 1.0 and Brain 2.0 keep me from experiencing all this until now?

That evening, I went back over my conversation with Lama Zopa Rinpoche that night outside the gompa and wondered if what he'd said to me had layers of meaning I couldn't grasp at the time. Was he making a prediction? Was connecting to this inner teacher what he meant by saying I was fortunate to "find" the Dharma? Was this blissful elevation what he meant by real happiness? Then I wondered: Could this experience be what the word "enlightenment" referred to? I had my doubts. First, I was disoriented and had more questions than answers. Second, I had not gained omniscience or any other special powers. Third, I wasn't certain that this was a positive development. Being blissed out felt wonderful; yet I was concerned about how destabilizing it was. I felt like I might have gotten in over my head and worried about becoming ungrounded. I was only in the first leg of my trip in India and had already unexpectedly gotten thrust from the known into the unknown, past the frontiers of normal everyday consciousness. I was in uncharted terrain and I had no map or idea of where to go or what to do next.

Thanks to that brief moment of elevated consciousness, I realized that any sense of being in control is actually an illusion created by the mind. However, that didn't make it easy to give up wanting to have a sense of control. Ever since I went to Harvard, I had become immersed in settings that emphasized a type A/Brain 2.0 approach to the world. For years, I was surrounded and shaped by overachievers who had their lives planned out, who knew exactly what they wanted to do and how they were going to become titans in their fields. During the course of my career, success had come from exerting control over situations and reducing risk and uncertainty. This was the way I had been brainwashed by my education and training to go about in the world to be successful and get what I wanted. But it had also become normal for me to feel anxiety and even panic whenever anything felt out of control. To alleviate the anxiety, I would assert even more control. Now that I saw through the illusion, I still didn't know how to break this hardwired habit of needing to feel in control and striving to control everything.

For a long time, I couldn't articulately talk about this experience because I didn't have the words to describe it. It would take years of searching through spiritual circles and reading the accounts of other people with similar experiences to eventually build up my vocabulary. For example, I learned the word "mysticism" months after I began having what could be called "mystical" experiences. I now understand "mystical" to be a catchall word that, according to the *Oxford English Dictionary*, means "having a spiritual symbolic or allegorical significance that transcends human understanding" and "inspiring a sense of spiritual mystery, awe, and fascination."[9] According to the

Merriam-Webster Dictionary, the word means "having a spiritual meaning or reality that is neither apparent to the senses nor obvious to the intelligence," and "involving or having the nature of an individual's direct subjective communion with God or ultimate reality."[10]

These definitions confirm what I can say from firsthand experience: these episodes are so mysterious that no one can be 100 percent sure about the significance or meaning of mystical experiences or visions, not even the person who has them. Further, they are often spontaneous, unexpected, and unpredictable. Mystical experiences often contradict one's own personal belief system and that of conventional religious doctrine.* Thus these experiences often lead to a period of disorientation and confusion as one tries to make sense of them. However, they are life changing precisely because they are so confounding. Besides, the disorientation and confusion I experienced were more than compensated for by a humbling and profound sense of awe, wonder, elevation, and bliss.

I have since learned from research conducted by Andrew Newberg, a pioneer in the field of neurotheology, and David Yaden, who collaborates with Newberg, that mystical experiences are more common than we realize. According to Yaden, about 20 to 30 percent of people have had mystical experiences, but usually only a couple of them in a lifetime.[11] The scientific term they use to describe these types of phenomena is "self-transcendent experiences," which they define as "transient mental states marked by decreased self-salience [a scientific word for self-importance] and increased feelings of connectedness." In a recent 2017 paper called "The Varieties of Self-Transcendent Experience," Yaden and Newberg propose that these tend to run along a "spectrum of intensity that ranges from the routine (e.g., losing yourself in music or a book), to the intense and potentially transformative (e.g., feeling connected to everything and everyone), to states in between, like those experienced by many people while meditating or when feeling awe."[12]

In his book *How Enlightenment Changes Your Brain*, Newberg uses

---

* During the Inquisition, many mystics who openly talked about their experiences were persecuted for heresy. One of the most famous of these cases was that of the mathematician and astronomer Giordano Bruno, who was depicted in the first episode of the 2014 television series *Cosmos: A Spacetime Odyssey* as a heroic martyr for science. In summary, Bruno claimed to see in a vision that the Earth revolved around the sun and that the universe was infinite. He was later burned at the stake for refusing to recant his beliefs, which contradicted the doctrine of the Catholic Church.

"Enlightenment" with a big "E" to refer to transformative, life-changing mystical experiences and "enlightenment" with a small "e" to refer to more common lightbulb moments and epiphanies. (So I will adopt his wording going forward.) Through this fascinating book, Newberg explains how brain imaging technology helped him uncover the neural mechanisms underlying a variety of spiritual practices that people use to bring themselves into a self-transcendent state of "Enlightenment." These studies validated that self-transcendent experiences do result in physiological changes in the brain that can be measured.

In addition, one of the books I found most helpful for processing my experiences was *After the Ecstasy, the Laundry* by Jack Kornfield, in which he presents insights from compiling the accounts of nearly one hundred people who experienced a spiritual awakening and what happened in their lives afterward. Since there isn't an equivalent word in the English language, Kornfield uses the Japanese word "satori" to refer to the first taste of Enlightenment, which he describes in this way: "In awakening, our whole sense of identity shifts. We let go of our small sense of self and enter the unbounded consciousness out of which we come. What becomes known with absolute certainty is that we are not and never have been separate from the world. It is as if our heart, our knowing, expands further and further until we contain everything, until we are the world."[13]

The main theme that emerged from analyzing these accounts is that the process of awakening is not a fairy tale with a "happily ever after" ending. The book reveals that it is not uncommon for people who have experienced satori to develop depression as they try to reintegrate into society. Thus, reading the book helped me come to terms with how anticlimactic my own experience with satori was. After being blown away by a sublime merging with an ineffable superconsciousness, I still had to be "Due" and take care of all the things she had to take care of, including herself. I still need to do the mundane chores of everyday life, like doing laundry, washing dishes, and cleaning the floors. I still need to sleep, eat, go to the bathroom, and take showers. I still get stomachaches and catch colds. Personal weaknesses and character flaws do not disappear. Difficult relationships are still difficult. None of life's necessities or challenges magically go away. Although my consciousness had shifted, the body or vehicle I had to operate to navigate living in this world was more or less the same. I was relieved that the book confirmed that this was, in fact, normal.

Kornfield's book helped me realize that satori is not a magical transformation that instantaneously turns someone into a saint or a buddha. Satori is only a beginning. Peak experiences spur us on because through them, we rediscover our true nature and see that our worldview and self-concept are illusions. Once I had a glimpse of being part of a much greater consciousness, I couldn't help yearning to experience it again, the way a moth flies to the light. Then I had to look deeply inside to investigate what keeps me from feeling that sense of elevation and connectedness all the time.

I had to find inside myself answers to questions such as: How does the mind construct the illusion of a separate and limited self? Why does the mind build this illusion? My efforts to dismantle this illusion have since taught me that it is the work of a lifetime because there are layers and layers of illusions to break through. These questions led me to further investigate and reflect on how activating Brain 1.0 and Brain 2.0 disconnects us from oneness and can still trigger people who have had peak experiences to behave like assholes every now and then (myself included).

5.

Guided by
an Inner Compass

Accept my word not out of respect, but upon analyzing it as a goldsmith analyses gold, through cutting, melting, scraping and rubbing it. —SIDDHARTHA GAUTAMA, the Buddha[1]

Enlightenment is man's emergence from his self-imposed nonage.* Nonage is the inability to use one's own understanding without another's guidance. This nonage is self-imposed if its cause lies not in lack of understanding but in indecision and lack of courage to use one's own mind without another's guidance. Dare to know! (Sapere aude.) "Have the courage to use your own understanding," is therefore the motto of the enlightenment.
—IMMANUEL KANT[2]

Enlightenment, joy, and peace can never be given to you by another. The well is inside you. —THICH NHAT HANH

A LINEAR THINKER might assume that since I'd had such incredible elevated experiences in the presence of Lama Zopa Rinpoche, I would subsequently convert to Buddhism and adopt him as a spiritual teacher. As I would see for myself, the path is hardly ever linear or predictable. My inner voice would instead guide me in a different direction—one that would lead me to discover the intersection between what spiritual teachings and scientific research reveal about human flourishing.

"When a student is ready, the teacher will appear" is one of the sayings so

---

* Nonage is the legal status of being a minor without the rights, obligations, and duties of being an adult and needing the protection and guidance of a parent or guardian.

often repeated in spiritual circles it has become a cliché.* When I first heard it, I couldn't help hoping it would be true, because wouldn't it be awesome to finally have a spiritual teacher? I wondered: What would my teacher look like, and how would I know when I met him or her? When I had this initial experience of satori, I genuinely considered the possibility that it meant my teacher had finally appeared!

Therefore, during the remainder of my time at Tushita, I made a point to read more books by and about Lama Zopa Rinpoche, including a compilation of letters he'd written to his students. Yet the more I read, the less his philosophy and worldview resonated with me. Lama Zopa Rinpoche seemed to adopt a shamanistic and ritual-based approach to Buddhism that I found incomprehensible. Since my wish to understand "reasoned faith" was what drew me to Dharamsala, my gut told me that this would be a bad student-teacher fit.

Since then, I have met various spiritual teachers and again and again, I did not feel any compulsion to adopt them as a guru. I realized the problem may be that I just can't buy into the concept of guru devotion. The best I can do is treat spiritual teachers with the same respect I have for college professors and thought leaders. The saying that is more in line with my experience would go: "When the student is ready, a teaching will be learned."

Nevertheless, the unexpected experience of satori at Tushita forced me to recognize that my concerns about Tibetan Buddhism were about form rather than substance, because this tradition does preserve an ancient wisdom that is authentic and genuine. Furthermore, experiencing satori also opened my mind to the possibility that every one of the world's religions, regardless of any of my misgivings toward religious institutions as political organizations, may be preserving and transmitting authentic and genuine truths inside packaging that is so distracting that it often serves as a decoy. To get at the core truths, or to purify gold, as Siddhartha Gautama put it, I had to learn to separate the essence from the external delivery mechanism. This meant I had to develop the patience and fortitude to resist giving in to my habitual tendency to throw the baby out with the bathwater.

---

* I have always found this line puzzling because the wording implies that a teacher would magically appear whenever a student is ready. However, many of the spiritual aspirants I met who said this seemed to have gone out of their way to find a teacher, so they made it a self-fulfilling prophecy.

Meeting More Mystics

Since that first experience of satori, what kept a skeptic like me from reverting to scientific materialism was my continued amazement at the unexplainable coincidences or synchronicities that continued unfolding. For instance, after the retreat at Tushita ended, the next planned stop on my itinerary was the Thosamling Institute, a small nunnery and retreat center for women nestled in the gorgeous foothills of the Himalayas. When I booked my retreat there in February, I had no idea that coming there would put me in the right place at the right time to see the Dalai Lama.

The day I arrived at Thosamling happened to coincide with a teaching given by the 14th Dalai Lama, Tenzin Gyatso, at a monastery within easy walking distance. While I arrived too late to attend the teaching, I got to attend the long life "puja" (the Sanskrit word for prayer ceremony) being held for him the next day. For me, it was an anthropological spectacle to see the Dalai Lama revered and venerated in a Tibetan environment. I had no idea how to understand the ceremonies and prayers, but it was obvious that all the Tibetan pilgrims, many of whom who had traveled very far to see him, were elated and euphoric. The atmosphere was so contagious, I absorbed their high spirits like a sponge. Among his people, the Dalai Lama is revered as a manifestation of the Bodhisattva of Compassion, named Chenrezig in Tibetan (and Avalokiteshvara in Sanskrit and Guan Yin in Chinese). Although I was thrilled to be in such close physical proximity to him, I found it impossible to put him on such a pedestal. For me, he was a great human being with extraordinary levels of wisdom and grace, but a human being nonetheless.

While I was thrilled to be able to see the Dalai Lama in a Tibetan environment, the highlight of my ten-day stay at Thosamling was becoming friends with a fellow American woman named Mary Reed, who was staying there on long-term retreat. Like me, she was a successful and nonreligious businesswoman who'd changed jobs to do work that was more meaningful in developing countries. In her thirties, she started having spontaneous, uncontrollable, mystical visions that, over time, eventually made it impossible for her to continue to work as a professional. Through personal networks, she was connected with the Karmapa, who is the second most senior lama in Tibetan Buddhism after the Dalai Lama, and was invited to come to see him in India to seek his guidance. During their meeting, he suggested that she stay in the area and take time to really understand who she is. In her search for a quiet sanctuary to follow these instructions, she found her way to Thosamling. When I met her, she had been there a few months and had only started trying to capture insights from her mystical experiences in writing.

Normally, when I encounter people who admit to having visions and conversations with Jesus and with angels, I either suspect mental health issues or an exploitative motive to profit from the naïveté of gullible people. However, Mary was clearly not crazy and not a con artist. She glowed and radiated a sense of peace, love, and warmth that magnetically drew people to her, myself included. Perhaps because of the commonalities we shared, it was Mary, rather than a formal "guru," who helped me most in making sense of and embracing the changes in my life.

One day, as we took a hike through the mountains above Thosamling together, I explained that despite how much joy and bliss came with having a spiritual awakening, in its aftermath, I was perplexed by how much anxiety I felt about the uncertainty of no longer having a clear path of progression in my life and career. I connected this anxiety to having been trained and indoctrinated in a type A approach to asserting control over uncertainty and having to grapple with a new awareness that I was (and am) not really in control. In contrast, she embodied so much joyfulness and presence, that I asked her how she had made this transition. She explained that she had come to recognize that *being* was easier than doing. She also shared that allowing things to unfold often led to better outcomes because there are too many factors, and we cannot control all of them. Besides, our egoistic views are so limited. There is more expansion and joy when we allow space for being surprised.

Several times during a lunch or dinnertime conversation, Mary shared with people at the table that she had recently had a vision that seemed to involve a symbolic meaning and asked people in the group what the image symbolized in their tradition. Hearing her talk about her visions and her writing made me curious. In our last conversation, I asked her about how the writing was going and what she was writing about. Mary then shared with me that in trancelike states, she encountered Jesus Christ, the Buddha, and angels (sometimes all together), who would transmit messages to her in pure energy rather than language. Then the challenge was for her to put what she received into words so she could share the message more broadly.

As I listened to her, I realized that what she described was how I had experienced my own realizations: an inner voice communicating with me, not in language, but in what I call thought forms. Eventually, months later, when I started to have visions, I also found them puzzling to interpret. As I grappled with the confusion and disorientation, it gave me comfort knowing that I was not alone. Mary had gone down this path ahead of me and was an e-mail away whenever I wanted to ask her for advice. What I appreciate most about Mary's wisdom is that she simply shares what she sees in a higher realm without the filter of any religion or ideology.

A couple years later, when Mary's book was published, I valued and respected her raw honesty and courage in sharing the long period of troubling confusion, self-torment, and destabilization she went through before she finally accepted and embraced her mystical awakening.[3] I also found in reading it validation for my own firsthand experience that mystical insights challenge what we hear from an orthodox religious establishment and lead us to question the beliefs and assumptions we hold in our minds about conventional reality. Most important, I learned from her journey to not torment myself about having confounding mystical experiences.

In hindsight, one thing I realized that Lama Zopa Rinpoche, the Dalai Lama, the Karmapa (whom I also saw as part of a large group visiting his residence), and Mary all have in common is extraordinary mystical abilities. I wondered if being in close proximity to mystics could activate dormant mystical abilities in someone who didn't have any reason to believe that he or she had any, like me. Getting to know Mary and meeting more mystics later in my journey confirmed my hunch that every mystic sees differently and has his or her own unique path. There is no absolute or one-size-fits-all mystical view. Rather than provide clear answers, I find that mystical experiences seem to open up even more questions. Plus, to my relief, awakening mystical abilities doesn't have to lead to a sudden erasure of one's individual character traits or personality. I have learned from my own evolution over the past few years that when a person is by nature grounded, rational, and practical, after becoming a mystic, that person is still grounded, rational, and practical.

Entering the Vipassana Retreat

My next stop after Thosamling was to enter the Himachal Vipassana Centre, which is located across the main road from the Tushita Meditation Centre on a mountain overlooking McLeod Ganj, the major Tibetan urban center of Dharamsala. There I was scheduled to do a completely silent ten-and-a-half-day meditation retreat in the Vipassana tradition, which is also known in the West as insight meditation (the word "Vipassana" is commonly translated as "insight" in English) and as mindfulness meditation.

I have to confess that initially when I signed up for this retreat a few months earlier in February, I thought of it as a bucket-list type of challenge, like running a 5k race or biking a hundred miles for charity. I thought the point was to see if I could survive ten days of silence without losing my mind or giving up. So after I signed up, I felt a sense of anxious trepidation whenever I thought about what I was getting myself into. Fortunately, many of the people I met at Thosamling, including Mary, had already done a Vipassana retreat and shared with me what their experience was like. Now after three weeks of being immersed in Buddhist teachings in Dharamsala, I came to see and appreciate this retreat as an opportunity to directly experience the wisdom transmitted by Buddhism in a traditional format. It seemed as close an approximation as I could possibly find as a layperson to experience how Buddhist meditation may have been originally taught in India. When the day came to leave Thosamling for the Vipassana retreat center, I arrived with a feeling of hopeful anticipation.

To share some context, the Vipassana retreat program I did was developed by S. N. Goenka, an Indian businessman born in Burma who trained under Sayagyi U Ba Khin, a highly regarded lay Buddhist meditation teacher who also held the position of accountant general of the Union of Burma. Goenka first learned Vipassana meditation to find relief from terrible migraines. He was so amazed by how effective the practice was in enhancing his overall health that he formally took Sayagyi U Ba Khin as his teacher. Coming from a devout Hindu family, he had to reassure his parents that practicing meditation did not make him a convert to Buddhism. After the military takeover of Burma, Goenka moved to India and, with the blessings of his teacher, began presenting Vipassana in India as a secular practice for calming the mind and improving well-being, which could benefit anyone regardless of their religious upbringing. His secular approach to Vipassana was so popular among Hindus in India, it sparked a revival of Vipassana in its country of origin. Goenka also attracted many students from around the world, and a number of these international students went on to establish Vipassana centers in their home countries under one overarching international umbrella.

Today, according to the organization's website (www.dhamma.org), there are over 170 Vipassana centers worldwide, 80 of which are in India alone. Like the one I attended in Dharamsala, all of the centers are run by volunteers and funded by donations. All offer the standardized ten-and-a-half-day retreat program that requires all participants to take a strict vow of silence. The teaching component is provided through video and audio recordings of Goenka guiding a past retreat, which are played in the evening as discourses. For the most part, Goenka is a very engaging speaker, so the videos are enthralling. Nonetheless, when Goenka asserted that his Burmese lineage is the purest transmission of Vipassana as originally taught by the Buddha (a claim that I could find no historical evidence to back up), I saw it as a sign that this leading authority on Vipassana still had a Brain 2.0 tendency to be strongly biased toward his own tradition. That meant a lifelong Vipassana practice by itself was not sufficient for understanding and taming Brain 2.0.

Although there are many forms of meditation in Asia, in the West, Vipassana or mindfulness has become the predominant form of meditation in the mainstream. Piecing together the history of the American mindfulness movement helped me understand why. In the 1960s, three of its key pioneers, Jack Kornfield, Joseph Goldstein, and Sharon Salzberg, traveled to Asia, where they became exposed to Buddhism through various teachers, primarily in the Theravada tradition, and engaged in intensive meditation practice mainly in the Vipassana style. Eventually, Kornfield, Salzberg, and Goldstein returned to the United States and began to share what they learned. In 1975 they cofounded the Insight Meditation Society, based in Barre, Massachusetts, and went on to write books on meditation that became widely popular.

One of the key people who studied with them was a researcher named Jon Kabat-Zinn, who had a PhD in molecular biology from MIT and worked at the University of Massachusetts Medical School. In 1979, he developed mindfulness-based stress reduction (MBSR), a more fully secular version of what he'd learned at the Insight Meditation Society. Thanks to his scientific research orientation and advocacy among the academic community, formal studies on MBSR were conducted to validate its potential health benefits in various clinical applications. As a result of the growing scientific evidence, the MBSR approach to meditation captured the popular imagination and has become almost as widespread as yoga classes.

The conflation of mindfulness meditation with stress reduction was initially very effective for generating popular interest in mindfulness. Unfortunately, this has also resulted in a limited Brain 2.0 conception of mindfulness as a stress reduction intervention when there are many other important ways in which mindfulness can positively affect one's life.

What I was about to learn from this Vipassana retreat was that the most powerful benefit of mindfulness meditation is how it activates and strengthens Brain 3.0 so that it becomes more effortless for me to be my best self, see the bigger picture, and bring myself into a higher state of consciousness. The practice helps me declutter my mind and makes it easier to tune in to my inner compass and connect to my inner wisdom. I have since continued this form of meditation on a regular basis to continue strengthening Brain 3.0 so I can tap into a deeper wellspring of purpose, energy, and strength.

As a brain geek, I consider the Goenka Vipassana program to be a powerful recipe for breaking habits and driving the rapid rewiring of the brain over a ten-day period. About one century ago, William James, widely hailed as the father of American psychology, laid the foundation for how we understand habits when he wrote: "All our life, so far as it has definite form, is but a mass of habits—practical, emotional, and intellectual—systematically organized for our weal or woe, and bearing us irresistibly toward our destiny." This conclusion was based on his observation that "nine hundred and ninety-nine thousandths [0.999, or 99.9 percent] of our activity is purely automatic and habitual, from our rising in the morning to our lying down each night" and that most of the activities we carry out in our daily routine are so fixed by repetition that they could be classified as "reflex actions."[4]

James also explained that no matter how strong habits are, there is a possibility to change them: "New habits can be launched, I have expressly said, on condition of there being new stimuli and new excitements. Now life abounds in these, and sometimes they are such critical and revolutionary experiences that they change a man's whole scale of values and system of ideas. In such cases, the old order of his habits will be ruptured; and, if the new motives are lasting, new habits will be formed, and build up in him a new or regenerate 'nature.'"[5]

In line with James's prescription, the Vipassana retreat forces participants to make a complete break with their normal habits and routines (which are "stored" in Brain 2.0), both in terms of observable behaviors and in terms of internal mental patterns, for ten and a half days. By reducing their reliance on Brain 2.0 neural networks, the retreat creates space for participants to build new neural pathways in Brain 3.0. The process begins at registration, when participants are required to make several commitments for the entire retreat period, which ensures that for people who are not monks or nuns, the experience is drastically different from ordinary everyday life.

First, I had to take a vow of "noble silence" from the evening of the first day until the morning of the last full day. The organization's website explains in its Code of Discipline: "Noble silence means silence of body, speech, and mind. Any form of communication with fellow students, whether by gestures, sign language, written notes, etc., is prohibited."[6] This also means that students refrain from reading or writing during this period. Participants are only allowed to communicate to a teacher or volunteer retreat manager about questions or problems and are told to keep these interactions to a minimum.

Second, I had to commit to follow the Buddhist Five Precepts in a strict manner:

1. to abstain from killing any being
2. to abstain from stealing
3. to abstain from all sexual activity
4. to abstain from telling lies
5. to abstain from all intoxicants

In alignment with these precepts, Vipassana retreat centers serve only vegetarian food so that no animals are killed to feed the participants. The Vipassana retreat centers are also designed to completely segregate men and women to ensure there is no physical contact between people of the opposite sex. There is a male instructor for the male participants, and a female instructor for the female participants. To minimize contact between people of the same gender, each individual gets his or her own privately partitioned sleeping quarters with only a bed and side table—which in my case was a cubicle smaller than a prison cell and about as sparsely furnished.

In addition, in Theravada Buddhism, there is also a practice of fasting by not taking any meals after noontime. Likewise, in the Vipassana retreat, dinner is provided only on the first evening. After that, participants are served only tea and snacks in the evening. So breakfast is literally what the word suggests: breaking the fast. I'm pretty sure that going without dinner for ten days forced my body to change its metabolism to adapt to the meal schedule.

Finally, the strict daily schedule was such a dramatic change from my ordinary routine, it must have facilitated my neural wiring to dramatically change. Altogether, about eleven hours of meditation were scheduled per day, excluding breaks. The wake-up bell was rung at four a.m., and the first meditation period started at about four thirty. We would have breakfast at six thirty and resume meditation until lunchtime at eleven. After lunch, we would resume meditation at one p.m. In the afternoon session, we would then be called in groups for quick check-ins with the teacher. There would be a tea break from five to six p.m., followed by an evening meditation session. Then at seven they would play videos of discourses by Goenka, followed by a Q&A session and further meditation. Finally, at nine thirty, the lights would go out.

While the retreat conditions and schedule may seem extreme, my understanding of neuroscience helped me to appreciate a method in the madness. For ten and a half days, my brain and body were forced to take a break from my normal mode of being externally oriented toward information coming in from the outside world through my senses. Therefore, this minimized my use of the major neural pathways normally activated in my daily life. This enforced sensory withdrawal created the opportunity for neural networks in Brain 3.0 to be exercised and strengthened by meditation so that new mental habits (meaning new "neural highways") could be built.

Understanding Vipassana as a Brain Geek

My experience and understanding of mindfulness during this retreat were greatly influenced by my having read Daniel Kahneman's *Thinking, Fast and Slow*, which integrates findings from behavioral economics, psychology, and neuroscience to shed light on how thinking happens. Kahneman writes, "Cognition is embodied; you think with your body, not only with your brain,"[7] and presents scientific evidence refuting the notion that our mental life, our mind, is housed entirely within our brain. Information is processed not only in the brain, but also throughout the entire body through physiological reflexes, facial expressions, and emotions, usually in a subconscious manner that comes into our mind as intuition and instinct. This is what Kahneman calls "fast thinking." Fast thinking unfolds through a set of neural mechanisms that cognitive scientists simply label System 1, which essentially corresponds to our "autopilot." In fractions of a second, System 1 somehow processes all the information coming into our senses at a level below conscious awareness and spits out an intuitive answer.

In contrast, System 2 is the shorthand label that cognitive scientists give our conscious attention system, which takes more time and effort to analyze information and generate a more informed answer. Kahneman describes the function of System 2 as "slow thinking." Given that System 1 corresponds to the autopilot, System 2 corresponds to the "pilot," or the conscious, rational self. The slow-thinking pilot consumes lots of energy to fuel the mental effort needed to focus, pay close attention to details, solve analytical challenges, and make plans. In contrast, the autopilot burns much less energy. Therefore, for the sake of energy efficiency, it makes sense to let the autopilot run the show and have the pilot take over only when necessary.

What people don't realize is that the autopilot is actually driving 99 percent of the time. To conserve energy, the pilot spends most of the time in "sleep mode" and wakes up only when the autopilot sends an alert signal that more effortful attention is needed to complete a task. However, if the autopilot manages to come up with a fast-thinking solution, the snooze button is pressed and the pilot returns to sleep mode. Only when it's clear that the autopilot cannot find a solution will the pilot fully wake up and expend energy on the slow thinking needed to address the situation.

Together, fast thinking and slow thinking work seamlessly together in generating the experience of "sentience," the scientific and philosophical term that refers to the subjective experience of being conscious in terms of feeling, sensing, perceiving, and thinking. The saying that "the whole is more than the sum of its parts" is especially true of the mind. We can't pinpoint exactly how the brain, nervous system, and body generate the conscious experience of mind. We can't pinpoint where fast thinking ends and slow thinking starts and vice versa. Therefore the experience of mind is best described as a stream of consciousness, a term coined by William James in 1890.[8]

Vipassana, or mindfulness, meditation only started to make sense to me when I thought about it through the lens of Kahneman's fast-thinking and slow-thinking framework. Basically, mindfulness trains us to use the slow-thinking system (the pilot) to watch the stream of consciousness that flows through the fast-thinking system (the autopilot). Since our fast thinking essentially reflects the mental habits and patterns that have been hardwired into our autopilot, if we want to change and rewire our autopilot, we need to first strengthen the ability of our pilot to observe and guide our fast-thinking system without getting swept into the fast-moving stream of consciousness. In this view, I see attention and awareness as the two core "muscles" of the pilot, and mindfulness meditation as a very effective workout for these muscles.

Being able to focus attention and to open awareness are vital to metacognition and interoception, two interrelated capacities that make up important aspects of self-awareness and executive functioning. Metacognition, which etymologically means thinking about thinking, refers to our ability to become aware of and understand our own thinking patterns and cognitive processes and to intentionally change our train of thought. Interoception can be broadly defined as the ability to sense signals originating within the body and also interpret them. Basically, interoception enables us to answer the question "How do I feel?" The interoceptive system is made of special nerve receptors that enable us to sense and assess our physiological condition by tuning in to internal vital signs, such as respiration, heart rate, hunger, thirst, and the need to use the bathroom, as well as our energy levels and emotional state.[9] The term "interoceptive accuracy" refers to our ability to correctly interpret these signals and take appropriate action—for instance, being able to tell that a headache accompanied by dryness of mouth is a sign of dehydration and then drinking water, or being able to tell that a growling stomach accompanied by irritability is a sign of hunger and then getting food.

From what I experienced, the Vipassana retreat was like a boot camp workout for developing metacognition and interoception. There were two main phases in the Vipassana retreat. In the first phase, over the first four days, we did a breathing meditation practice that enhanced my ability to focus attention and my metacognitive awareness. In the second phase, which involved the remaining six days or so, we did a body scanning practice that built up my interoceptive awareness and further enhanced metacognition by enabling me to see and feel how the mind is fully integrated within the body, such that there really is no biological separation between body and mind.

In the first phase, we were merely told to pay attention to the sensation of breathing at the small area of skin between our nostrils and upper lip. And of course, what I observed was how impossible it was to pay attention to something as simple as the breath for more than a few seconds before my mind started wandering. Since I have known for a long time that I have a racing mind and a sometimes wild imagination, I didn't consider this to be a major revelation. Plus, scientists had discovered that mind wandering is completely normal and had proposed that it may have evolved as a means to give our executive functioning neural pathways a break.[10] According to a 2010 study by Daniel Gilbert and Matt Killingsworth at Harvard, people's minds tend to wander about 47 percent of the time.[11]

Intriguingly, through brain imaging studies, neuroscientists learned that the areas of the brain active during mind wandering correspond with the "default mode network," which got its name because these areas remain active even when people are resting and not focused on any activity. The primary function of the default mode network appears to be integrating past experiences and interactions into our maps of the world so we can construct meaning and simulate scenarios for how to interact with the world and other people. This may explain why mind wandering is often focused on our personal lives and concerns that are most important to us. That said, Gilbert and Killingsworth gave their journal article on the study the title "A Wandering Mind Is an Unhappy Mind" because they found that people are less happy when their minds are wandering than when they are fully engaged in what they are doing in the present moment, regardless of the activity. Therefore, it couldn't hurt to learn how to help my mind wander less.[12]

During the Vipassana retreat, as I reflected on how I could even be aware that my mind had wandered, I realized I was using my slow-thinking pilot (System 2) to observe my fast-thinking autopilot (System 1) racing and wandering all over the place. I then saw that my fast-thinking mind was (and is) very much like a monkey, constantly jumping from thought to thought through what seemed like endless random associations. In the absence of external sensory stimulation, I realized how much noise and commotion there was inside my head, which I gave the name the "mind-track."* My mind-track was often such a loud, chaotic, cacophonous theater that I often found it impossible to follow the instructions to pay attention to the breath.

Then slowly, it began to dawn on me that perhaps the point of the meditation practice wasn't so much about forcing my attention on the breath as it was about becoming familiar with my mind-track and the specific patterns by which my own monkey mind jumped around. Like lifting the hood of a car to

---

* Similar to the soundtrack of a movie, my mind-track accompanies and amplifies my experience of life; the difference is that my mind-track is raw, unedited, and unpolished.

see how it is built, observing my mind-track enabled me to see how neurons fired and wired together in my brain to create stories and epic dramas as well as the dynamic hologram of an identity and sense of self.

The daytime instructions were frustratingly minimal, but we fortunately got much more explanation in the evenings when the videos of Goenka's discourses were played. Goenka shared that the aim of Vipassana is to directly experience in our bodies the Buddhist principle that all phenomena are impermanent and subject to change. All our physical sensations and discomforts and all our thoughts and emotions arise and pass away like bubbles surfacing in water. If we don't grasp at, get attached to, or resist these phenomena, they pass away naturally.

To further understand this principle, we were instructed to practice "sittings of strong determination," in which we were to try not to move or adjust our position for at least an entire hour. And sure enough, I saw that the discomfort I experienced, as well as the chatter of stories that played out about this discomfort in my mind-track, really did arise and pass away. This realization enabled me to neutrally observe and accept physical sensations, thoughts, and feelings without having to judge or label them as good or bad. This then helped me develop a stronger sense of equanimity, which is described on Wikipedia as "a state of psychological stability and composure which is undisturbed by experience of or exposure to emotions, pain, or other phenomena that may cause others to lose the balance of their mind."[13]

In the second phase of the retreat, the meditation instructions evolved. We were told to create a focused "laser beam" of attention, equivalent in size to that of the small area between the nostrils and lips (approximately the size of a nickel) that we had been focusing our attention on to feel the sensation of breathing. Then we used this beam of attention to scan our bodies in a continuous fashion. We started by slowly scanning the exterior of the body, then after a day or two, we were instructed to scan faster and then to scan the interior of the body. Then, near the end of the retreat, we were instructed to use two beams to simultaneously scan the left and right sides of the body in a symmetrical fashion.

Again, I couldn't help trying to understand what was happening from a scientific perspective. What I realized was that this second phase of continuous body scanning was like a long endurance workout for the interoceptive system. The scanning forced me to tune in to the interoceptive nerve receptors throughout my organs, muscles, and bones and to learn to override the tendencies to emotionally react to these signals and sensations, categorize them as good or bad and pleasant or unpleasant, and cling to them or resist them. Instead, I learned to just feel them as raw sensations with an attitude of curiosity and acceptance.

At the point of transition from the first phase into the second phase, having spent four days paying attention to the sensations in that tiny area between my nostrils and my lips, my sensitivity in that area was so refined, I could pick up the subtlest and most minute changes in temperature between the inhale and the exhale, the movement of the tiny hairs, and the pulsing of blood through the capillaries. So when I began to move a "laser beam of focused attention" throughout my body, it was like spreading that degree of sensitivity to the rest of my body. In time, I could feel the beating of my heart, the pulsing of blood through my veins and arteries, and the movement of the smooth muscles that make up my digestive system.

By the ninth or tenth day, I could feel subtle vibrations throughout my body that seemed like electromagnetic waves—like I was a river of subtly vibrating, pulsating energy. It was like being able to feel my body dissolve into the subatomic particles coming together to form dynamic structures that lasted only nanoseconds at a time. Feeling my body dissolve into energy in that way was surprisingly life-giving rather than scary. I felt whole, integrated, alive, and blissful. It was a mind-blowing experience. It was incredible to finally see how much there was to sense in a universe within my body that I had never noticed before. Until this retreat, I had never developed my interoceptive system enough to be able to tune in to these subtle sensations.

The other thing I noted during this second phase was how much harder it was to keep my mind steady. The sensations kept drawing my mind away from the meditative exercise into the past. As I scanned my body, memories associated with sensations in different body areas would rise into conscious awareness, often for the first time in years. As these long-forgotten scenes from my life resurfaced and played out in my mind-track, I kept getting pulled into the quicksand of emotional catharses I could not control. The sounds of people crying all around me in the meditation hall suggested that I wasn't the only person experiencing this. Wrestling with the monkey mind to resist or suppress these memories and the accompanying emotions was of no use. The only way out of the quicksand was to surrender into it by letting the memories

surface and letting the emotions run their course until they passed away. Ironically, it was from surrendering to the mind-track that profound insights began
to arise.

Insights on Brain 1.0, Brain 2.0, and Brain 3.0

By watching my mind-track during these endless days of meditation sits, I
came to recognize that I had three different "fast-thinking" personas inside
me. How I thought, felt, and acted was very different depending on which of
these personas was in charge. During my meditations, these personas began
to come to life, so I gave them names: the Inner Godzilla, the Inner Teen Wolf,
and the Inner Sage. Eventually, I connected these three emotional states with
three patterns of brain activation I simply called Brain 1.0, Brain 2.0, and Brain
3.0, according to the sequence by which the corresponding neural networks
and brain regions develop and mature.

After days of watching my mind-track for hours and hours, I became much
more intimately familiar with my three personas and how they seemed preprogrammed into my brain.

- The Inner Godzilla is the insecure part of me that is afraid—of losing,
  of being left behind, of being treated unfairly, of never having enough to
  live on, of going hungry, of being alone, of failure, of never being com-
  petent enough, of not being enough, of not belonging, of being rejected
  or not included, of not being protected, of not feeling safe, etc. It is the
  part of me that can get overwhelmed and react by smashing things or
  disappearing. By tuning in, I eventually connected my Inner Godzilla
  to the neural structures of the freeze-flight-fight system. Since these
  structures are fully formed at birth, I named this emotional state
  Brain 1.0.
- The Inner Teen Wolf is the restless part of me that anxiously wants to
  have the external world give me whatever I crave but also lets the external
  world define my self-worth and determine my happiness. It is the part of
  me that wants to have all my desires fulfilled immediately, to win every
  competition, to always have control, to be at the top of a hierarchy, to be
  powerful, and to have maximum advantage in every situation. I eventually connected my Inner Teen Wolf to the neural structures of the

dopamine, or reward, system. Since these structures develop in adolescence, I named this emotional state Brain 2.0.

- The Inner Sage is the centered part of me that can wisely see the bigger picture, connect to a greater purpose, act in the service of the greater good, and keep picking myself up and persevering even when situations seem hopeless. It is the part of me that resourcefully manages to find a way out of a dead end. It is the part of me that is compelled to move toward continuous growth and self-actualization. I eventually connected my Inner Sage to my frontal lobes, where neuroscientists believe executive functioning, higher abstract processing, and optimism reside. Since the frontal lobes are the last parts of the brain to mature (in our midtwenties), I named this emotional state Brain 3.0.

As scenes from across my life arose and played out in my mind-track, I felt regret for many of the things I did and said in Brain 1.0, when the Inner Godzilla took over. The scenes reminded me how I pushed people away and burned bridges to protect myself. In Inner Godzilla mode, I felt like I was contracting like a turtle retreating into its shell. By spending so much time in this state, I built strong armor and thick walls, and ended up imprisoning myself inside them. I also saw how being in this state was very tiring because I exhausted a lot of energy carrying armor and walls everywhere I went. Tuning in to what it felt like in my body to be in Inner Godzilla mode made me see that my body was very agitated. My breathing was shallow and my heart beat fast. I felt fear, helplessness, and a sense of desperation and exhaustion. In this emotional state, my mind was naturally drawn to the negative and consumed by worry, anxiously playing out all the things that could go wrong.

I also felt regret for many of my actions in Brain 2.0, when the Inner Teen Wolf was in charge. My mind-track showed me how I could get stuck in carrot-chasing or reward-seeking mode for years at a time. The scenes showing times when I was in Brain 2.0 reminded me of how often I sacrificed the present moment in order to jump through hoops and maximize opportunities to achieve success in the future. Yet when the successes did come, I merely shifted my focus to the next major milestone on the horizon. Tuning in to what it felt like in my body to be in Inner Teen Wolf mode made me realize that I was like a junkie who continuously needed another fix of achievement and success to fill an anxious void inside me. The hunger was physical, mental, and emotional. My heart raced. I fidgeted and became restless as my arms and legs filled with energy to run after what I wanted. Whenever I didn't get what I wanted, the Inner Teen Wolf quickly devolved into the Inner Godzilla.

It was like my mind-track was showing me that the chase for more, more, more had consumed me and given me such extreme tunnel vision that I completely neglected, discounted, and ignored the things that bring meaningful joy to life, like family, friendship, and making a difference. Worse, my body was burned out from the endless chasing. When I finally understood that craving only fueled more craving plus burnout, I realized my body (via my mind-track) was asking me to finally break out of this dopamine-and-adrenaline-fueled way of living.

My mind-track also showed me scenes of how different I was in Brain 3.0 because my Inner Sage would guide me. The Inner Sage enabled me to love learning just for the sake of learning, even when there were no external rewards. I saw that as a child, my Inner Sage helped me persevere through the many crises that took place at home. In Brain 3.0, I could calmly meditate on a challenging problem until I found a solution. My Inner Sage naturally looked to understand the human condition and examine universal issues shared by all people. It helped me to generously forgive misunderstandings and failures by seeing them as part of a learning process. The scenes showed me that in Brain 3.0, I could be fearless about pursuing growth, wisdom, and self-actualization. I saw that over the course of my life, my Inner Sage moved me forward in my development. My Inner Sage helped me to earn a seat at Harvard. Then, when PTSD caused me to get stuck in Brain 1.0, even during the darkness of that period of my life, my Inner Sage guided me to reconstruct my mind and rebuild Brain 3.0. And now my Inner Sage was using my mind-track to finally get me to realize that my Inner Sage is my natural state of being—not the Inner Godzilla or the Inner Teen Wolf.

I also observed in real time that when I was in Brain 3.0, it was far easier to do the meditative exercise of scanning my body than when I was emotionally in Brain 1.0 or 2.0. When my Inner Sage was in charge, it was easy to detach from my stream of consciousness and watch my thoughts the way I appreciated waterfalls and waves on a beach. My heart rate calmed and I felt filled with a sense of peace and serenity. I felt my muscles relax and my body open and release tension. It felt like I was expanding and embracing the world with curiosity, awe, and hope. I often felt a swelling of joy and bliss in my heart, which automatically lifted my shoulders, neck, and head and straightened my spine. In this mode, insights kept surfacing that helped me to understand the meditation practices and teachings.

Eventually, I noticed that my System 1 (the fast-thinking autopilot) and my System 2 (the slow-thinking pilot) had different qualities depending on whether I was in Brain 1.0, Brain 2.0, or Brain 3.0, which I summarize in the table below. By observing my mind-track, I saw that System 1 and System 2 functioned in an optimal way only when I was in Brain 3.0.

Brain State	System 1 (Fast-Thinking Autopilot)	System 2 (Slow-Thinking Pilot)
Brain 3.0	Calm, open, curious, creative, present, optimistic, and a sense of abundance	Attuned to people, environment, and senses; can see bigger picture and interconnections
Brain 2.0	Restless, anxious to do or get something, a sense of deprivation and scarcity	Tunnel vision focus on getting carrot/reward; tuned out from surroundings, disconnected from present
Brain 1.0	Closed, defensive, self-protective, pessimistic	Can't focus or concentrate; experience brain fog

Furthermore, meditation created the mental space my pilot needed to step back from my fast-thinking monkey mind and take note of which emotional state I was in. This space also allowed me to make a conscious choice to shift my pilot more fully into Brain 3.0 and connect more strongly with my Inner Sage. This then enabled me to detach from whatever story or trigger was activating Brain 1.0 or Brain 2.0 and to shift my autopilot into Brain 3.0. Observing this shifting unfold was utterly fascinating. As I started to spend more time meditating in Inner Sage mode, I began to unpack what was happening in my mind during the body scanning. The process went like this: One, I noticed raw sensory input (sensations). Two, I noticed the activation of associated thoughts, images, or memories. Three, I noticed Brain 1.0 turn on and stimulate rumination or Brain 2.0 turn on and stimulate planning to get something I wanted. Four, I noticed my Inner Godzilla or Inner Teen Wolf build up elaborate stories. Five, I noticed coaching from my Inner Sage to detach. Six, I returned my attention to body scanning.

I began to watch with amusement and curiosity how raw sensory input and whatever associations they activated triggered Brain 1.0 and Brain 2.0 to weave stories, and eventually I was able to watch this happen without my mind (and body) fully going into Inner Godzilla or Inner Teen Wolf mode. Then I created an experiment by playfully observing how changing the way I perceived the same sensory input or stimulus could activate different neural circuits and create different inner experiences.

For instance, one day, as I started one of the sittings of strong determination, I noticed a spider crawling on the rug near me. Up until this point, the sight of a spider instantly triggered Brain 1.0. My typical reaction included freaking out, shrieking in fear, and getting far away or grabbing something or someone to kill the spider. This time, I calmly observed Brain 1.0 activate. I felt the impulse to freak out course through my body as my heart rate increased. I heard my Inner Godzilla shriek in my head and whine: "This sucks! Of all the places this icky spider could be, why the hell is it near me?! Please, please, please, leave me alone! Please don't be poisonous." It was very hard to focus on the body scanning because I kept imagining that every sensation I felt on my skin could be that of the icky spider crawling on me. I felt like a helpless victim.

Then my Inner Teen Wolf took over and the story in my mind immediately changed. My Inner Teen Wolf turned the situation into a battle between me and the spider. It said something like this: "Little spider, you are not going to ruin my meditation. You better watch out, little guy. Don't mess with me. I'm much bigger than you." I then watched my Inner Teen Wolf scheme about ways to get rid of the spider without killing it because the Five Precepts I had promised to keep during the retreat forbade me from deliberately killing anything. I felt this strong impulse to blow it really hard to get it far away from me. However, the possibility of blowing it onto someone else kept me from actually carrying out that impulse. I continued to be too agitated to focus my mind on the body scanning.

Then I asked myself: why am I making such a big deal of this? When it occurred to me that the spider was actually very vulnerable because anyone could inadvertently step on it, I was overcome with compassion. My Inner Sage naturally took over and the story in my head transformed. I started to see the spider as my meditation buddy and wanted to protect it. I didn't feel bothered by the spider anymore. I was even okay with the idea of it crawling on me so I could look after it. That gave me the peace of mind to finally focus on the body scanning. At the end of the session, I picked up the spider and took it outside where it would be safe from being inadvertently squished.

This experiment of observing my mind view the same stimulus in Brain 1.0, Brain 2.0, and Brain 3.0 helped me to see that there is a degree of free will in activating neural circuits to prime our mind's interpretation of unfolding events and how we respond to them. With enhanced metacognition and interoception, it finally became possible for me to unpack the process of perception and interpretation and to preemptively activate my Inner Sage so that the Inner Godzilla and Inner Teen Wolf would be less able to hijack my mind.

On the last day of the retreat, I noticed that the noisy commotion that characterized my mind-track at the beginning of the retreat had settled down. My mind-track had become quiet, calm, and peaceful. I felt joyful and lighthearted. Outside the meditation hall, the world had never looked more beautiful and awe inspiring. I admired details in everyday objects I hadn't noticed before. The colors of the leaves, the trees, and the ground were more vivid. The sky was spectacularly gorgeous. Everything seemed like a miracle. I lost track of time, appreciating everything.

I guessed that part of the reason I felt like a different person was because my brain had dramatically changed. It was like my neural superhighways were now more connected to the Inner Sage, and this gave it the ability to calm and quiet the Inner Teen Wolf or Inner Godzilla. The anxiety I had felt before coming into the Vipassana retreat about the looming uncertainty in my future was now drastically dialed down. Although nothing in my external life situation had really changed, there was a noticeable internal shift. It felt like inner peace and calm had become my new normal way of being. Somehow in ten days, my brain had rewired so much that I shifted my default state from being in Brain 2.0 to being in Brain 3.0. I finally felt like my true self! Equanimity was the natural by-product of a stronger Inner Sage turning down the volume of the Inner Godzilla and the Inner Teen Wolf. A less noisy mind-track, in turn, made it even easier for me to tune in to my inner wisdom and abide naturally in a state of inner peace, harmony, and joy.

What I learned during the Vipassana retreat enabled me to finally understand the Four Noble Truths, the core teachings of Siddhartha Gautama, in a way that is grounded in neuroscience. These foundational teachings emerged from his multiyear reflection and meditation on the universal human experience of "dukkha," a Pali word that is often translated as suffering but also means stress, dissatisfaction, discontentment, and unhappiness. The First Noble Truth simply states that dukkha exists. The Second Noble Truth states that the causes of dukkha are aversion, craving, and ignorance. The Third Noble Truth states that dukkha can cease when one liberates oneself from aversion, craving, and ignorance. The Fourth Noble Truth states that the cessation of dukkha can be achieved by following the Noble Eightfold Path.

This teaching made much more sense to me after I saw how aversion, craving, and ignorance can be overlaid onto the brain's three emotional states: aversion corresponds to Brain 1.0, craving corresponds to Brain 2.0, and ignorance corresponds to not having sufficiently developed the Brain 3.0 neural pathways needed to calm and tame Brain 1.0 and Brain 2.0 when they are triggered. I then realized that in the First Noble Truth, Siddhartha Gautama was simply observing that no matter what human beings do to be happy, unhappiness is the norm because happiness is very elusive and ephemeral. Then in the Second Noble Truth, he demonstrated that he must have figured out that human beings are wired to avoid pain and unpleasant sensations in Brain 1.0, crave and approach pleasure in Brain 2.0, and, in these states, ignore information that is not immediately relevant to self-preservation and to rewards that give us pleasure. In the Third Noble Truth, he explained further that the antidote to aversion and craving is to develop equanimity, and the antidote to ignorance is to gain wisdom into the true nature of reality. This demonstrated that he had somehow strengthened Brain 3.0 to such a degree that he could calm and deactivate the neural circuits for aversion and craving, thereby transcending Brain 1.0 and Brain 2.0. Then in the Fourth Noble Truth, he provided a recipe for self-directed neuroplasticity that explains how to live in a manner that strengthens Brain 3.0 and reduces the activation of Brain 1.0 and Brain 2.0.

In the Sallatha Sutta ("suttas" are teachings in the Pali Canon), Siddhartha Gautama used the metaphor of being hit by two arrows to help his followers understand the cessation of dukkha. The first arrow is the actual incident, which can indeed cause pain and injury; but the second arrow is self-inflicted—it comes from how our minds react to the first arrow to make the experience more negative. The second arrow represents resisting and resenting what has happened to us through aversion (wallowing in self-pity and bemoaning our pain) and craving or desire (wishing this hadn't happened to us and wishing to escape the pain). Those who are ignorant of the Four Noble Truths shoot the second arrow because they are attached to desirable states. His point is that if we don't shoot the second arrow, we will minimize our experience of dukkha.

While the metaphor made sense to me, before the Vipassana retreat, I didn't see how I could stop shooting the second arrow. It seemed impossible to overcome hardwired instincts to avoid pain and grieve about misfortune or to crave pleasure and rejoice when I get what I want. During the Vipassana retreat, I realized that the Inner Godzilla and the Inner Teen Wolf are the parts of my mind that shoot the second arrow. I suffer more when Brain 1.0 hijacks my mind and the Inner Godzilla exaggerates the negativity of my experiences, or when Brain 2.0 hijacks my mind and the Inner Teen Wolf exaggerates the pleasure or happiness I'd get from attaining the object of my desire, so I cannot be content with what I have. I also experienced for myself that even when life's arrows trigger Brain 1.0 and Brain 2.0, the neural pathways of Brain 3.0 can calm Brain 1.0 and Brain 2.0 so that the Inner Godzilla and Inner Teen Wolf don't take over the mind-track. That means that when Brain 3.0 is fully developed, the slings and arrows of life can still cause pain, but I can choose not to unleash the Inner Godzilla and Inner Teen Wolf.

As for the Noble Eightfold Path, the eight elements are often translated as Right View, Right Thought, Right Speech, Right Action, Right Livelihood, Right Effort, Right Mindfulness, and Right Concentration. However, these English words do not adequately capture the original meaning. Thanks to the Internet, I was able to find transliterations of the original terms in Pali. I also learned from examining online Pali glossaries that the word "samma," which gets translated as "right," is related to the English word "summit." There is no precise English word that captures the meaning of "samma," but rather than "right," it means something more like highest, best, complete, whole, perfected, wise, and sacred[14]—essentially what Buddhists believe are the qualities of our innate Buddha Nature. By closely examining the original Pali wording, I came to understand that the Noble Eightfold Path is essentially a call to live in alignment with our Buddha Nature by intentionally breaking habits that activate Brain 1.0 and Brain 2.0 and replacing them with new habits that enable us to spend more of our lives in Brain 3.0.

In case there is interest, I offer my translation and interpretation of the elements of the Noble Eightfold Path in the table below.

Element	Common translation	My perspective[82]
Samma Ditthi	Right View	Higher/Wise View—cultivating perspectives that activate Brain 3.0 and enable us to see through delusion and discern the true nature of reality
Samma Sankappa	Right Thought	Higher/Wise Intention or Aspiration—exercising and cultivating Brain 3.0 by generating an intention of metta (loving-kindness) and karuna (compassion) for all beings [see the compassion meditation in chapter 8] that is aligned with our Buddha Nature
Samma Vaca	Right Speech	Higher/Wise Speech—exercising and cultivating Brain 3.0 to speak in ways that uplift and bring other people into Brain 3.0
Samma Kammanta	Right Action	Higher/Wise Conduct or Behavior—exercising and cultivating Brain 3.0 to act for the greater good without causing harm, in alignment with our Buddha Nature
Samma Ajiva	Right Livelihood	Higher/Wise Livelihood—pursuing work that activates Brain 3.0 and sustains our lives without exploiting or causing harm or cruelty to other sentient beings
Samma Vayama	Right Effort	Higher/Wise Diligence—diligently cultivating Brain 3.0 through daily practices so we can more easily embody our higher values and aspirations and reduce our susceptibility to get triggered into Brain 1.0 or Brain 2.0
Samma Sati	Right Mindfulness	Higher Consciousness, Awareness, and Remembering—exercising and cultivating Brain 3.0 to observe and become familiar with the mind, distinguishing between raw sensory input and mental phenomena in response to sensory input, such as feelings, emotions, thoughts, and interpretations
Samma Samadhi	Right Concentration	Higher/Complete Absorption—activating Brain 3.0 to connect with our Buddha Nature, transcend our limitations, and see and sense how things are interconnected in a cosmic bigger picture

Following the Footsteps of the Buddha

When I first conceived the idea to visit the Buddhist pilgrimage sites honoring Siddhartha Gautama's major life milestones, I saw myself more as an archeological investigator than as a spiritual pilgrim. I needed to see the historic evidence with my own eyes before I could believe that this mythical figure called the Buddha actually lived on this planet. Very rarely have historical figures received the reverence and honor that has been heaped on Siddhartha Gautama for 2,500 years. To enjoy this type of legacy, he must have been an extraordinary human being when he was alive. I also hoped that perhaps after seeing these archeological sites, I might be able to understand the deeper meaning of his honorific title, the Buddha, which is often translated as the "Awakened One" or the "Enlightened One."

The magical legends about the Buddha forced me to wonder: was Siddhartha Gautama an extraordinarily inspiring but biologically normal human being whose wisdom and accomplishments became exaggerated in the retelling by later generations of Buddhists who never met him in person, or could he possibly have been a superhuman being as the myths around him imply? As it turned out, visiting the sites did not answer this question. In fact, I have come to realize that nothing short of becoming enlightened like the Buddha could ever allow me to truly understand what "Buddhahood" really means.

The classic Buddhist pilgrimage itinerary is called the Astamahapratiharya, a Sanskrit word that means the "Eight Great Sites of Wonders" (or "Miracles"). The four most important sites commemorate major events in the life of Siddhartha Gautama: Lumbini, where he was born; Bodhgaya, where he became enlightened; Sarnath, where he gave his first sermon, in which he taught the Four Noble Truths; and Kushinagar, where he died. In addition, four more sites were added because they are where the Buddha is said to have performed great miracles: Sravasti, where he spent much of his life teaching and where it is said the Buddha performed the great Twin Miracle, in which he emitted flames from the top half of his body and a stream of water from the lower half of his body and then ascended into heaven to visit and give teachings to his mother's spirit (she had died shortly after giving birth to him); Sankhasya, where the Buddha supposedly descended to Earth after visiting his mother; Rajgir, where Siddhartha Gautama gave many teachings at a nearby mountain called Vulture's Peak and where the Buddha miraculously subdued a wild, demented elephant that was sent to kill him; Vaishali, where Siddhartha Gautama established the very first monastic order for women in the history of India and where supposedly a monkey miraculously offered a bowl of honey to the Buddha and then a band of monkeys dug a water tank for him to use. Vaishali is also where the aged Siddhartha Gautama preached his last sermon before his death.

For practical purposes, I visited only six of these eight sites. I didn't visit Lumbini because it is actually located in Nepal and I didn't want to deal with the hassle of crossing the border and coming back into India. I also didn't visit Sankhasya because it is in a more remote area that is harder to get to. For this reason, Sankhasya is not on many Buddhist pilgrimage tour itineraries. Further, I was skeptical about its historical authenticity because I had heard mentioned that neither Sankhasya nor the story for which it is famous is actually mentioned in the Pali Canon, the collection of authoritative historical teachings of the Buddha preserved in Theravada Buddhism.

Visiting these sites was a sweeping history lesson in the rise and fall of empires. What I saw were the ruins of once grandiose monuments, neglected and forgotten from the time Buddhism was stamped out during the Muslim conquest of India until the nineteenth century, when a team of British archeologists led by Alexander Cunningham rediscovered and excavated them.* These monuments were built, more than two centuries after Siddhartha Gautama passed away, by King Ashoka (circa 304–232 BCE), one of the greatest and most revered emperors in India's history, who conquered and ruled one of the largest empires ever assembled on the Indian subcontinent. According to historians, Ashoka converted to Buddhism after growing tired of the destruction and suffering caused by years of warfare and empire building. When Ashoka made a pilgrimage to these eight sacred sites, he had pillars erected to commemorate his visit and also funded the construction of stupas and monasteries on their premises. I inferred from Ashoka's erection of monuments to commemorate the miracles at Sravasti, Sankhasya, Rajgir, and Vaishali that the legends of the Buddha's miraculous powers had already become widely popularized during Ashoka's time.

Under Ashoka's patronage, Buddhism prospered and spread throughout

---

* Cunningham was the first director-general of the Archaeological Survey of India, which was established in 1861 after he spent many years lobbying the British Indian government to excavate and safeguard the antiquities of India. Under his guidance, the team carried out a detailed survey of Buddhist sites at a time when India's Buddhist heritage had been largely forgotten by the local population.

India and nearby countries. Buddhism continued to flourish in India under the patronage of various kings until the tenth century, when a series of invasions by foreign armies began. The invaders pillaged monasteries, burned vast libraries of scriptures, destroyed monuments, and massacred monks and nuns. Survivors fled southward to Sri Lanka, northward to Tibet and Nepal, and eastward to Bhutan and Burma. By the thirteenth century, except for a few pockets in the east and in the south, Buddhism vanished from its mother country. Thus, the sacred sites of the Buddha's life also fell into neglect and oblivion.

To locate the sites of the Astamahapratiharya, Cunningham's archeological team had to rely on the detailed accounts written by Chinese monks Fa Xian (337–422 CE) and Xuan Zang (602–664 CE), who traveled to India to bring back scriptures to China. In the mid to late 1800s, when his excavations gave proof that Siddhartha Gautama was an actual historical figure, it unleashed a great surge of interest and fascination with Buddhism in the Western world that has lasted until the present. In 1879, Sir Edwin Arnold's *The Light of Asia*, an epic poem about the life of the Buddha, became an immediate success and was translated into many languages. Arnold's poem helped to further popularize Buddhism and turn the Buddha into a household name known around the world. These pilgrimage sites have since been partially restored by the Archaeological Survey of India and reopened to the public as displays of India's great historical legacy.

As a nonreligious pilgrim, I originally planned to arrive at these sites, walk around, read signs, hire a guide to share more detailed information, and take photos. But the "Advice on Pilgrimage" that I received from Lama Zopa Rinpoche via e-mail from Venerable Sarah made me reconsider. Here is an excerpt of his instructions:

> *Pilgrimage needs faith. More faith, more happiness. Otherwise, you are just like a tourist looking at ruins. Pilgrimage is not just going to holy places and taking pictures. You go to holy places for blessing. So it's good to do practices at the different holy places. . . . Pilgrimage is good if you know how to do pilgrimage. Otherwise it's just travelling like a tourist. Also, when you go to these holy places it reminds you of impermanence. Once these places were great cities but now there are just stones. A thousand years ago these places were quite different. But even though there are just stones now these stones are so precious. Amazing. Can you imagine how blessed these places are? They are places where the Buddha was. . . .*

At the beginning do prostrations. . . . Do different prayers at these holy places and make prayers to have realizations.

Once again, what he said cut straight to the bone. I thought about what he shared and realized it couldn't hurt to take the pilgrimage journey a bit more seriously and treat these archeological sites as sacred. The pragmatic part of me wondered: why waste a precious and rare opportunity for receiving and sending blessings? To follow the spirit of his instructions, I came up with a set of simple practices that I learned at Tushita and Thosamling. When I arrived at a site, I would begin by doing three prostrations to show respect. Whenever I came across a sacred monument such as a stupa, I would circumambulate it three times while chanting simple mantras that I could easily memorize, such as "Om Mani Padme Hum," the mantra of compassion. Then if there was time as well as shade, I would sit in meditation, do a few more mantras, and say prayers, such as the Four Immeasurables, to bless all living beings to have happiness and health. Before leaving, I would then do three more prostrations to say farewell to the site.

During the pilgrimage journey, I found it was easy for me to feel reverence and respect for the sites that marked where Siddhartha Gautama as a human being lived and shared teachings. However, I found it hard to process the sites that supposedly marked magical miracles performed by the Buddha. I also found it impossible to rely on history to separate fact from fiction because the mythology around the Buddha as a superhuman figure had already been integrated by Ashoka into the archeological evidence in the third century BCE and then echoed and further embellished in all the historical records that came after. So what I really got from the pilgrimage tour was a more nuanced understanding of the history of Buddhism and how myths about the Buddha evolved over time and got dressed up in different ways by the countries they traveled into.

Nonetheless, it was stunning to be in Bodhgaya for the celebration of Vesak, a holy day that commemorates the anniversary of the Buddha's Enlightenment. Because of the Vesak festival, Bodhgaya was crowded with pilgrims. The main attraction in Bodhgaya is the Mahabodhi temple complex built on the site where the Buddha was enlightened. When I visited the complex, I was puzzled that many signs marked the occasion as the Buddha's birthday. By asking around, I learned that people believed that Siddhartha Gautama was born, became enlightened, and passed away into parinirvana (a state of complete liberation from the cycle of rebirth) on the full moon of Taurus. Thus, Vesak is a triple holiday.

Tens of thousands of pilgrims representing Buddhist traditions from all over the world conducted ceremonies at the sacred site all throughout the holy day. At night, the entire complex was lit by the surreal light of the full moon and by thousands of candles and lanterns. Sitting there in meditation and prayer among the devotional Buddhist pilgrims, I couldn't help but feel awe and wonder for how I'd ended up in Bodhgaya on their holy day by happenstance. Was it possibly true that I was there because of karma? Was I possibly there for a reason I could not comprehend?

Sitting there among the multitudes, I wondered: Does Enlightenment have to involve superhuman abilities? If we stripped away the legends, couldn't the unembellished story of a normal human being who learned to tap into a source of wisdom and peace within him and transmit that wisdom and peace to others be enough to inspire people to apply his teachings? Did the legends arise because people needed to turn the Buddha into a superhero who could perform miracles? Or did these legends grow from actual occurrences? There was no way to discover the answers to these questions.

While in Bodhgaya, I stayed at the Root Institute, another center under the FPMT umbrella organization founded by Lama Zopa Rinpoche. I learned from talking to folks at the Root Institute that modern historians in India portray Siddhartha Gautama as a spiritual leader who initiated a reformation of Hinduism by challenging the caste system. This portrayal of Buddhism gained prominence at the time of India's movement for independence from Britain when the well-known social reformer Bhimrao Ramji Ambedkar called for all Untouchables to abandon Hinduism and convert to Buddhism to free themselves from the oppressive caste system. According to this reinterpretation, the historical movement that Siddhartha Gautama started would have been seen as radical because it welcomed people from all castes and said that all people could become enlightened on their own through meditation; Buddhism was also the first spiritual school in India to ordain women as nuns, a decision that greatly upset the religious and cultural establishment. I learned that unfortunately, Ambedkar's strategy to free the Untouchables from oppression by converting them to Buddhism didn't entirely work because many Hindus simply associated the new Indian Buddhists with a lower caste and continued to discriminate against them. Learning more about their background and their suffering gave me more motivation to greet and honor the many thousands of Indian Buddhist pilgrims I saw in Bodhgaya.

The pilgrimage was a wonderful experience, but at the end of the celebrations, I still wasn't confident I understood what Siddhartha Gautama stood for. It then dawned on me that I may also need to become more familiar with the tradition he was brought up in to more fully appreciate his teachings and philosophy. I wondered if learning about yoga in the Hindu tradition could shed more light on Siddhartha Gautama's teachings. I had no idea that by diving into yoga, I would come to see that Siddhartha Gautama was only one of many spiritual teachers in India who made efforts to retransmit the essence of India's ancient spiritual teachings to the people of their times, and that many of these revered yogis are believed to have had extraordinary superhuman abilities.

From Bodhgaya, I took an early morning train into Kolkata to meet Shailesh, who had graciously arranged for his assistant to pick me up at the train station and for me to stay in a spare apartment he specifically kept to house out-of-town guests. I barely knew my host and I didn't want to outstay my welcome or impose too much on his hospitality, so my plan was to play it by ear and be ready to leave after taking the course he was arranging for me that upcoming weekend. Fortunately, there were so many budget airlines in India that it would be easy to book a one-way ticket at the last minute. Nevertheless, I spent most of the train ride with butterflies in my stomach because I had no idea what to expect. Going to Kolkata meant taking a leap of faith that whatever was about to unfold would be fine, or that if anything did go awry I would be able to sort it out.

Yoking into the Incomprehensible

Looking back, I will always remember Kolkata as a surreal portal between two worlds: the normal physical, material reality I had grown up assuming was all there is and a more ephemeral, elusive realm that defies comprehension and analysis. It was in Kolkata that I first learned that the word "yoga" comes from a Sanskrit root, "yuj," which means to "yoke" or "join together" and that in its original context, yoga refers to a mystical state of "divine union" when one's consciousness is merged with what sages often referred to as the Paramatma, which can be translated as the "Supreme Spirit," the "True Self," the "Oversoul," or "Source." It was in Kolkata that I had a series of unexplainable encounters with states of consciousness that would forever change my view of life and the universe.

My train ran about five hours behind schedule so I arrived at the Kolkata train station at night. As it was too late to meet Shailesh, he arranged for his personal assistant to pick me up at the station and bring me to his guest apartment so I could rest. Shailesh would then meet me in his office the next morning. As I mentioned earlier, when I first met Shailesh, I was very intrigued to learn more about the way he incorporated spirituality into his life as a businessman.

When I arrived at his office, I learned that he had actually carved out part of the floor space of the office of his brokerage company to operate a Pranic Healing clinic and used the training rooms to teach courses in Arhatic Yoga and Pranic Healing. As Shailesh enthusiastically showed me around the healing clinic, it was clear that this was his passion project. He introduced me to the healers and told me that they would take care of me and host me while I stayed in Kolkata as his guest. He instructed them to tell me about what they did as professional Pranic Healers, explain how the clinic operated, and answer my questions. He also asked them to give me a tour of Kolkata, and together, they brainstormed a list of sacred sites to take me to visit. At that time, I was unfamiliar with the famous spiritual teachers who had lived in Kolkata, so I didn't recognize any of the names or places they listed except for Mother Teresa. I was astounded because I had not expected this degree of hospitality or to be surrounded by such an enthusiastic welcome committee.

I initially came to Kolkata thinking I would maybe stay one week. But in total, I ended up spending about three weeks there to take more courses at Shailesh's invitation. Shailesh had arranged for me to take the introductory-level Basic Pranic Healing course with another instructor during the upcoming weekend. He then invited me to stay another week because he was offering the same course I had originally taken in Singapore with him and wanted me to review it. After that, he then invited me to stay yet another week because he decided to offer the Advanced Pranic Healing course so I could experience it. But unfortunately a last-minute family emergency came up so he had to cancel the last course.

Altogether, there were about eight full-time healers who worked at the clinic. They were all my age or younger. As I got to know them, I learned that many of them were college graduates who had given up jobs in Kolkata's booming tech sector in order to work as full-time healers in the clinic. Like

Shailesh, they all took Master Choa Kok Sui (MCKS) as their spiritual guru and were earnest and sincere about devoting their lives to spirituality and carrying on the legacy of MCKS by spreading his teachings. All of the healers practiced Arhatic Yoga as well as Pranic Healing. The degree of devotion and faith they displayed was completely beyond my comprehension and thus very intriguing.

Spending time at the Pranic Healing clinic was like stepping into another realm where normal physical laws concerning space and time didn't quite apply. Watching the healers at work was a bit like finding myself in a Harry Potter movie. They used crystal wands, invocations, and prayers to do distance healing on their clients. The most incredible part was that after the healings, the clients would consistently text or call to let them know that they could instantly feel the effects of the healing session and then schedule another session. I couldn't help wondering if it was possible to explain all of this using the placebo effect. It was mind-blowing because it defied my notion of what was possible and forced me to further reconsider my view of reality.

The first site they arranged for me to visit was the ashram (a spiritual retreat center) established by Paramahansa Yogananda (1893–1952), a world-renowned spiritual leader and author of the bestselling classic *Autobiography of a Yogi*.* However, at the time, I had only heard of the book in passing as one that Steve Jobs had revered. (I later heard that Jobs had even prearranged for the book to be given out during his memorial service as a gift to those who came to pay their respects.) Nevertheless, I did not know the name of the person who wrote this book and, therefore, I couldn't link the book to the ashram I was visiting. In hindsight, I find it amusing how innocently clueless I was wandering around the buildings and gardens without any idea why they were special.

Luckily, one of the healers had a connection with someone who was part of that spiritual community, who kindly agreed to serve as our guide. It was a special tour because he had a key to the crypt where Yogananda meditated and brought us in there so we could feel the energetic imprint that remained— which the healers promptly started scanning and describing. Oddly, I didn't

---

* Formally called the Yogoda Satsanga Math in Dakshineswar, this site also serves as the headquarters for the Yogoda Satsanga Society of India, the spiritual organization founded by Yogananda in 1917.

think I would be able to sense anything, but I did feel something like the up-lifting warmth and joy I felt around Lama Zopa Rinpoche at Tushita. So I guessed that whoever meditated there must have been as extraordinary as Lama Zopa Rinpoche.

It was only when we entered the gift shop and I saw the book *Autobiography of a Yogi* being sold there that I realized the significance of the place we were visiting and appreciated how special it was to have been inside the crypt where the author used to meditate when he was in Kolkata! Right at that moment, one of the healers, let's call him Gupil,* spontaneously gave me a copy of the book as a gift. It was a clear sign that my adventure in India was about to get much more interesting here in Kolkata.

By reading *Autobiography of a Yogi*, I learned that the word "yoga" means "divine union" in Sanskrit and that being a "yogi" means being someone who experiences this mystical state of divine union. I found it to be a very hard book for someone with a scientific mind to digest because in it, Yogananda shared many fantastical stories of incredible, miraculous deeds performed by great yogis. For instance, he shared that Lahiri Mahasaya, the founder of Kriya Yoga, the spiritual lineage that Yogananda practiced and brought to the United States, could not be photographed without consent. Whenever anyone tried to photograph Lahiri Mahasaya without his permission, his image would not appear in the developed film. Yogananda also wrote about his experiences meeting a yogi who could bilocate (which means to be present in two locations simultaneously), a yogi who could make things manifest out of thin air, and others who could perform miraculous healings. He also talked about encounters with saints who no longer had to eat food to live. Furthermore, there is a chapter in the book about the resurrection of his own guru, Sri Yukteswar, whom he said visited him several months after his death, not in spirit but in flesh-and-blood form.

Despite my skepticism, the book came to life for me in a unique way because I got to see several of the sites where stories in the book took place, such as the Kolkata train station, the Kali temple, and Sri Yukteswar's ashram where Yogananda lived and trained for many years, in addition to the ashram Yoga-nanda founded. My hosts also brought me to places associated with spiritual

---

* "Gupil" means secret. This is not his real name.

teachers mentioned in the book, such as Ramakrishna and his disciple Vivekananda. Strangely enough, as I read the book, it felt like the spirit of Paramahansa Yogananda was present with me trying to reassure me that the stories he shared were authentic and true. Yet I also discounted such a notion as being a figment of my imagination—how could such a thing possibly happen? Besides, even if it were possible, why would his spirit care so much about whether I believed his stories that he would visit me? Wouldn't his spirit have more important matters to attend to?

Yet even though I didn't fully believe the stories in the book, I also had a hard time dismissing them because I was surrounded by earnest aspiring yogis who believed without a doubt they were true. The more I got to know the healers hosting me, the more I realized that their devotional faith was grounded in their own mystical encounters. Reading the book in Kolkata in the Pranic Healing clinic (rather than in the United States) made a profound impression. It was disorienting to find myself immersed with people who were so enthusiastic about embracing and interpreting the spiritual experiences shared in the book as confirmation of their own beliefs and intuitions, whereas I soberly viewed these experiences as inconclusive anecdotes that could possibly be explained or interpreted in a variety of ways.

At that time, I was still processing my experiences at Tushita, which had given me a transformative reminder of being a superconsciousness that was much larger than "Due." That experience of satori had already forced me to acknowledge that it is possible to experience a higher level of consciousness and a spontaneous blissful joy that, unlike ordinary happiness, is decoupled from external circumstances. So I began to wonder: Could what I experienced at Tushita be an example of divine union? What if the stories in the book were true? What would this mean? Did it imply that there really is much more to life than the physical, material world we see with our eyes—that other spiritual realms really exist? Did it imply that all humans have the potential to experience divine union? Else, how could a skeptical nonbeliever like me have such mystical experiences?

As I spent time with the Arhatic yogis in the clinic, my mind struggled to accept the validity of the Pranic Healing practices. I had always assumed the placebo effect was behind people's belief in faith healing. Previously, whenever I heard yoga teachers or spiritual people talk about energy healing, I reflexively dismissed it all as woo-woo, New Age, wishful thinking that lacked evidence.

However, after hearing so many testimonials from people who seemed sane about how they'd recovered from serious conditions because of Pranic Healing, I was intrigued and curious to learn more and to experience it for myself. Yet even as I took the Basic Healing course in Kolkata and learned the key concepts, I couldn't fully buy in.

I learned in the course that MCKS's approach to energy healing was based on the premise that diseases and pain could be treated by cleansing the human energy body using techniques that focused primarily on restoring the healthy functioning of key energy centers in the body called chakras, removing diseased energy from the affected area, and infusing the area with fresh spiritual energy. A lot of it seemed like an exercise of visualization and imaginary make-believe. In the class, when we did healing exercises with a partner, I couldn't tell whether the healing made any difference because I was already in good health and nothing was bothering me.

The turning point came later at a food court at a shopping mall where one of the more senior healers took me to hang out for fun. The food was so spicy, it burned my entire mouth so much that I was in tears. Seeing my suffering, she did a quick healing for me under the table, and in seconds, my mouth and tongue stopped burning. This left me stunned because I didn't have any faith that it would work, but it still did. So didn't this mean that what happened was not really faith healing? While I wasn't convinced enough to change my life and become an energy healing evangelist, like Shailesh and his healing team, I did decide it was worth investigating further.

In parallel, as I finished reading the book *Autobiography of a Yogi*, I began to have mysterious experiences I had never had before and, in some cases, haven't had since. For example, when I retook the "Achieving Oneness with the Higher Soul" course in Kolkata, there was an exercise in which all participants reverently greeted a life-sized photograph of MCKS as a way of paying respects to the teacher. When I did this during the course in Singapore in February, nothing out of the ordinary happened. This time in Kolkata, the image of MCKS instantly came to life and winked at me with a smile. When I blinked in startled astonishment, the image reverted to a normal photograph. I was so freaked out, I couldn't talk the rest of the day. Again, I wondered if I had imagined it. Yet logically, if my imagination was capable of doing something like that, it should be able to do it again. I went back to stare at the photograph several times and found that my imagination was not capable of re-creating that uncanny event.

Such a thing has not yet happened again since—but I haven't made a habit of staring at images of spiritual teachers either.

I also started to experience for the first time a phenomenon called hypnogogic hallucinations. Hypnogogia refers to a transitional state of consciousness between being awake and being asleep, a state in which lucid dreaming, sleep paralysis, and hallucinations are often reported. It is also a state that can be intentionally or inadvertently induced by practices like meditation. In the mornings in Kolkata, I often sat meditating alone in my bedroom in the guest apartment. A couple of times, when I was in a deep state of meditation, I heard a fleeting audible voice speak to me as if it were coming from a person in the same room. Each time, it was able to transmit only one or two words, like "home" and "come home," before I snapped out of the meditation in astonishment. Strangely, it was a male voice that spoke in an American accent, a voice I didn't recognize. It was clearly not the voice of MCKS, which I had become familiar with by listening to recordings of his guided meditations.

When I shared this experience with my new friends, the healers in Kolkata, they speculated that it could be the voice of my inner teacher, "Higher Soul," or spirit guide. We debated whether it meant for me to literally go home to my parents in Philadelphia or to metaphorically return home to "Source." No one could say for sure. I seemed to be the only person who wondered if I was experiencing auditory hallucinations as a result of something going awry with my ear, auditory nerve, and/or auditory cortex. The hypochondriac in me also hoped this was not a sign that I was developing some type of brain tumor.

In another meditation, I experienced what would be best described as a life review. In my mind's eye I saw many of the experiences and lessons of my life, which I had earlier thought of as being separate and disjointed, become dots that interconnected and led me to Dharamsala, Bodhgaya, and Kolkata. Further, I saw that all these dots were meant to converge and push me forward to fulfill my life's purpose. The message seemed to be that everything I'd ever learned was going to be useful as I moved forward. Yet exactly what my life purpose was and even what I should do in the immediate future were still elusive to me. Later, in another meditation, I had a vision of myself on a stage in front of a huge multitude of thousands or perhaps tens of thousands of people, giving what seemed like a teaching on meditation. This seemed very odd to me because I had only started meditating a few months earlier, and I could not imagine what I could possibly say about meditation that anyone would listen to. At the time, it hadn't yet occurred to me that teaching meditation could be part of my life purpose.

Then in another meditation, I had an experience of satori very similar to what happened at Tushita. I felt bliss welling up in my heart and shooting up through my head and into the sky. I then felt this unexplainable sense that the prayers of everyone in the entire world were coming to "me," and that was when I realized that "I" was no longer "Due." All the prayers were coming to "me" as the Collective Consciousness, yet in this form the prayers couldn't be directly answered. The energies of these prayers had to be redirected back to embodied human beings, who were all physical manifestations of this unified Collective Consciousness and had to take the initiative to respond to these prayers in the physical world. I saw streams of energy go to many, many people, because all of us had a role to play in answering these prayers. As I came back into "Due," I realized a stream of these energies was being directed to me to manifest something that would serve the greater good. While I still didn't quite know what that would be, I had the sense that it was all going to unfold naturally in due course and that I would know what to do when the moment came.

As someone who originally dismissed praying as a waste of time and assumed that God either didn't exist or didn't care about human affairs, I was stunned. Did this mean that prayers were not bullshit? If this was an experience of "divine union," that meant that whatever the divine is, it is not like anything I was ever told by a religious authority figure. What I experienced clearly did not align with how my Catholic upbringing had depicted God as a bearded white man on a heavenly throne who passes judgment on the lowly mortal sinners he created to see fail and then punish for an eternity. Nevertheless, I couldn't make up my mind as to whether to take this experience seriously or to question it as a figment of my imagination. It continued to stump me as to why a skeptical nonbeliever with very limited exposure to and training in spiritual practices could even have mystical experiences. The rational part of me was frustrated that I seemed to be lost in a field experiment with so many variables outside of my control that I could no longer draw any reasonable conclusions from my observations or explain what was happening using science. However, the intuitive and curious parts of me were mesmerized and wanted to go deeper.

From then on, the lines seem to blur between yoga and normal life. It became a regular occurrence for me to slip into trancelike hypnogogic states, even when I wasn't meditating, and to my surprise, my new friends embraced this development with excitement and enthusiasm. One moment, I would be present with them, following the group conversation, and the next moment, I would be spellbound by what was unfolding in my mind-track. Whenever this happened, they would let me be until it was time to move to the next event we had scheduled. Then they would gently alert me that it was time to go. I would have no idea how much time elapsed. Sometimes I would bliss out and the world and noise around me would fade away into a joyful silence. Other times, a voiceless inner teacher would explain esoteric subjects to me using thought forms rather than words. However, as I didn't have faith that this was a reliable way of getting information, I wondered whether perhaps this was just my imagination running wild.

Although I was weirded out by my experiences, nothing seemed too weird for my new friends, who were very curious and often asked me to share what I was experiencing. I learned that they too had experienced many mystical phenomena and that these experiences fueled their sense of spiritual devotion. In fact, the urge to deepen their connection with the divine was what motivated them to practice Arhatic Yoga and to work as energy healers. Therefore, they reveled in sharing their deeply personal experiences with someone who was now experiencing a similar shift.

Eventually, it occurred to me that Gupil could read my mind. I slowly pieced it together because he often responded to questions or statements I thought in my mind but did not actually express aloud. When I asked him about it, he confessed that it was true that he could catch some thoughts, but not from everybody. Upon my further questioning, he explained that this gift, or "siddhi," as yogis liked to refer to these types of abilities, developed spontaneously after he started practicing Arhatic Yoga. Then he asked me not to talk about it to other people because the Arhatic yogis were supposed to keep these gifts private (taking great care to not show off).

Even with all the unusual things that were unfolding, the most magical aspect of my time in Kolkata was the extremely generous hospitality I received from Shailesh and his team, something I didn't feel that I had earned. I found it incredible that they would take in a stranger in this way. Eventually, when Shailesh invited me to have lunch with him, I mentioned that during his course, I saw the photo of MCKS come to life and was very shaken by that experience. He responded by sharing that he had intuitively sensed when we first met in Singapore that I had a special karmic connection to MCKS and that he saw this as proof that the hunch was right. Apparently that hunch had prompted him to invite me to Kolkata and host me indefinitely.

After that lunch conversation, I asked the healers how many other people Shailesh had hosted in this manner. They explained that I was actually the very first person in the two or three years since they opened the clinic that Shailesh had ever welcomed like this. They also shared that my visit was a gift for them because many of them had never had the occasion before to visit the sacred sites they took me to see. They were grateful that my visit created a unique opportunity for us to go on what they considered a sacred pilgrimage together. Learning all of this left me even further confounded. Could it possibly be true that a karmic connection to MCKS was behind everything that had unfolded for me in Kolkata?

Karmic connection or not, I got a clear gut feeling that MCKS was most likely not my "guru" when the healers played a video of MCKS giving a talk because they wanted me to get a better feeling for who he was when he was alive. In the talk, I was taken aback by the comments he made about Vipassana. I was surprised that he seemed to disparage it as a practice because it could result in an emotional catharsis that people weren't able to handle. To me, his statements presented a superficial and lopsided view of what could happen when people dive into a Vipassana retreat without adequate preparation. My own experience showed me that practicing Vipassana on top of a strong foundation in Buddhist teachings is a powerful means of becoming familiar with one's own mind and developing a deep level of self-awareness and self-acceptance that is an integral part of self-mastery and self-actualization. I was also concerned that many of the Arhatic yogis, who didn't have any exposure to Vipassana, seemed to blindly accept all of MCKS's opinions as truth without questioning them.

After watching the video, it occurred to me that qualities like self-awareness and equanimity were relatively strong among Buddhist practitioners because the Buddhist tradition strongly emphasizes these qualities. However, in contrast, these qualities seemed less developed among the Arhatic yogis I befriended in Kolkata, perhaps because MCKS's approach didn't seem to emphasize these qualities. Thus, even though my new friends in Kolkata believed that MCKS must have brought us together for a reason and that we were meant to work together to spread his teachings, I was not so sure. The way his students seemed to look at the world was too divergent from my own worldview. I could appreciate, respect, and value his teachings and legacy, and was still interested in taking more of his courses, but I couldn't see myself as a disciple or "foot soldier" in the movement, as the healers liked to refer to themselves.

Even as my experiences in India cracked open my shell and rendered me more open to spirituality, they also showed me that my hardwiring was not compatible with following a conventional spiritual path. Whereas most of the spiritual aspirants I had met so far—whether they were Tibetan Buddhists, Theravada Buddhists, Arhatic yogis, or students of other traditions—seemed to choose one path as the one and only spiritual path for them, I preferred to explore and understand many spiritual traditions to look for overall patterns and see how they aligned with scientific research without becoming attached to or biased toward any particular path as being "right."

While many of the people I encountered seemed to choose one guru or spiritual teacher and then devote themselves as disciples to that person by putting their guru on a pedestal and proactively spreading their guru's teachings, I found that type of mindset and behavior just didn't resonate with me. The most I could feel was the reverence and respect I held for an outstanding professor at Harvard or Wharton, or a great humanitarian like Gandhi or Mother Teresa. Whereas many aspirants seemed to think of their guru as an infallible great master and appeared to soak in all of their teachings and the wondrous stories about him (or her) without any doubt, my fact-checking instincts could not be suppressed.

Eventually I accepted that I may simply not be capable of experiencing that type of devotion. To me, all gurus are human beings who learned from trial and error just like the rest of us. I instinctively take what any guru said and what is said about this person with healthy skepticism. It's just not possible for me to put a human being on a pedestal and follow him or her with mindless obedience, no matter how many miracles the person is said to have magically performed. The way I'm wired, my right prefrontal cortex (the analytical, reality-checking, braking system in Brain 3.0) is just too strong to be deactivated by exuberance or blind faith.

In the end, even though I hesitated to fully embrace the belief systems of the various spiritual schools I encountered in India, by the time I left Kolkata, my confidence in the worldview I held before 2012 was shattered. The fact that a person like me, an atheistic skeptic without any faith, could have a series of mystical experiences turned my life upside down. Too many things I had once assumed were impossible, I had experienced as being possible after all. I had seen and felt too much to turn back and pretend that none of this had ever happened.

Like a moth drawn to a flame, I felt compelled to go deeper. So I decided to extend my sabbatical for the rest of 2012 to investigate spiritual practices further. I wanted to experience a traditional yoga teacher training and compare what I would learn from studying classical yoga scriptures such as the Bhagavad Gita and the Yoga Sutras with the more New Age views presented by Yogananda and MCKS. I made plans to take a couple months to see a few more places in India, then visit family in the United States and friends in Europe, before returning to Singapore in mid-July 2012 to look into yoga teacher training programs there.

Yoga Teacher Training

When the time came to go back to Singapore, I started to have second thoughts. Did it really make sense to take off more time to go through yoga teacher training when I had no intention to ever work as a yoga teacher? How long was I going to put my career in limbo? Didn't it make more sense to start planning my relocation back to the United States?

Shortly after my arrival in Singapore, my partner in the February course at the Centre for Inner Studies reached out to share that Shibendu Lahiri, the great-grandson of Lahiri Mahasaya, the founder of Kriya Yoga, would be in Singapore that very weekend to give a Kriya Yoga initiation, and it was open to all who were interested. I was very curious to see how Shibendu Lahiri carried on his great-grandfather's lineage so I decided to attend.

From the very beginning of his talk, it was obvious that Shibendu had a very different perspective from that of Paramahansa Yogananda, whose branch of Kriya Yoga came to be called the Self-Realization Fellowship. The literature of his organization states: "One of the essential goals of Paramahansa Yogananda's mission was 'to reveal the complete harmony and basic oneness of original Christianity as taught by Jesus Christ and original Yoga as taught by Bhagavan Krishna; and to show that these principles of truth are the common scientific foundation of all true religions.'"[15] In contrast, Shibendu's talk revealed a strong patriotic bias toward India's spiritual traditions. He even made several comments that he found rituals in Christianity to be rather strange and counterintuitive. He clearly did not share Yogananda's appreciation of Christianity or see its underlying "oneness" with India's yoga tradition.

Thus, what stood out most for me from attending Shibendu's talk was learning that over four generations, a striking degree of variability and subjectivity in the preservation and transmission of Kriya Yoga had emerged. Moreover, there was no way to distinguish which transmission of Kriya Yoga was more "pure"—that of Lahiri Mahasaya's own great-grandson or that of its much more famous spokesperson, Paramahansa Yogananda. Observing this variation helped me to see and appreciate how the original teaching of any spiritual school naturally changes as it is transmitted through different disciples over successive generations.

The Kriya Yoga initiation was fascinating. After we received a "shaktipat" (a blessing from a guru that transmits spiritual energy to a student) from Shibendu, in which he touched the chakras along each person's spine and head, he and his assistants instructed the new initiates on how to perform the Kriya Yoga breathing techniques, such as drawing out the inhale and exhale each to over twenty seconds and how to focus on specific chakras while doing the breathing technique and repeating a mantra. As someone who cannot multitask, I found the techniques very difficult and cumbersome to practice. I tried them regularly for about a week and then eventually dropped them completely. Although I chose not to practice Kriya Yoga, the initiation reignited my curiosity to understand more about classical yoga philosophy and history.

Therefore, rather than give in to second-guessing myself, I decided to stick to my plans to look into yoga teacher training programs. I followed the recommendation of a friend and yoga teacher in Vietnam to connect with an Indian Singaporean master yoga teacher named Paalu, with whom she had studied and whom she felt was authentic and knowledgeable. From looking at the website for his yoga studio, Tirisula Yoga, I learned that Paalu had studied yoga in Mysore under B. N. S. Iyengar and one of his highly distinguished students, V. Sheshadri. He had also studied a different yoga style created by B. K. S. Iyengar (no relation to B. N. S. Iyengar), which uses tools and props to enable students to perform asanas with more optimal alignment for their specific bodies. In addition, Paalu had studied Sanskrit and various forms of meditation and taught Reiki.

When I met with Paalu for a consultation, it didn't take long to intuitively "know" that he was also a mystic. Being in his presence, I could palpably sense that he was connected into a higher consciousness and that these experiences informed the way he taught. My conversation with him confirmed that his Registered Yoga Teacher (RYT) training program was a unique opportunity to learn from a master yoga teacher who was Indian by heritage, could read Sanskrit and thus had a deeper direct understanding of the Yoga Sutras and other ancient Indian teachings, was fluent in English, and was still thoroughly modern because he grew up in Singapore. Therefore, I concluded that this chance to learn from an authentic yet modern Indian yogi wasn't an opportunity I would easily come across again, especially after I returned to the United States, so I decided to go for it.

I informed Paalu that I was planning to relocate to the United States before the end of 2012 and wanted to complete the training before the move. Therefore, we decided it would be best for me to take the upcoming one-month intensive yoga teacher training, which would start soon in August. Paalu explained that while the two-hundred-hour yoga teacher certification would be for hatha yoga (the overall umbrella name for the style of yoga that emphasizes asana practice), he had designed the yoga asana part of the training around the Ashtanga Primary Series as taught by B. N. S. Iyengar in the Mysore tradition created by Krishnamacharya. Given that my visit to the Mysore Palace the year before had kindled a fascination with Krishnamacharya, this suited me just fine. He also explained that there were three main components to the training: yoga philosophy classes that would be taught by Paalu; yoga anatomy classes that would be taught by his teaching partner, Satya; and asana practices that would be cotaught by Paalu, Satya, and their larger team of yoga teachers.

During the teacher training, I found it spooky how Paalu could sense and say things about people that they had never told him. For instance, he intuitively sensed where people had past injuries and on more than one occasion, he casually mentioned health conditions people were experiencing. Nevertheless, having him as a teacher could also be frustrating because he was not particularly interested in structure or time management.

In the yoga philosophy lectures, it felt like Paalu's monkey mind would frequently jump around and digress into long, unrelated tangents—as my Inner Critic fumed at the lost time and loss in productivity. Because of this, for better or worse, the course didn't fully cover all the materials we were supposed to know to take the Yoga Alliance Yoga Teacher examination, and what we did cover often lacked a sense of coherence and depth. Then again, a part of me also wondered if Paalu digressed on purpose to force me to confront my continued attachment to productivity and acing exams—two things that enlightened yogis are probably indifferent toward. In hindsight, the underlying cause of my frustration was really a hunger to learn as much about yoga philosophy from Paalu as possible, since I knew my time with him would be short.

In spite of his many detours and ramblings, I found I had to try to listen attentively because every now and then, Paalu would say something that would hit me like a bolt of lightning—pearls of wisdom that have guided me through the present day. For instance, he opened the first class by explaining that although we were there to take a course called "yoga teacher training," yoga could not actually be taught—yoga had to be shared. He also explained that we could only share yoga when we were overflowing with it. If we felt empty or depleted, then we would only further exhaust ourselves by attempting to share what we didn't have enough of. If we are not overflowing, we should instead acknowledge it and practice self-care.

Another time, during a private conversation, he helped me to understand that the true path is pathless. He also told me that the period of time for me to travel around searching for answers in the external world had come to an end, and that I had to now see for myself that "spiritual" paths taught by spiritual organizations eventually become decoys. He explained that the true path is in the heart. That meant I had to learn to turn inward and to trust my own inner guidance. He also helped me to understand that sacred teachings are hollow if they are treated as secondhand knowledge like academic learning. Sacred teachings must be grounded in firsthand experiences and direct realizations to come to life as wisdom. Thus he encouraged and challenged me to experience the meaning of the teachings and put them into my own words. His advice gave me the courage to break free from feeling dependent on spiritual organizations and gurus for guidance and direction, and instead start trusting my own heart to show me the path.

Thanks to his Indian heritage, knowledge of Sanskrit, and mystical and clairvoyant gifts, Paalu had a unique perspective on how to interpret the Sanskrit teachings in the Yoga Sutras, a collection of verses summarizing the practice of yoga first written thousands of years ago by a sage named Patanjali.* "Ashtanga Yoga" literally means the "Eight Limbs of Yoga," which Patanjali

---

* Exactly when the Yoga Sutras were written is widely debated. The range seems to be between 500 BCE and 500 CE.

recorded and described in the Yoga Sutras. For reference, I have listed the Eight Limbs in the following table.

	Eight Limbs of Yoga
1	**Yama:** Restraints, of which there are 5: ahimsa (nonviolence), satya (truthfulness), asteya (nonstealing), aparigraha (nonavarice), and bramacharya (sexual innocence/purity)
2	**Niyama:** Virtues/observances of which there are 5: sauca (purity and cleanliness), santosa (contentment), tapas (discipline), svadhyaya (contemplative study of self), and isvarapranidhana (concentrated attunement to the True Self)
3	**Asana:** Steady poses, the training and disciplining of the body to align with and surrender to the True Self
4	**Pranayama:** Expansion of life force (usually cultivated today in the form of breathing techniques)
5	**Pratyhara:** Withdrawing from the senses or withdrawing from worldly orientation and turning inward to sense the True Self
6	**Dharana:** Focused, one-pointed concentration
7	**Dhyana:** Contemplation/awareness (Sanskrit root for the Japanese word "Zen")
8	**Samadhi:** Absorption/union/oneness with the True Self

Prior to taking the yoga teacher training course, I had mainly experienced yoga as a fitness class at the gym that was entirely focused on the third limb, asana. However, during the yoga philosophy classes, I learned that in its original context, asana is really seen as a preparatory practice for meditation. The aim of asana is to strengthen, balance, and align the body so one has the physical capacity to practice the higher limbs, which are progressively advanced forms of meditation and mind training. While it is not clear exactly how many asanas exist, the Traditional Knowledge Digital Library, a master digital database in India that serves as a repository of its traditional knowledge and ancient cultural heritage, has documented at least 1,500 asanas so far.[16]

As anyone who has taken a yoga class has experienced, performing asanas on a regular basis increases flexibility, balance, and strength, as well as calmness and centeredness. After experiencing these benefits, I naturally wanted to learn more about the underlying physiological mechanisms. During the teacher training, I was told that holding various asanas promotes well-being by improving the circulation of blood in vessels and plasma in the lymphatic system, activating the glands of the endocrine system, balancing the autonomic nervous system by stimulating the parasympathetic nervous system, and cleansing and detoxifying internal organs. Therefore, we were told, asana practice, in addition to being a physically challenging workout, provides therapeutic healing.

In glancing at research articles on yoga as a clinical intervention, I found an overall concern that many of the studies that were done were not designed to meet the gold standard for rigor, probably due to the high costs of running double-blind, randomized clinical trials. Further, there is so much variability in the yoga protocols tested in the various studies that I found it hard to justify blanket statements about yoga interventions. Nonetheless, as someone familiar with the amount of effort, perseverance, and funding it takes to design and run scientific studies, I also appreciated the challenges and structural obstacles for reaching a higher burden of proof. Generally, researchers have to apply for grants in a very competitive market, so researchers are naturally incentivized to pursue topics that have a higher likelihood of getting funding. The readily available funding and prestige for studying yoga interventions are probably tiny relative to what's available for innovative cures for cancer.

Nonetheless, even if conclusive evidence for the benefits of yoga had not yet been collected, I remained open to exploring through logical induction how these mechanisms could possibly work. Naturally, I found the most interesting part of the yoga teacher training was the anatomy portion, in which Satya walked us through how yoga practices affect the body. We covered the respiratory system, the circulatory system, and the nervous system in broad strokes. Then we dived into more detail on the musculoskeletal system to understand what muscles, ligaments, and bones are involved with specific asanas. I especially loved how this knowledge came to life as my classmates and I performed the asanas and practiced adjusting each other. It completely changed the way that I related to my body by helping me further develop my sense of proprioception, which is the "ability to sense stimuli arising within the body regarding position, motion, and equilibrium."[17]

Proprioception involves an accurate perception of the spatial positioning of one's own body parts with respect to each other during the performance of a particular motor activity. This includes sensing muscular exertion, force, movement velocity, size or mass, and weight or heaviness. Proprioceptive awareness arises from specialized sensory nerve endings called proprioceptors located in the muscles, tendons, and joints, which provide information to the brain about skeletal muscular activity so that a person with eyes closed can still sense what various parts of his or her body are doing in space.[18] Proprioception is essential to properly fine-tune the movement and coordination of different parts of the body to sit, stand, walk, climb, turn, dance, swim, jump, and catch, grab, hold, lift, or throw something. It allows us to maneuver around obstacles in the dark and to reach for or manipulate objects without looking at them (such as a musician playing an instrument just from feeling it). Scientists also believe that the deterioration of proprioception with age may explain the increased risk of falls among the elderly.[19]

As someone who was not exposed to athletics, dance, or gymnastics as a child, I did not have many opportunities to develop my proprioceptive faculty. As a result, I was extremely clumsy throughout my childhood until this point in my life. I often had unseemly bruises on my knees and elbows from regularly bumping into furniture and doorways. I often hit my head on cabinets. I could even occasionally close car doors on my own foot, which I learned to compensate for by wearing thick protective shoes so it wouldn't hurt so much. So I knew firsthand that undeveloped proprioception equals dismal body intelligence.

This completely changed with the yoga teacher training. By intensely practicing asanas and learning the underlying anatomy for getting my body into these positions, I became much more tuned in to my proprioceptive system. As I developed a better sense for how gravity exerted its force on my body and how my muscles, bones, and ligaments adjusted to counteract gravity so I could hold these various postures, I actually felt like I had more gravity as a person. My posture also naturally improved as the muscles that support weight and balance got stronger. In addition, practicing yoga also made me more tuned in to my interoceptors, the sensory receptors in internal organs that provide information on vital functions such as heart rate, blood oxygen level, hunger, thirst or dehydration, bladder fullness, etc.

Yoga training taught me that to develop body intelligence, we have to tune in to both proprioceptors and interoceptors to get a fuller picture of how our body feels as we move. Mindful movement exercises, like yoga or mindful walking, are very effective practices that create mental space to develop greater proprioceptive and interoceptive awareness. As I became more tuned in to my body, I also became more grounded and less lost in thoughts inside my head, which led me to be more present. As I began to feel more integrated physically, I noticed that I also felt more integrated emotionally and mentally. I felt like my gut feelings got stronger and more reliable. Thus, it became easier to listen to my gut and have faith in its wisdom.

One day in yoga philosophy class, I was intrigued by Paalu's explanation of the many layers of meaning for the term "hatha" yoga. He shared that several hundred years ago, a sage had explained that "hatha" means the union of sun ("ha") and moon ("tha"). At a mystical level this is a metaphor referring to the union of the Higher Self and the lower self. In Sanskrit, the word "Atma" refers to the Spiritual Self (the divine soul), and the word "Jivatma" refers to the incarnated soul, that is, the part of the Atma that gets incarnated in human form and has the tendency to forget and become separated from the Atma until the Jivatma "awakens."* Thus, hatha is a metaphor for the reuniting of the awakened Jivatma with the Atma. At an energetic level it refers to the balancing and integrating of masculine (sun) and feminine (moon) energies in the major energy centers and meridians of the body. At a more concrete level, it could be understood as balancing and harmonizing the two arms of the autonomic nervous system: the sympathetic nervous system and the parasympathetic nervous system.

I couldn't come up with a scientific way to validate the more mystical concepts, but what Paalu shared about the autonomic nervous system inspired me to apply my knowledge of anatomy to observe how my practice of yoga might be affecting my own physiology. Basically, scientists see the human nervous system as an aggregate of many subsystems. The first high-level classification is the central nervous system (CNS), which consists of the brain and spinal cord, and the peripheral nervous system, which consists of everything else. The peripheral nervous system has two main components: the somatosensory nervous system, which controls voluntary motor movement and transmits signals from the senses to the CNS (this includes proprioceptors and interoceptors), and the autonomic nervous system, which controls the critical functions performed by vital internal organs, all of which do their jobs without the need for

---

* Master Choa Kok Sui referred to the Atma as the "Higher Soul"; however, in my view, the word "higher" is an imprecise adjective to describe a soul, because it implies a soul can be classified or divided at multiple levels. Instead, my Inner Sage tells me that "Atma" refers to a soul in its entirety.

direct conscious control. A key touchpoint between the somatosensory nervous system and the autonomic nervous system is the interoceptive system, which consists of the interoceptors distributed in our internal organs that enable us to tune in and monitor our internal vital functions.

The autonomic nervous system consists of two main branches: the sympathetic nervous system that arouses the body with energy to immediately react to stressors, and the parasympathetic system, which relaxes and calms the body to maintain vital life functions. In short, the sympathetic nervous system governs the body's freeze-flight-fight functions, and the parasympathetic system governs the rest-and-digest functions.*

The key nerve of the parasympathetic system is the vagus nerve, which connects the brain to the heart, the enteric nervous system, and other vital internal organs inside the rib cage and abdomen. The vagus nerve is the largest cranial nerve in the body and plays a key role in helping the autonomic nervous system maintain homeostasis, the catchall scientific term for the internal self-regulating two-way feedback processes by which our body keeps our organs functioning optimally. The fact that about 80 percent of fibers in the vagus nerve are afferent, meaning they carry signals from the organs to the brain, allows us to infer that the vagus nerve's main job is to let the organs tell the brain what is going on with them.[20]

According to Robert Sapolsky, a leading expert on stress and author of *Why Zebras Don't Get Ulcers*, stress causes the body to switch into sympathetic mode by activating the freeze-flight-fight cascade, which by design, turns down parasympathetic functions, deactivating the vagus nerve and disrupting digestion so the body can direct more energy to the arms and legs to either run or fight.[21] This means that in high states of stress, when the vagus nerve is taken "offline," the relay of information from our vital organs to the brain is disrupted. In that way, our head and our bodies become disconnected and out of sync, and this renders our vital organs vulnerable to developing chronic disease.

The vagus nerve is so important to our well-being that scientists have come

---

* Some researchers now propose expanding the autonomic nervous system to include the enteric nervous system (which is made of over 500 million neurons that are embedded along the full length of the gastrointestinal tract, beginning where people swallow food from the mouth into the esophagus and ending where people excrete waste at the anus) and the heart nervous system (which contains over forty thousand neurons). Furthermore, some people argue that because the enteric nervous system and heart contain so many neurons, they should be considered mini-brains.

up with the construct of vagal tone as an index of heart rate variability and respiration patterns to measure the functioning and health of this nerve. This is how the concept works: When we inhale, we turn on the sympathetic nervous system and slightly increase heart rate. When we exhale, the parasympathetic nervous system turns on, activating the vagus nerve to slow heart rate. So the heart beats faster when we inhale and slower when we exhale. High heart rate variability equals strong vagal tone. Minimum variability, or low vagal tone, means that the parasympathetic nervous system has trouble putting the brakes on stress. According to Sapolsky, when a person's heart rate doesn't slow down during exhales, this is a marker of someone who not only turns on the stress response too often, but also has trouble turning it off.[22]

Chronic stress, which means experiencing elevated stress levels for long periods of time, wreaks havoc on our cardiovascular system. It increases blood pressure and puts our immune system on extended red alert, which leads to high inflammation. Over time, chronic stress makes us more susceptible to disease because the body wasn't designed to handle high states of stress for more than short bursts. In order to return to homeostasis, the body needs to spend the majority of time in parasympathetic mode to regenerate. The challenge for human beings is that once we are stressed (when Brain 1.0 or Brain 2.0 takes over), our minds tend to continue to fixate and ruminate over our challenges so that we continue to push down the "on" button for the sympathetic nervous system instead of returning to parasympathetic mode. Thus, as vagus nerve function remains compromised, we tend to stay lost in our heads and fall even more out of sync with our bodies.[23]

This is where I found yoga to be very beneficial—it is a way to dial down the sympathetic nervous system and restore the parasympathetic nervous system. To confirm this, I found several articles referring to research findings that show that taking slow, deep breaths stimulated the vagus nerve and thus helped enhance parasympathetic functioning. One article on the NPR website described the vagus nerve as the "brake" for the stress response and explained that slow, deep breathing engages the brake.[24] This mechanism is definitely at work in yoga when we do ujjayi breathing, which involves taking deeper breaths to fill the lungs while slightly contracting the throat muscles. This form of mindful deep breathing could be why yoga, even in the form of power yoga, feels much more calming and stress relieving than other forms of exercise I have tried.

I decided to experiment by using myself as a test subject. I did a number of breathing exercises that were taught in the yoga training to see the impact on my heart rate. Ujjayi breathing definitely slowed my heart rate. The alternate-nostril breathing exercise also seemed to have a similar effect. When I did the slower breathing exercises, feelings of stress and tension seemed to melt away, leaving me feeling more calm and centered. From my observations, the effects of practicing asanas also seemed to get enhanced by pairing it with ujjayi breathing. Just holding stretching poses without the breathing wasn't nearly as calming or centering. Furthermore, ujjayi breathing seemed to make it easier to stay focused on the present moment, reducing the wandering of my mind during yoga practice. The breathing also made it easier to tune in and pay attention to the information coming from my proprioceptive and interoceptive systems. It was fascinating how interconnected all these bodily mechanisms seemed to be.

As much as I enjoyed the yoga teacher training, after I got certified, I still could not see myself teaching yoga as a profession. My main motivation for doing the training was to understand traditional yoga teachings and the history of yoga and to empower myself to guide my own yoga practice and be more selective in how I consume yoga offerings in the marketplace. The course gave me what I wanted and more. I now had a deeper understanding of yoga teachings and a high-level understanding of the evolution of yoga in modern times—that is, how the practice of yoga has been co-opted into a quasi-spiritual wellness lifestyle consumer industry. In the end, the most important benefit I got from learning yoga, and why I continue to practice it, was to be better grounded and integrated within my own body.

A Deepening Metamorphosis

Toward the end of the yoga teacher training program, after intermingling many different practices from various spiritual traditions and teachers, ordinary life became a little bit less ordinary in ways that I could not easily rationalize away through my knowledge of science, no matter how much I tried to do so. Instead, I had to look at science as a way of making sense of and understanding why these changes were all of a sudden happening at this point in my life.

As I shared earlier, I had started to see halos around people following meditation ever since February 2012. Nonetheless, the saying "seeing is believing" wasn't the case for me. I wanted to keep an open mind that there were many possible explanations. Perhaps this was a result of some sort of scrambling of electric activity in my retinas, optic nerves, or visual cortex that happened whenever I opened my eyes after a meditation. As I was aware there is a scientifically documented phenomenon called the corona effect, I didn't consider seeing halos as conclusive proof that there is such a thing as a human energy field or aura. The truth is that I wasn't ready to believe I could see energy. Even now, long after all that was about to unfold happened, I'm still not sure I'm ready to believe.

Sometime in late August 2012, I started to see these halos even without meditating—though they would appear much more clearly after a meditation. I noticed I could also see them vividly when I was in groups discussing spiritual topics. In September, I started to notice a prism or rainbow effect in that, on one side of a person, the halo would be tinged with stripes of red, orange, and yellow, and on the other side of a person, the halo would be tinged with stripes of blue, indigo, and violet, just like a rainbow divided in two. It baffled me that what I saw followed the sequence of the colors in the visible spectrum that I had learned in high school using the acronym ROYGBIV (red-orange-yellow-green-blue-indigo-violet), with the exception that green, the color in the middle, was missing.

Furthermore, at the top of people's heads, at the supposed location of the crown chakra, I began to see an electric violet orb of light that looked sometimes like a lightbulb or star and other times like a funnel. I also noticed a consistent pattern that this "lightbulb" on top of people's heads seemed to radiate more brightly after a meditation, while doing yoga, or during a discussion on spiritual topics. It was such a stupefying and stunning sight to behold that whenever I noticed it, I continuously rubbed my eyes and cleaned my glasses just to see if it would go away. It didn't.

Around that time, after months of enhancing the Brain 3.0 neural networks of the "observer" by intensely exercising metacognition, interoception, and proprioception, I began to regularly experience mystical states of meta-awareness (awareness of awareness) in everyday life in which being "Due" temporarily melted away—an experience that is very hard to capture in words. It's like entering a core state of vast awareness in which there is no observer, yet also realizing that the human form is a vehicle for a much higher consciousness, the "True Self," to extend into the physical world. Whatever I saw, this higher consciousness saw. Whatever I touched, this higher consciousness touched. Whatever I thought, felt, or did, this higher consciousness was aware of it. There was absolutely nowhere to hide from this higher consciousness. This revealed to me that humans can hide things from each other and sometimes from ourselves, but it is impossible to hide anything from the True Self. The True Self is aware of everything. This revelation was also a call to surrender and merge into the True Self, to let my identification with being "Due" completely melt away. But, to be honest, "Due" wasn't quite ready for this and asked for a rain check.

Since February 2012, it had also become normal for me to see phosphenes such as blue swirls or blinking or spinning orbs of white light whenever I closed my eyes. It happened while meditating, while resting, while closing my eyes in the shower, and while falling asleep. In the dark, I could even see phosphenes with my eyes open. Then, while I was in Dharamsala, it started to become normal for me to see what appeared as black-and-white television static in the moments before I fell asleep or as I woke up. I didn't know what to make of these phosphenes, so I regarded them as a subset of hypnogogic hallucinations that didn't necessarily signify anything.

Starting in mid-September 2012, the phosphenes morphed into a series of mystical experiences that my rational mind could neither explain nor dismiss. Again, coming out of sleep, I would see what looked like television static, but now it was like my consciousness was flying toward the static. As I approached the static, I saw that it was actually an infinite field of rotating globes of white light on a black background, and as each sphere of light rotated, it would reveal either a "1" or a "0" in black. It seemed like the field of globes was dynamically creating some sort of colossal and indecipherable binary code so quickly it was impossible for my mind to process or remember any pattern. It was stunning to behold this field stretching to infinity in all directions.

In some of these episodes, I would fly into one of these spheres and then see visions that were beyond my imagination and shook me to the core. They lasted only a few moments before I would be drawn back into my body in the exact reverse sequence my consciousness seemed to travel away from it. What made these experiences different from dreams was that everything appeared even more vividly than they do in high-definition screens at the movie theater. My dreams are never like that. Also, I wasn't asleep. I was fully conscious and alert with my eyes closed.

Even to this day, I still feel unable to adequately describe what I saw using words. For example, in the last of these visions, upon entering a sphere, I found my consciousness in the entryway of what looked like a lodge or temple facing the door. Then the door opened to reveal a magnificent bearded "Christlike" figure dressed in what looked like a radiant blue ball of light holding over his left shoulder a little Asian girl with her back toward me, dressed in what looked like a radiant red ball of light. Beyond the door was a celestial, ethereal type of light. This figure immediately made "eye contact" with me, and it was as if he was telepathically communicating with me using energy. If I had to put the message that came through into English, it would go something like this: "Know that I have been carrying you throughout your entire life. You and I are yoked together for all eternity. We are joined in oneness."

I immediately came back into my body in a state of shock with these lines from the Bible in my mind-track: "Come to me, all who labor and are heavy laden, and I will give you rest. Take my yoke upon you, and learn from me, for I am gentle and lowly in heart, and you will find rest for your souls. For my yoke is easy, and my burden is light."[25] My body felt waves of ecstatic bliss, and it was like my consciousness or energy was so large, there was no possible way it could ever fit into my little human body. My mind reeled in shock. As someone who had abandoned Catholicism, didn't have any faith in Jesus Christ being the Messiah, and doubted the scientific possibility of a virgin birth—it made no sense for my mind or brain to fabricate such a vision of a Christlike figure!

Even as I looked for ways to explain it away or pretend it was just my imagination or a vivid dream, I also had to ask: What if it was a genuine vision or a message from a higher spiritual realm? Was a great Being trying to help me see that this Earth is not a God-forsaken planet? Then it also hit me that this was the same being whose voice I heard in hypnogogic trances while I was in Kolkata. Could it be possible that my Inner Sage is not only a metaphor but actually a higher being who had just found a way to miraculously summon me to meet him face-to-face? Could this Christlike figure be the "True Self" that sages and mystics talk about? Was it possible that this Christlike figure is continuously incarnating through humanity to express itself in the physical universe and to experience life in physical form? Was it possible that all of humanity is really one great, unified Collective Consciousness, symbolically represented by this Christlike figure?

As I processed what happened, I began to wonder: Do out-of-body experiences imply that consciousness is not limited to the brain? If consciousness is not a mere by-product of neural synapses firing, does this mean that consciousness is possibly eternal? This light that I could now see enveloping people and shining as an orb above their heads—was this the force of an "eternal" consciousness animating flesh and blood into life? I had more questions: How common is it for people to see this etheric light? Does this explain why across the world, in the religious art of just about every culture I have come across, holy figures are depicted with radiant halos and brilliant auras?

If my vision was reliable, and if I dared to believe the message it conveyed, it would force me to look back at my life with a completely different lens. It meant I was never alone, am never alone, and will never, ever be alone. Was it possible that all those times in my childhood in which I was able to gather the inner strength and fortitude to overcome terrible, overwhelming ordeals—my resilience arose from the support of this great Being who had just revealed that He is always carrying me? It was further disorienting to consider that the little girl being held and the majestic figure holding the girl were (are) somehow *one Being*. How could I possibly make sense of the idea that my "True Self" is not "Due"—that "Due" is more like a temporary role that the True Self plays in something like a highly sophisticated virtual reality video game? I (as "Due") found it much easier (and more comforting) to look up symptoms for brain tumors online and try to reassure myself I didn't have one.

In early November 2012, I had a completely different type of hypnogogic trance. It began as usual with seeing blue-violet swirls of light. Then suddenly I saw in my mind's eye a vividly clear upside-down dark blue triangle, which proceeded to spin so quickly it created a vortex that pulled my consciousness into it. Meanwhile, I felt a huge amount of pressure in the space between my eyebrows, so much so that my ears popped and hurt the way they do during scuba diving whenever I descend underwater too quickly. As I came to the other end of the vortex, I saw a magnificent iridescent circular portal open and beyond the portal, darkness so thick there was no hint of light. Again the experience ended as suddenly as it began, with me pulled back through the vortex into my own body. Again I was tempted to dismiss it as a very vivid and bizarre dream. However, I couldn't because it was clear to me that I was fully awake and the feeling of pressure between my eyebrows and inside my ears remained for an entire week before it finally went away. I have never had any dream leave such physiological traces afterward.

When the vision ended, I immediately remembered that I had seen diagrams of an upside-down blue triangle during the yoga teacher training in a class on chakras. I quickly looked up depictions of the chakras, and eerily, I saw that the diagram for the "ajna" chakra, the one that lies between the eye-brows, consists of an upside-down indigo-blue triangle. I vividly remembered that during that class, I had been very skeptical about the quality, validity, and reliability of the information presented on chakras. I wondered: How did people even come up with this knowledge in the first place? As far as I understood, it was impossible to verify or validate any information on the chakras. Who could say whether it was not misinformation like old wives' tales being mind-lessly passed on from one generation to another? I even rolled my eyes when I learned that for the exam, we needed to be able to identify each of the major chakras by its symbolic illustration and provide its Sanskrit name. I was peeved at having to memorize information I considered irrelevant and useless.

After seeing the upside-down triangle and feeling very strong pressure where the ajna chakra is supposed to be located, I was forced to reconsider my previous stance. The fact that this chakra is often also called the "third-eye" chakra had much more significance to me now. I had to ask myself: What if these illustrations weren't bullshit after all? What if they originally came from ancient mystics who had visions even more powerful than what I had just experienced? Regardless, that doesn't imply that all of the information being passed on about chakras is authentic and undistorted, but I now had to ac-knowledge the possibility that surviving remnants of ancient wisdom are contained within this esoteric body of knowledge that has been handed down by yogis for thousands of years.

I had learned from my exposure to various yoga traditions and teachers that ancient sages in India had developed a three-part framework for understanding the evolution of the soul and how aspects of it coexist across different planes or realms of consciousness. In Sanskrit, the word "Jivatma" refers to the portion of the soul that incarnates into human form and becomes identified with each human incarnation, the word "Atma" refers to the full soul or individuated essence that mysteriously evolves through successive physical incarnations, and the word "Paramatma" refers to the Supreme Spirit, True Self, or Source, of which the Atma is a microcosm. Once this vision occurred, I had a "know-ing" that this framework could be trusted. I also somehow "knew" that just as the Atma proceeds from, is sustained by, and eventually returns to the Para-matma, the Jivatma also proceeds from, is sustained by, and eventually returns to the Atma. As the Sanskrit terms are a bit unwieldy to use in conversations with people who are not yogis, I use the words "embodied self" interchangeably with "Jivatma," "Higher Self" interchangeably with "Atma," and "True Self" interchangeably with "Paramatma." In line with this framework, my intuition was telling me that the vision was a form of notice from the Higher Self to "Due," the embodied self, that the return process had started.

I have to confess, I found experiencing that level of pressure inside my head to be so jarring that I decided to hit the brakes. I stopped doing intensive practices, in case they were directly connected to my having these unexpected "out-of-body" types of experiences or hallucinations. I had had enough mysterious visions to get the message that there is much more to human consciousness than science can account for. The more I saw, the more unanswerable questions I had. Honestly, I wasn't ready to go through the portal or to discover what lay beyond it. I wanted to stay grounded. I wanted to stay in my body. I also didn't want to cause a tumor, aneurysm, hemorrhage, or stroke to occur inside my brain. I had gone further than what I had ever imagined possible. I had no desire to push the boundaries any further.

Later, when I read *Descartes' Error* by Antonio Damasio (in which Damasio provides neurobiological insights from studying patients recovering from brain tumors and other types of brain damage), I was dumbstruck by these excerpted lines from René Descartes' *Discourse on Method, Part IV*:

> *From that I knew that I was a substance, the whole essence or nature of which is to think, and that for its existence there is no need of any place, nor does it depend on any material thing; so that this "me," that is to say, the soul by which I am what I am, is entirely distinct from body, and is even more easy to know than is the latter; and even if body were not, the soul would not cease to be what it is.*[26]

I was amazed by how Descartes had, four centuries earlier, beautifully articulated what my mystical experiences had revealed to me! These lines prompted me to wonder whether Descartes may have also had mystical experiences.*

Ironically, Damasio's take on these lines also demonstrated that people reading Descartes' writings through the lens of scientific materialism tend to

---

* The Wikipedia page on René Descartes actually states that according to his biographer, Adrien Baillet, Descartes did have visions and believed he had communed with a divine spirit. "René Descartes," Wikipedia, accessed July 18, 2017, https://en.wikipedia.org/wiki/Ren%C3%A9_Descartes.

filter out the mystical insights about the nature of the soul and interpret his writings as a philosophical justification for mind-body dualism, that is, the separation of mind/brain from body. This led me to wonder: what if people who have not had mystical experiences, regardless of how brilliant they are intellectually, are unable to understand the writings of Descartes and other philosophers describing their mystical experiences the same way that some-one who has had mystical experiences would intuitively understand them?

By the end of 2012, I had to acknowledge that my year of exploration and investigation had dug up way more than I had bargained for. I couldn't deny that my mind-track had shown me glimpses of mysterious higher realms be-yond my comprehension. I still could not fully grasp what "Enlightenment" (in the Buddhist, Hindu, and Jain mythological sense) actually means, but I couldn't deny that with my own eyes I could see people light up and glow when they were meditating, discussing spiritual topics, or doing any activity in which they seemed to connect to their soul, like singing or dancing, or speak-ing on a topic that they were passionate about. Like the law of gravity that bounds us regardless of whether we know about it or believe in it, I can literally see that "Spirit" sustains us and flows through us regardless of whether people know about or believe in Spirit.

It also dawned on me that I started to have mystical experiences only after I began shifting from Brain 1.0 and Brain 2.0 into Brain 3.0 and that these experiences had become more frequent and intense after I learned how to use meditation to proactively strengthen Brain 3.0 and quiet Brain 1.0 and Brain 2.0. Therefore, I began to connect the ability of Spirit to animate and flow through a person to the individual's development of Brain 3.0. Now that I was spending more time in Brain 3.0, I could feel more and more Spirit was flow-ing through me and it didn't seem to matter that I still did not consider myself religious nor spiritual and that I tended to question everything that spiritual organizations taught. Furthermore, I found that reading ancient texts such as the Bhagavad Gita, the Yoga Sutras, and the Dhammapada, as well as the Bible, seems to activate Brain 3.0, which then enables more Spirit to flow through me and bring me into a state of calm elevation in which an inner teacher trans-mits into my mind-track wisdom contained in between the lines of these texts.

This made me wonder about the cause-and-effect relationship between emotional states and "spiritual connectivity." I observed that whenever I got triggered into Brain 1.0 or Brain 2.0, I would feel contracted and disconnected; but whenever I paused and created space to recenter myself in Brain 3.0, I felt reconnected with my Inner Sage and naturally uplifted and rejuvenated. That meant it is definitely possible that living predominantly in Brain 1.0 and Brain 2.0 mode can result in people becoming so misaligned and blocked that Spirit can barely trickle through them and that by revving up Brain 3.0, people can open the floodgates for Spirit to flow through them. I found again and again that getting swept away by Brain 1.0 and Brain 2.0 resulted in my becoming temporarily deaf to my Inner Sage, and in hindsight, whatever it was that triggered Brain 1.0 and Brain 2.0 was usually not worth losing this higher connection. Actually, nothing is worth losing this higher connection. Thus, my wish to stay connected with my Inner Sage/Higher Self became my main motivation to be in Brain 3.0 as much as possible.

What continues to intrigue me most about my Higher Self is that what it shares rarely conforms to any conventional religious teachings. It defiantly points out flawed statements made by famous spiritual teachers and guides me away from dogmatic obedience to any guru, doctrine, or set of rituals. Furthermore, it is logical, rational, and curious. It embraces science. It is confident in its wisdom, yet humble and patient. When I'm too slow to follow, it keeps trying different ways to get me to understand something until I finally grasp it. There is a complete absence of harsh criticism, judgment, or manipulative control.

Looking back, my spiritual journey in 2012 was akin to rediscovering and finally appropriately valuing a precious life-giving gem that I had carried inside me all along but had discounted and taken for granted. Even if embracing this gem shattered my worldview and identity as an atheistic agnostic, there was no point picking up the pieces of this old identity and gluing it back together. The best thing to do was to say good-bye to the way my life used to be and let go of the stories I once used to define my sense of self so I could accept the unfolding of a new path and rejoice in the rediscovery of my Higher Self. It was like realizing that my life had been a black-and-white television show and then finally experiencing life in full vivid HD color. It was like being a barren tree sapling pushing through a long drought and, after receiving rejuvenating rains, finally getting to bloom into full life. Why not cherish the change and flow with the emergence of a new, more integrated and connected way of living and being?

6.

Creating and Sharing Calm Clarity

Your vision will become clear only when you can look into your own heart. Who looks outside, dreams; who looks inside, awakes.
—CARL JUNG

The further the spiritual evolution of mankind advances, the more certain it seems to me that the path to genuine religiosity does not lie through the fear of life, and the fear of death, and blind faith, but through striving after rational knowledge.0.
—ALBERT EINSTEIN[1]

If you do follow your bliss you put yourself on a kind of track that has been there all the while, waiting for you, and the life that you ought to be living is the one you are living. When you can see that, you begin to meet people who are in your field of bliss, and they open doors to you.
—JOSEPH CAMPBELL

IT'S HARD TO explain this, but the truth is, even though I am credited with being the founder of Calm Clarity, how it unfolded was more like Calm Clarity came to me and told me how to build it, and I reluctantly went along. Faith is not something that comes naturally to me so it was not easy to trust my Inner Sage. At many times, I was so filled with doubt and anxiety that I felt like abandoning this uncertain path altogether to find more stability and security in a normal job. Uncertainty still triggers my Brain 1.0 to take over and demand that I go back to the way life used to be before I awakened—when I lived wholly within my comfort zone of reason, science, and analysis and the illusion of certainty and control. But somehow, my Higher Self keeps breaking through the noise and reminding me that I am not doing this alone. Then, once I recenter myself in Brain 3.0, I see yet another interconnection between spirituality and science and the next step emerges. All these little steps somehow added up to bring Calm Clarity to where it is today.

Calm Clarity Is Born in Philadelphia

The real test of any personal transformation that happens in a sheltered space is whether it lasts when one returns to the "real world." In my case, that involved relocating to the United States and moving into my parents' home in a rough neighborhood in Philadelphia at the very end of 2012. My first priority upon arriving was to help my parents manage all the health issues they had left unaddressed because they didn't speak enough English to understand what their doctors were telling them and to help them navigate the obstacle course that is health insurance coverage. Back in my childhood environment, I learned that all the buttons my parents could push to trigger my Inner Godzilla were very much still there, but to my surprise, it was like my Inner Godzilla was much slower to wake up and get aroused. Furthermore, the calm I brought with me seemed to also have a contagious impact on my parents.

My whole life, my parents have argued and bickered loudly on a near daily basis, and in my younger years, I often added to the cacophony and drama in the house by getting into screaming matches with them. When I first moved back home, I often woke up to the noise of them losing their tempers at each other. Rather than get sucked into the drama, I coached my parents to help them understand each other's perspectives and the role each of them played in creating the situation. As my parents became more empathetic and patient with each other, the house also became much more calm and quiet.

At first, my mom couldn't believe how much I had changed. She remarked continuously about how I had always had the worst temper in the family and how impressed she was that I had become so calm and centered. Eventually, my mother became so curious that she asked me to teach her how to meditate. Within months, her blood pressure normalized and her doctor said she would no longer need hypertension medication. After seeing the benefits to my mother and myself, my father also started watching YouTube videos on meditation to learn on his own.

After getting my parents' health issues under control, I began to really think about what I would do next for work. I wasn't yet sure that I was ready to start a social enterprise because I had no experience or familiarity with how the nonprofit sector operated. I had no idea how to raise funds. Therefore, I decided to explore opportunities in the nonprofit space and accepted the invitation to join the local board of a youth development organization I had supported for years.

However, the more I learned about the nonprofit space, the more concerned I became that so many people working in this space were either burned out or about to burn out. Many of the people I met seemed to have gotten stuck in Brain 1.0 because they and their organizations didn't prioritize self-care or proactively address the secondary trauma they experienced while serving their beneficiaries. Since they were continually facing the risk of a budget deficit, organizations also tended to get trapped in Brain 2.0 chasing funds. This made them myopically focus on satisfying the whims of donors, even if it derailed them from their mission. Many of the organizations I encountered were so caught up in survival mode and weighed down by toxic stress that they lost the ability to shift into Brain 3.0 to pioneer creative and innovative solutions.

Soon after my return, I was shocked to learn of an impending $300 million budget shortfall at the Philadelphia School District for the upcoming school year, 2013–2014.[2] This crisis eventually forced the school district to give pink slips in June 2013 to all nonessential staff, which included the majority of assistant principals, teaching aides, nurses, guidance counselors, and cafeteria workers. It broke my heart to learn that life had not gotten any easier for young people growing up today in low-income communities in Philadelphia. In fact, with most of the guidance counselors laid off, it had become even harder for rising seniors to find the information and help they would need to get into college.

The crisis made me think back on my own situation and mindset as a teenager and how I would have dealt with this situation. I remembered that I never wanted to feel dependent on handouts or charity. I always preferred to stand on my own feet. What I would have valued was training and advice to become more effective at navigating through the chaos in my world, at advocating for support when necessary, and at making a positive impact on my environment and the people I care about.

So I began with a thought experiment: if I could meet and mentor my fourteen-year-old self, what would I share with her? With the benefit of hindsight, knowing that she would eventually have to resolve the damaging effects of developmental trauma and toxic stress, I would want to expose her to a top-notch life skills training that was grounded in brain research that could help her build up resilience and immunity to toxic stressors and potentially reverse the negative impact of trauma on brain development. I also realized that for this type of training to be embraced by the public school system, it had to be unquestionably secular and evidence based.

Since to my knowledge, such a training program did not yet exist, that meant someone would have to create it. However, as a self-taught brain geek with no formal neuroscience training, I lacked confidence that I was qualified to create or deliver such a training program. Schools seemed to want to collaborate only with prestigious university research programs run by renowned professors funded with multi-million-dollar grants. Why would school leaders take me seriously? I was an outsider without deep pockets, connections, or an organization to back me up. I couldn't see an obvious or feasible path forward.

In spite of my reluctance, while I was meditating, the words "Calm Clarity" surfaced in my mind-track. I recognized it right away as the name of the program that I had been thinking about. Yet, this wasn't enough of a sign for me to move forward. Having a name emerge did not address any of my doubts, so I continued exploring other options.

Then one night I had a very unusual dream, unlike any I had ever had before, and it was even more striking because I rarely dream or remember my dreams. In this dream, I saw that I was giving birth to a baby in some sort of lodge or cabin in the Himalayan Mountains. To my surprise, a Tibetan monk in traditional robes, whom I don't recall having ever met in person, served as the midwife. He gracefully handed me the baby, a boy, and told me: "His name is Namgyal." Next, I was at a U.S. Consulate office getting the paperwork filed to bring Namgyal back to the United States with me. To my shock, the baby was already a toddler who could run around, and I became anxious that the consulate officials would reject the application because Namgyal didn't look his age. Next, I was at the airport on the U.S. side, waiting to get through immigration, and somehow Namgyal had already grown into a young boy, making me anxious again that the authorities would reject his paperwork. Next, as we moved from the baggage claim into a car that picked us up, Namgyal had grown into a teenager. Finally, as we arrived home, Namgyal was already a young man with Christlike features. He simply got out of the car and rolled away on what seemed like a skateboard or hoverboard with complete confidence that he knew exactly what he was doing and where he was going.

When I woke up, I remembered that the name of my retreat instructor at

Tushita was Venerable Namgyel, but I had never learned what his name meant. I quickly did an Internet search and learned that the meaning of the Tibetan name Namgyal (also spelled Namgyel) is "victorious one who conquers all obstacles to achieve the supreme victory." Coincidentally, Namgyal Monastery is the name of the personal monastery of the Dalai Lama. Then it dawned on me that the baby in the dream—whom I had brought back with me from India and whom the consulate and immigration officials could not actually see—was a symbolic baby: it was Calm Clarity. Through the dream, I was being told that Calm Clarity, as a concept in my subconscious (or superconscious) mind, was "born" during my travels in Dharamsala, in the foothills of the Himalayan Mountains, and came back with me to the United States. Although the delivery was facilitated by Tibetan Buddhism, Christ was somehow involved. The dream assured me that Calm Clarity would unfold, grow, and thrive. The message seemed to be: stop hesitating, move forward.

To get a better idea of what to actually do, I had to meditate to tap into inner guidance. Similar to the meditation I had done in India where I saw all the experiences in my life become interconnected dots, in this meditation I saw that everything I had learned from science, from my business career, and from my journey in 2012 had to be combined and synthesized to create what would be Calm Clarity. Furthermore, Calm Clarity would be differentiated from academically created programs because it had to be deeply grounded in the wisdom and compassion that comes from directly transforming pain, suffering, and trauma. My Inner Sage then explained that before I was born, my soul had made the choice to undergo these experiences so that I could create Calm Clarity—which was a very hard thing for my mind to process. My mind-track showed me that Calm Clarity would be built from science and from spiritual wisdom and that this combination would make Calm Clarity so compelling that leadership teams at major companies would also value it. Then I saw it would be a social enterprise serving people across the socioeconomic spectrum. I also saw that the Higher Selves of the participants would collaborate with Calm Clarity to bring them into Brain 3.0. Once people could tune in to their own Inner Sage, their Higher Selves would be able to guide them further.

Through that meditation, my life purpose finally crystallized. Furthermore, my soul/Inner Sage/Higher Self also gave me a high-level road map or vision for how Calm Clarity would unfold over time. The question was: Could I believe it? Could I take a leap of faith that this road map really would unfold somehow? Could I really trust in what my Higher Self seemed to be calling me to do? Could I trust that I was really connected into Spirit and that I was interpreting these messages accurately? My skeptical, analytical mind (thanks to a very, very strong right prefrontal cortex) wanted to protest, but by this point, my Higher Self had already built a solid track record for guiding me through the most incredible journey, one I could never have imagined experiencing. My Inner Sage asked: Didn't I have enough experiences to be confident that this universe is not random? Didn't I have enough mystical experiences to see that all life and all form are unified into a larger Collective Consciousness, which, for lack of a better word, people often refer to as "Spirit" or "Source"?

My Inner Sage encouraged me to trust that Spirit was backing me up and that everything would turn out fine. I didn't have to understand how or know what the exact steps were; I just had to trust my intuition and follow its guidance. The steps would emerge along the journey. As I analytically assessed my options, I realized that I did not actually have a better alternative and probably could never come up with a better alternative than the path that Spirit placed before me. So why not give it a try and see what might happen? The worst thing that could happen was failure—and I had already proven my resilience to handle setbacks. The best thing that could happen was that the vision and road map would come true. Wasn't that worth the risk of failure?

In addition, while I was building my career, I had observed that high-quality leadership programs were marketed primarily to affluent audiences and that, for the most part, none of these leadership firms made any effort to reach people in low-income communities who could really benefit from leadership skills training to overcome very stressful challenges. From my own experience, I saw that people in low-income communities really only had access to programs created by well-intentioned nonprofit staff or policy makers in a vacuum, which were usually terrible in comparison to the training programs provided to the elite. I felt that people who were facing life-and-death challenges on a regular basis deserved better training. So I made a clear intention to create a program tailored to helping young people facing extreme adversity at the same standard of excellence that I would offer to a CEO. Thus, I decided to pour my heart and soul into creating an effective, meaningful training that, had I received it as a high school student, could have transformed my life.

Since then, as the vision that emerged in my mind somehow began manifesting unpredictably in bits and pieces, it began to be easier to have faith. Yet the path has not always been smooth. It was full of dead ends and closed doors that wouldn't budge open. Nevertheless, whenever I hit obstacles, I couldn't stay disappointed for long because, as I recentered myself in Brain 3.0, my Inner Sage helped me to distill the lessons I needed to learn from the experience and then guided me to move in another direction that was often much better than the original one. This sense of flowing and living in grace has become part of the magic, wonder, and mystery of living in Brain 3.0 and sharing it with others.

Building Calm Clarity

The journey of building Calm Clarity involved several major interconnected streams: my ongoing inner transformation and deepening understanding of the convergence between science and spirituality, supporting people as individuals to steadily shift from Brain 1.0 and Brain 2.0 into Brain 3.0 in many different aspects of their lives, and collaborating with institutions such as businesses, nonprofit organizations, and schools in order to nurture and cultivate Brain 3.0 among larger groups and change their culture to be more in line with activating and sustaining Brain 3.0 (rather than Brain 1.0 or Brain 2.0).

After the vision came to me, I got started by taking the most logical first step. I reviewed the science-based books and articles that I had found most inspiring and insightful. Then I used their citations and references to compile and track experts whose studies were relevant to understanding and explaining how to direct brain development toward enhanced well-being and effectiveness. I read the books that those experts wrote, watched the TED Talks and other lectures they gave on YouTube, followed mass media articles and blog posts announcing findings from their latest studies, and then tracked the experts that they cited, and so forth. As I compiled the research, key themes emerged that enabled me to connect the findings to various essential life skills.

Compiling the research on meditation led me to make the practice of "metta" (often translated as compassion or loving-kindness) meditation a cornerstone technique thanks to recent studies on its benefits and potential to reduce the physiological effects of toxic stress and trauma. Earlier in 2013, David Kearney at the University of Washington School of Medicine had published the results of a pilot study that involved training veterans with post-traumatic stress disorder to practice metta meditation for twelve weeks, which concluded, "Overall, loving-kindness meditation appeared safe and acceptable and was associated with reduced symptoms of PTSD and depression."[3] In addition, research led by Barbara Fredrickson at the University of North Carolina at Chapel Hill found that metta meditation led to an increase in positive emotions, enhanced social connection, and improved the functioning of the vagus nerve.[4] In parallel, Cendri Hutcherson, Emma Seppälä, and James Gross at Stanford found that even just a few minutes of metta meditation led to increased feelings of social connection and positivity toward meeting new people.[5]

Investigating further, I learned that vagus nerve activity and social connection are associated with oxytocin, a natural chemical circulating in the human body and brain that got nicknamed the "cuddle hormone" because hugging someone is linked to the release of oxytocin from the pituitary gland, although it is not exclusive to cuddling. Increases in oxytocin are linked to a wide range of affectionate physical gestures, including holding someone's hand and petting a dog or cat. Oxytocin also gets released when we fawn over cute babies and watch adorable animal pictures and videos on the Internet. This is significant because higher levels of oxytocin are associated with enhanced trust and feelings of connectedness, prosocial behavior and cooperation, and higher resilience to stressors.[6] In summary, the science pointed toward increasing oxytocin as a possible mechanism for healing the adverse impact of trauma.

Next, I decided to apply insights from my research to "hack" traditional meditations, such as metta meditation, to enhance their physiological benefits while making them more convenient and easier for beginners to practice without getting frustrated like I did when I first tried to learn meditation. The versions I created aimed to rev up Brain 3.0 by activating specific neural networks and to turn on certain biochemical cascades in relatively short periods of time, say ten to fifteen minutes. I began by testing them myself, and when I was satisfied with what I created, I also asked friends to user-test them and give me feedback.

At that point, metta meditation was the only key meditation practice that I had not yet practiced consistently over a period of time. So I decided to experiment with practicing the metta meditation I created every day for three months to see what changes could take place within me. Metta meditation simply involves wishing good outcomes, like happiness, health, and peace for oneself, for other people in one's life, and then for all people. In my version, I ended the meditation with a gratitude journal practice to really amp up the activation of Brain 3.0, the activation of the vagus nerve, and the release of oxytocin in the brain and body. (See chapter 8 [page 227] to learn more about the Calm Clarity Compassion Meditation.)

Just as Fredrickson's research predicted, over that period of time, I did find that I became more positive and optimistic; that it was easier to feel connected with people, even complete strangers; and that I wasn't as reactive when people were rude or mean to me. As a mind-hacker, I attributed these changes to the following mechanisms: (1) regularly activating and strengthening the compassion and gratitude neural circuits in Brain 3.0 also activated and strengthened the neural circuits for positive emotions and resilience, (2) elevating levels of oxytocin led to my feeling more connected with others and to my behaving more altruistically, and (3) strengthening the functioning of my vagus nerve, which connects my brain to my heart as well as other key internal organs, enhanced the functioning of my parasympathetic system and reduced my reactivity to Brain 1.0 triggers, meaning it gave me the ability to hit the brakes on the freeze-flight-fight cascade.

In parallel with the scientific reading and research, I also continued to spiritually explore. I read and reviewed many ancient texts for inspiration, such as the Dhammapada, the Bhagavad Gita, the Bible, the Tao Te Ching, and the Stoic teachings of Epictetus and Marcus Aurelius. I also read poems by Rumi to get a sense for the Sufi mystic tradition. After seeing an exhibit on the Dead Sea Scrolls at the Franklin Institute (the major science museum in Philadelphia), I found and read translations of the Gnostic Gospels discovered at Nag Hammadi, which were available online. That exhibit also inspired me to take a fresher look at Christian traditions, so I returned to my childhood Catholic church for Mass and observed Quaker meetings (after all, Philadelphia is the central hub of the worldwide Quaker community). I also continued to explore and learn about Tibetan Buddhism by attending teachings at the Tibetan Buddhist Center in Philadelphia, whose spiritual director, Lama Losang Samten, is a renowned sand mandala artist and a disciple of the Dalai Lama. Although I don't consider myself a Buddhist (I still don't buy into many aspects of Buddhist ideology), having an authentic lama from Dharamsala right here in Philadelphia has been extremely beneficial as an anchor. In the end, since I could go to only one of these places on Sundays, I chose to prioritize hearing Lama Losang's teachings.

Even though I had stopped doing intensive spiritual practices and was instead combing through the science of well-being and testing the science-based meditations I created for beginners, I continued having mystical experiences that my analytical mind struggled to process. Not long after I read the Gnostic Gospels, it began to happen occasionally during moments in which I was doing nothing but sitting quietly in peaceful awareness with a sense of joyful contentment; I would see the air and empty space around me turn visible as vibrant luminous mist. This would last anywhere from a few seconds to a bit more than a minute if I let myself just peacefully appreciate the beauty and wonder of it without trying to analyze it or make sense of it. No matter how many times it happens, it's not the type of thing I could ever get used to. Like watching a sunrise or solar eclipse, it is one of the most captivating and uplifting sights to ever behold. Whenever it happens, it calls from my memory these passages from the Gnostic Gospel of Thomas:

> Jesus said, "If your leaders say to you 'Look! The Kingdom is in the sky!' Then the birds will be there before you are. If they say that the Kingdom is in the sea, then the fish will be there before you are. Rather, the Kingdom is within you and it is outside of you." . . .
>
> They asked him: "When is the kingdom coming?" He replied, "It is not coming in an easily observable manner. People will not be saying, 'Look, it's over here' or 'Look, it's over there.' Rather, the kingdom of the father is already spread out on the earth and people aren't aware of it [do not see it]."[7]

I had to wonder: Could what I was seeing be somehow tied to these lines? Could this luminous mist possibly be hints of a higher realm that interpermeates our physical universe? And of course, by asking these questions and trying to make sense of what I was seeing, it would all fade away.

Unfortunately, there is no way of knowing for sure if these strange mystical phenomena I began seeing weren't a result of faulty functioning in my retinas, optic nerves, and/or brain. Around that time, one of my friends had shared on social media a quote from an outspoken atheist named Dan Barker that said: "Insanity is believing your hallucinations are real. Religion is believing that other peoples' hallucinations are real." This, of course, struck a nerve. At that point in time, I still found my own mystical experiences too weird for me to talk about openly. To reassure myself that I was still sane and grounded, I felt I really needed to double down on science.

Despite my wish to emphasize the science, Spirit continued to deepen what I learned from science. For instance, my understanding of metta meditation changed one Sunday morning, when I decided to visit the Quaker meeting in the historic Arch Street Friends Meeting House in downtown Philadelphia. Since no one provided any instruction on how to sit there quietly during the meeting, I practiced the metta meditation by wishing good things for all the people in the room with me (see instructions on the compassion meditation in chapter 8 [page 236]). Somehow, I started to feel the benevolent presence of the many spirits whose bodies were buried beneath the land, which had served as a cemetery in the 1700s before the meetinghouse was built on it in the early 1800s. In a manner I cannot explain or understand, in my mind's eye, I sensed that they were forming a sacred circle around us, blessing all of us. Then as I sent loving-kindness, compassion, and peace to everyone in the room and throughout the world, I had the sensation that I was holding a baby in my arms, whom I was bathing with love and tenderness.

Then I realized my consciousness was somehow merging with that of "Mother Earth," and that this "Divine Mother" is the "Soul" of our planet continuously holding and nurturing all of humanity as a baby in Her loving and soothing arms. Somehow, this great Being telepathically explained to me that no matter how many mistakes human beings have made and will make in our evolution, She always embraces and nurtures us, whether or not we are consciously aware of it. She also explained that by embodying unconditional love and sharing it with the world, people connect with Her. The more we act like the Divine Mother—the more we transmit Her nurturing and healing energies into our world and the more we manifest Her love, compassion, and wisdom—the more we heal ourselves, each other, and our planet.

I had to wonder: Was this merely my imagination or was I really tuning in to a divine consciousness embodied in our planet? If the latter were possibly true, then is it also possible that whenever humans share unconditional love, we are channeling and embodying the love of this Divine Mother? I felt so physically, emotionally, and spiritually uplifted, nourished, and fulfilled that it was hard to dismiss this as just a figment of my imagination. So I decided to incorporate what I learned from this experience into the Calm Clarity version of the compassion meditation and see what happens and how people respond.

When I guide the compassion meditation, as I instruct people to self-generate the pure, unconditional love that an ideal mother has for her child, I also set an intention to connect to this Divine Maternal energy and transmit Her unconditional love to everyone in the room. I often find that by the end of guiding this meditation, for a brief period of time, I see a pink halo around everybody's heads and a pink glow suffuse the entire room. If I'm looking at a piece of white paper, the paper also temporarily appears pink to me. I've come to wonder if this pink light is associated with the flowing of unconditional love and with the opening of people's hearts. Yet how could I or anyone ever validate this for sure? What I can share is that people who I guide through this meditation consistently tell me that it made them feel really good and relax in a way that they didn't believe was possible and that they want to continue to practice the meditation on their own.

After creating a set of Calm Clarity meditations, I started to prototype the curriculum, which I intentionally designed to be secular and science-based so that the public school system would be able to implement it. I began by creating modules that explain how the meditations work by drilling down on the science and intertwining the meditations with key life skills. I also saw a clear opportunity to differentiate Calm Clarity from other offerings in the "spiritual marketplace" by making it as scientific, concrete, and practical as possible. Ironically, being a skeptical, logical, data-driven mystic seemed to put me in a unique position to present meditation in a way that skeptical, logical, data-driven people would find convincing.

I focused the training on explaining how the brain functions, helping people understand and observe their own emotional states (Brain 1.0, Brain 2.0, and Brain 3.0), and providing sound, evidence-based tools and techniques to activate various aspects of Brain 3.0. The first prototype had eight modules to develop different life skills. Each module was structured like a TED Talk that interweaved science with interactive exercises, guided meditations, and group discussions. By the fall of 2013, the prototype curriculum was ready to be pilot tested with an audience, but I had no idea how to find young people to user-test the program. After living abroad in Asia for seven years, I didn't have much of a social network in Philadelphia. As it turned out, I spent only a couple weeks scratching my head wondering how to move forward because the universe stepped in.

Pilot Testing Calm Clarity

One day in September 2013, out of the blue, Laura, a former colleague from Vietnam, reached out to me on an instant messenger chat platform—something she had never done before (and hasn't really done since). As a graduate of the University of Pennsylvania, she had lived in Philadelphia for a number of years and still had many friends in the area. Once she heard me explain what I was working on, she offered to make a few introductions. Then, one of these friends, Brett, a venture capital investor who would later join the Calm Clarity advisory board, connected me with Rich, the founder of Schoolyard Ventures, an organization that teaches entrepreneurship to high school students. During our first phone call, Rich offered to immediately recruit students in his program to be part of a focus group that would experience the modules and provide feedback. Since two of these student volunteers preferred to go through the Calm Clarity Program in a manner that was more convenient for them, they pitched me on the idea of running a pilot program at their school, Masterman High School (an elite magnet school that is consistently ranked as the top public high school in the state of Pennsylvania), and connected me to the teacher who ran the enrichment program there.

Because Masterman was such a pressure cooker and many of the students did not have healthy coping mechanisms for stress, the teacher jumped at the idea of running a Calm Clarity pilot there and offered to do everything she could to make it happen. She recruited students and helped collect before-and-after surveys consisting of validated instruments I wanted to test to confirm whether they were relevant to measuring the impact of the program. Thus, within a few months of drafting the curriculum, I found myself testing it with inner-city students at the top high school in Pennsylvania. After running the pilot with students during the winter trimester, I incorporated their feedback to refine the program and then tested the improved version with a second batch of students in the spring trimester.

Although I had put so much time and energy into making the Calm Clarity Program entirely secular and science based, there was a part of me that wondered: Would the spiritual wisdom embedded in the program still get transmitted? If it was true that Spirit more readily flows through people when they are in Brain 3.0, then would the participants experience a sense of spiritual elevation as they activated Brain 3.0? To my amazement, as I facilitated the training with the students, I could literally see the energy in the room shift as their minds and hearts opened. The students would radiate brightly like pillars of light with brilliant halos. For short periods of time, the space in the room would even turn into bright luminous mist. It amazed me that a workshop on the scientific mechanisms for shifting into Brain 3.0 could lead to the same energetic changes I saw when people engaged in discussions on explicitly spiritual topics.

Thus, in spite of my wishes to build a secular curriculum that emphasized only the science, my eyes forced me to acknowledge that Spirit was giving me vividly beautiful visual reminders that consciousness encompasses much more than the brain. Of course, I kept this observation to myself. Moreover, I had to really ground myself and concentrate so as to not let what I was seeing distract me from covering the lesson plan and keeping the course moving on schedule. Seeing these unusual things still made me wonder if I possibly had a brain malfunction. As far as I know, other than these strange sights, I had no other symptoms of a brain tumor.

Interestingly, around this time, I was also forced to learn about the human eye in more detail because my father and then my mother developed eye conditions that required urgent treatment. In late 2013, my father had a retinal tear that caused blood to contaminate the macular fluid in his eyeball. I had to take him to a retinal specialist to have the tear sealed with a special laser. Then as a precaution, after realizing that my mother had never had her retinas examined before, I arranged an examination for her. The optometrist discovered that she had oversized optic nerves that were a possible sign of glaucoma and urged me to take her to a specialist right away. Fortunately, the specialist found that my mother's optic nerves were still healthy but warned that she was at very high risk for developing narrow-angle glaucoma. As the lenses of her eyes hardened with age, it was causing the angle between her cornea and iris to narrow. As the angle narrowed, the drainage system for the eye could get blocked, which would cause the internal pressure of the eye to spike quickly. If this happened, in a matter of minutes, the increased pressure would irreversibly damage the optic nerve of that eye. Therefore, she recommended my mother immediately undergo a preventative outpatient procedure called a laser iridotomy to create a small hole in the edge of both of her irises that would provide another means for the fluid in her eyes to safely drain.

To be on the safe side, I also had my eyes and my retinas checked. To my great relief, the optometrist found that my eyes and retinas were healthy and normal—aside from being nearsighted and having astigmatism since I was a teenager. However, I was still left with the baffling mystery of why I had begun seeing things I had never seen before. My rational mind knew that by process of elimination, it was possible that whatever strange sights I was seeing could be a result of unusual activity in my optic nerves, visual cortex, or some other area of my brain. However, it is not as easy or affordable to have these internal structures examined. Besides, how could I even explain to my doctor, my insurance company, and the specialist the reason I had for wanting to have my optic nerves and brain examined? To the best of my knowledge, I did not (and do not) have any symptoms or signs of a brain tumor, aneurysm, hemorrhage, or stroke. Furthermore, I am not aware that there is any type of medical specialist who could figure out the brain mechanisms that could possibly account for or demystify "clairvoyant" phenomena. So this remains for me a mystery on which science has not yet shed light.

During this period of time, I also began to build an advisory board and board of directors through what seemed like a series of serendipitous encounters. As an introvert, I have never been a particularly social person and my natural inclination is to mostly live like a hermit. However, on several occasions, I felt compelled to accept an invitation to attend a workshop, gala, or party, and then got inadvertently placed next to someone who was naturally passionate about the type of work I was doing and wanted to stay in touch and support Calm Clarity.

For example, I met Dave Denious, a partner at a large law firm who now serves on the Calm Clarity board, by sitting next to him at a gala for the nonprofit organization where I had served on the board. He had taken a class with Robert Thurman, a leading scholar of Buddhism, when he was a student at Amherst College and was both fascinated by Eastern philosophy and passionate about youth development. We stayed in touch and eventually, as Calm Clarity took shape, Dave agreed to serve on our board and to help advise me on legal matters.

Next, I met Rick Bellingham, a former HR executive who pioneered the first corporate wellness initiatives in the 1980s and now works as an executive coach, when a mutual friend, also a Harvard graduate, invited both of us to participate in a workshop where he shared a new form of qigong movement he'd invented. Rick loved hearing about my vision for Calm Clarity and offered to support and mentor me in any way he could.

Then I met Tom Tritton, a former president of Haverford College and the Chemical Heritage Foundation, at the holiday party hosted by CultureWorks, the organization that I chose as Calm Clarity's fiscal sponsor (having a fiscal sponsor enabled us to start formally operating as a nonprofit project under their 501(c)(3) umbrella). Tom was intrigued by my taking a scientific approach to demystifying spiritual practices and my focus on creating a trauma-informed resilience training specifically for low-income high school and college students. After I showed him the draft curriculum, he could see that the content was solid. So, even though he barely knew me and had earlier shared that he was too busy to take on a new project, he decided to take a chance on helping me build Calm Clarity.

Finally, after we realized we needed help from a technology expert to understand how to leverage technology to think about how to grow and scale, Dave reached out to his network for recommendations for technology experts in the area and got connected with Chris Kohl, the chief information officer at Vertex, who responded enthusiastically. Chris had been on his own journey of self-improvement and was very interested in the concept of self-mastery. When he saw the materials, he was floored by the robustness of the Calm Clarity Program and volunteered to help build the organization.

Serendipitous connections also led to unexpected collaborations to pilot test the program with our various target audiences. Someone I met at a breakfast networking session connected me to her former supervisor at the University of Pennsylvania's Netter Center for Community Partnerships. The discussions led the director of the Netter Center's programs at Sayre High School in West Philadelphia to invite me to conduct a pilot during their upcoming summer-bridge program called "Leaders of Change," which would include students from high schools across West Philadelphia.

I have to admit that I was a bit nervous about this particular pilot because the students in the spring Masterman pilot had shared that they thought the program was great for them because they love learning, but warned me that students from West Philadelphia, which has a reputation for being one of the toughest areas in the city, might not embrace it. Sure enough, on the first day, many of the students in the summer pilot arrived with pent-up anger and attitude. The students were drawn from a half dozen different high schools, and quite a few of them really didn't want to be in the same room together. I overheard someone declare that a student from another school, who had given her a look she didn't like, better watch out.

I clearly got the students' attention when I kicked off the program by sharing my own story of growing up in violence and dealing with trauma, getting into Harvard, overcoming PTSD, building a career, and coming back as a social entrepreneur to share what I had learned with them without anyone paying me to do so. They were entranced as I walked them through the science behind the Calm Clarity framework and explained how people shifted between Brain 1.0, Brain 2.0, and Brain 3.0. One of the students insightfully shared during an exercise that he felt like who he is in Brain 3.0 is the real him, and that when he is in Brain 1.0 and Brain 2.0—that's not really him. When I guided them through their first meditation, the room was so silent, I could hear a pin drop. The students had never experienced anything like Calm Clarity before, and when the session ended, many of them came up to me to say thank you and to share their excitement about applying what they learned and sharing it with their families.

As the summer unfolded, I realized that many of the students in the West Philadelphia pilot were even more attentive and hungry to learn this information than the students at Masterman had been. Although I had to run a condensed version because the Netter Center allocated only two hours a week for a period of five weeks for me to run Calm Clarity training, the staff remarked that there was a clear before-and-after difference. At the end of the program, I asked the students to fill out a feedback survey. What the participants shared confirmed that the program had made a positive impact on their lives.

For example, one of the young women who came in ready to pick fights wrote: "I just want to say that Calm Clarity calmed me down. . . . It helps me/us focus more on what we need to get done. I had a lot of anger before Calm Clarity and I feel less angry now. It made me a better 'leader of change.' I participate more in everything. I don't hold back."

Similarly, another young woman wrote: "You actually motivated me to do well in life. I'm more calm. I appreciate everyone. I want more people to support me. I loved the 'forgiving' lesson and learning about what you cannot control. . . . I now know how to forgive people and try to have a positive attitude. I forgive a friend of mine who hurt me in the past. I tried putting my feet in his shoes to understand. I forgive him for the most part and learned not to hate him."

A young man who shared that his family was evicted from their apartment and spent the summer homeless wrote: "The program has benefited me because it helped me realize a big mistake that I was making in my life. Instead of playing video games, I realized I have to come back to reality. . . . When I first joined Calm Clarity, I was too hyperactive and never stayed calm. But later in Calm Clarity, I've learned to calm down. . . . I've found out that I was very stressed and that meditations actually help keep me calm. My mom has noticed that I don't always look mean now."

I was particularly amazed by what was shared by a young man who was covered in tattoos and seemed to intentionally project a tough-guy front. What he wrote confirmed my hunch that he really was paying careful attention: "I come home more calm on Mondays [after Calm Clarity] and I always share what I learn from Calm Clarity because this type of info anyone can use. They keep what I teach them in mind and think differently now. The learning for me was to think when I'm going through obstacles and how to overcome the stages of being successful. I'm more welcoming to people and networking with others."

In the fall of 2014, I was invited by the faculty at Cabrini College (now Cabrini University) to run the Calm Clarity Program as a credit course. At that point, I had expanded the program into ten modules based on participant feedback and suggestions to give more time to dive deeper into some of the topics using interactive exercises. Also, I had finally worked out an approach for collecting meaningful data using relevant validated instruments.

Once again, testimonials shared by the college students confirmed that they found the program life changing. For example, a senior who had shared in class that she'd decided to take the course because she was having so many panic attacks that she couldn't sleep, and could barely function, wrote in her survey at the end of the program: "As a senior, I considered withdrawing from college because I was too stressed out. Since being engaged in Calm Clarity, I feel excited, more prepared and less stressed when it comes to graduating."

A freshman who shared in class that she was grieving the recent loss of her mother and had seriously considered taking time off from college wrote: "Since my mother had passed away right before I started college, I felt like giving up on life. But I knew that she wouldn't want me to do that.... Honestly, this class not only encouraged me to stay in school but to also focus on thinking positive and being a positive impact on others."

In early 2015, when I completed the data analysis from the Cabrini College pilot, I felt like I had hit a big milestone. To my great joy and relief, the analysis showed that there were statistically significant changes as measured by key instruments, such as the Brief Resilience Scale, the General Self-Efficacy Scale, and the PERMA-Profiler (a multidimensional instrument developed by researchers at the Positive Psychology Center at the University of Pennsylvania to measure well-being). I finally had the preliminary data and testimonials to show the program could make a significant impact on low-income students and potentially increase college success rates.

However, I soon learned the hard way that no amount of compelling data could convince schools with tight budgets to invest in piloting a new program such as Calm Clarity. Regardless, the real value of the data was that it gave me the validation I needed to have confidence that the program was effective. Since I couldn't indefinitely bootstrap programs for schools without earning income, my advisory board suggested that it was time to shift gears and focus on building a self-sustaining business model. I already knew from the guiding vision that the next step was to pilot the training with professionals and refine it so that the experience would motivate executives and businesses to shift into Brain 3.0.

Making the Case for Enlightened Leadership

To think about how to apply Calm Clarity to the business world, I found myself meditating on enlightened leadership, the concept that had originally intrigued me when I walked through the halls of the palace at Mysore learning about the legacy of Krishna Raja Wadiyar IV. Now that my own inner transformation had given me a different way of seeing the world, I wondered whether Wadiyar's experience of yoga (meaning "divine union") may have influenced the way he saw the world and the decisions he made as a leader. Did a mystical sense of interconnectedness and oneness with all life lead him to act for the greater good? Did mystical experiences also apply to great leaders like Mahatma Gandhi, Martin Luther King, and Nelson Mandela? Did their Higher Selves guide them and give them strength and energy? How else could a human being persevere through so many obstacles with the patient faith that his or her cause would eventually emerge victorious? How else could a person steadfastly believe that light would triumph over darkness?

Yet I also wondered: is the business world ready to explicitly talk about leaders connecting with and embodying their Higher Selves? I sensed that I would be more effective at creating a dialogue by subtly presenting these concepts using neuroscience. Once people understood how the emotional states of their brain affected leadership, they would understand the negative impact of being in Brain 1.0 and Brain 2.0. After they had a direct experience of what it felt like to be in Brain 3.0, then they would be open to going deeper. Being a veteran of the business world, I knew that to get companies to pay attention, they had to see a link between shifting people into Brain 3.0 and top- and bottom-line financial results.

I found inspiration in the work of Ellen Langer at Harvard University, who is widely regarded as "the Mother of Mindfulness" because she pioneered research into the concept of mindfulness in the 1970s. Interestingly, Langer's definition of mindfulness has nothing to do with meditation or yoga. What she focuses on is people's nuanced ability to discern and actively notice new things in every moment and keep themselves from falling into a mindless autopilot mode. In the Zen tradition, this corresponds to a notion called "beginner's mind"—the ability to experience any activity in life, no matter how mundane, with childlike curiosity, as if you are doing it for the first time. It's the ability to keep yourself from letting past conditioning and preconceived notions (which are "stored" in your autopilot) interfere with your engagement with the present moment.

In particular, one statement she made during an interview with Krista Tippett on the *On Being* website really stood out to me: "Mindlessness is the application of yesterday's business solutions to today's problems. And mindfulness is attunement to today's demands to avoid tomorrow's difficulties."[8] This observation captured what happened to BlackBerry, Yahoo, Kodak, Blockbuster, Sony, and American automobile manufacturers as a group—all of these companies didn't respond early enough to signals that their once dominant products and services were no longer relevant to consumers. Unfortunately, too many companies discover only after it's far too late that yesterday's solutions have an expiration date. In today's fast-changing economy, organizations can no longer afford to let mindless patterns and ingrained ways of thinking and doing things drag them down.

Nevertheless, the challenge for companies to apply the wisdom of Langer's observation is that the human brain can get so overwhelmed by change and uncertainty that people cannot think straight, create, or innovate. Therefore, there is a pressing need to help companies overcome mindlessness by showing leaders, as well as employees at all levels, how to shift into Brain 3.0.

Almost two decades ago, when I first started my career as a management consultant, it was possible to write a five-year strategic plan and anticipate how the world would evolve. But today, it is getting harder and harder to predict and forecast even six months to one year into the future. No one can tell when a new technology will revolutionize their particular field or industry. No one knows how long their expertise and skills will remain relevant. No one can anticipate when what they do for work will become obsolete.

Unfortunately, human beings are hardwired to perceive change, uncertainty, disruption, unpredictability, and loss of control as threats to be defended against. These threats trigger Brain 1.0 to unleash huge amounts of stress hormones that put us into a state of freeze-flight-fight, which reduces blood flow to Brain 3.0, and without Brain 3.0, we become "mindless." When we are in Brain 1.0, we are too cognitively impaired to think clearly and respond effectively to the challenges at hand, so we cannot even consider the challenges on the horizon. In the book *Scarcity*, two behavioral economists, Sendhil Mullainathan and Eldar Shafir, explain that in a state of stress, people's IQ may drop by almost an entire standard deviation.[9] Without Brain 3.0 fully functioning, we can't take in new information. So instead, we tend to resist change by defiantly throwing past solutions at the problem, which of course are not effective, or by going into Brain 2.0 to escape the stress and anxiety through denial, suppression, or self-medication.

Therefore, the key aim of Calm Clarity's Mindful Leadership Program is to help people understand how much more effective we can be if we proactively strengthen Brain 3.0 and build up our emotional immune system, so that when stressful challenges arise, we will not be hijacked by Brain 1.0 and Brain 2.0 and mindlessly react. By being able to keep Brain 3.0 "online," we increase our neurobiological capacity to respond effectively to a complex situation. See figure 1 below.

Shifting into Brain 3.0 enables us to respond effectively

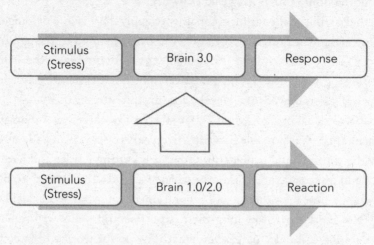

Figure 1: This diagram illustrates how stressors normally bring us into Brain 1.0 / 2.0, in which we react instinctively without being thoughtful. As we strengthen Brain 3.0, even when stressors occur, we are able to remain centered.

I was lucky to have Rick Bellingham as a sounding board to talk through my ideas. He told me about a "leader-to-detractor" scale that we saw could overlay beautifully with the three emotional states of the brain. I created a new scale (see figure 2 below) to illustrate how when people are in Brain 1.0, no matter what level or position they're in, they often detract from a team's performance because they create a toxic environment inside and around them. In Brain 2.0, people do what they are paid and rewarded to do in a transactional, self-interested manner. In cases where the rewards are scarce and the stakes are high, people may viciously compete to win in ways that undermine the organization's values and even cut corners and make unethical decisions that jeopardize the overall organization for the employee's own short-term benefit. It is only when people are in Brain 3.0 that they can see a bigger picture, go above and beyond their job description, and take the initiative to create new strategies and projects that serve the organization's greater vision and mission.

Along with shedding light on how brain states affect people's behaviors

How emotional states impact behavior

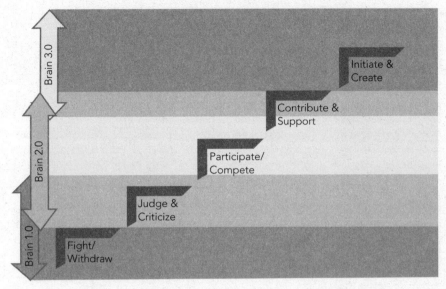

Figure 2: We can discern what emotional state people are in through their behavior.

and motivations, I also wanted to help companies understand that emotional states are contagious. People in Brain 1.0 can easily trigger Brain 1.0 in everyone around them and hijack the larger group into Inner Godzilla mode. People in Brain 2.0 can also easily trigger Brain 2.0 in everyone around them and hijack the larger group into Inner Teen Wolf mode. The upside about emotional contagion is that people with enough "horsepower" in Brain 3.0 can also lift up people around them and serve as the group's Inner Sage. Leaders who inspire us as a society—great humanitarians such as Mahatma Gandhi, Martin Luther King, and Nelson Mandela—are essentially individuals who have so strongly cultivated Brain 3.0 in themselves that they contagiously activate Brain 3.0 at a national and international level, inspiring countless multitudes to come together to work toward a vision that is much bigger than themselves. I capture these ideas in figure 3 below.

How emotional states impact interpersonal interactions

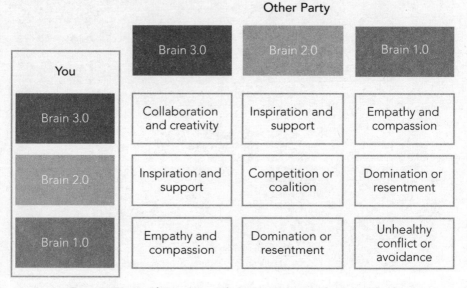

You	Other Party		
	Brain 3.0	Brain 2.0	Brain 1.0
Brain 3.0	Collaboration and creativity	Inspiration and support	Empathy and compassion
Brain 2.0	Inspiration and support	Competition or coalition	Domination or resentment
Brain 1.0	Empathy and compassion	Domination or resentment	Unhealthy conflict or avoidance

Figure 3: Emotional states impact the way we interact with others and are often contagious.

It's easy for people to see how being in Brain 1.0 is extremely toxic for a company's culture and performance. However, most people don't realize that high activation of Brain 2.0 also contributes to an unhealthy political and toxic culture. When the leadership team and supervisors operate primarily in Brain 2.0, people get locked into tunnel vision pursuing goals to win rewards. The people working under these leaders do not feel it is safe to question their choices or to be the messenger of bad news. As a result, the leadership team falls out of touch with what is really happening in the business. In a Brain 2.0 culture, people fight to get credit for good performance, but no one takes accountability for bad performance. People secretively hoard information and knowledge to increase their advantage, even when other coworkers really need the information to do their jobs effectively. In really bad cases, teams and entire business units may give in to the temptation to manipulate figures and reports to cover up bad performance or to exaggerate good performance. Wherever Brain 2.0 is given free rein, the risk of unethical behavior and scandals increases.

If this observation hits close to home, the good news is that it doesn't have to be like this. We can break out of being hostages to Brain 2.0 by shifting into Brain 3.0. In Brain 3.0, we no longer feel like we are scrambling or have to scramble to get through the day; yet we may feel more energy and actually produce higher-quality work in less time. By making more conscious choices about what we commit to in our schedules, we can get back more time to spend in ways that refresh and rejuvenate our spirits. This, in turn, enhances our creativity and our ability to connect emotionally and to be present with other people.

Figure 4 below illustrates how the amount of time that people within an organization spend in Brain 1.0, Brain 2.0, and Brain 3.0 can create a toxic culture, a myopic culture that trades off long-term performance for short-term gains, or a high-performing culture with the capacity for honestly diagnosing problems and concerns and finding innovative solutions.

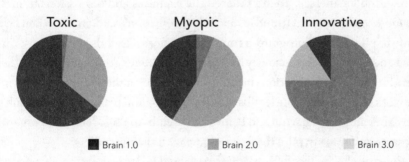

Is your company bringing out the best in your people?

What percent of time does your organization spend
in Brain 1.0, Brain 2.0, and Brain 3.0?

| Toxic | Myopic | Innovative |

■ Brain 1.0   ■ Brain 2.0   ■ Brain 3.0

All companies impact and are impacted by
the brain activation of their employees.

Figure 4: How much time an organization spends in Brain 1.0, Brain 2.0, and Brain 3.0 has an impact on its performance.

For organizations to truly fulfill their vision and mission, to truly maximize their contribution to the economy and society overall, to truly unlock the potential of their employees, they need to shift into Brain 3.0. They need to support their leadership team and staff to exercise, develop, and strengthen Brain 3.0 to the point where they are not easily swept into Brain 1.0 and Brain 2.0 by change, by uncertainty, or by the behaviors of other people with whom they interact.

For people to really change, intellectually understanding the concepts is only a starting point. People need to have a direct visceral experience of what it feels like to shift from Brain 1.0 and Brain 2.0 into Brain 3.0. They need to feel for themselves how beneficial, yet easy, it is to take a few minutes to activate Brain 3.0 through a meditative exercise. When I started running pilots with professionals, the participants confirmed that once they activated and shifted into Brain 3.0 using the compassion meditation, they could immediately experience how much more clarity they had, how much more centered they felt, and how much more effective they could be. Once that happened, even the most skeptical people were willing to carve out time to meditate on a regular basis to activate and prime Brain 3.0.

To develop a mindful leadership program that could be delivered as a two-day training, I simply chose four modules from the full ten-module program we pilot tested with college students (which we now refer to as the Calm Clarity Essential Life Skills Program) that I thought would be most relevant to professionals and businesses. Then I tailored the exercises and discussions to make the topics more relevant to and resonant for professional audiences. When I was ready to pilot test the program, a couple of my closest friends in New York City, Irina and Sujata, each graciously hosted informal focus groups by inviting their friends who worked in various professions to gather at their homes, experience the workshop, and give me feedback. I am also grateful to the Wharton Club of New York and the Wharton and Penn alumni clubs in Philadelphia for organizing sessions for me to pilot the first module with their members.

Overall, the feedback from our professional pilots was extremely positive. People who had already been exposed to the diverse research I incorporated into the program shared that they were impressed by how the Brain 1.0, 2.0, and 3.0 framework interweaved findings from different fields together in a very intuitive and easy-to-understand manner. Many people also loved the Calm Clarity Compassion Meditation. In particular, engineers raved about the way I designed the arc of the first module to explain the underlying scientific mechanisms of the meditation, so when it was time to do the meditation, they already understood why and how it worked. They were amazed by how quickly the ten-minute Calm Clarity Compassion Meditation helped their bodies relax and release stress and tension. Many shared that they immediately began to practice it regularly on their own and found that they could sleep better at night.

After incorporating the feedback from the professional pilots, I started offering the Calm Clarity Mindful Leadership Program as a two-day weekend retreat in Philadelphia in the summer of 2015. After I worked out the logistics of running the weekend retreats, I started experimenting with a buy-one-give-one model. In early 2016, I started the Calm Clarity College Scholar Program, in which the tickets purchased by participants sponsor scholarships for low-income, first-generation college students to also attend the same retreat at no cost. In time, as companies began requesting that I run the mindful leadership training for their staff, I also turned these into buy-one-give-one opportunities for companies to sponsor a Calm Clarity collaboration with a college or non-profit organization to deliver training to low-income students.

What I learned from sharing Calm Clarity with corporations is that people at all levels of an organization are hungry for Brain 3.0 and tired of the burn-out that comes from the roller-coaster ride of operating in Brain 1.0 and Brain 2.0, no matter how successful they have been in doing so. Once people under-stand the emotional states of their brain, they readily admit that they can feel the wear and tear that swinging between Brain 2.0 and Brain 1.0 for long periods of time is having on their bodies, and that they intuitively know that living this way is not healthy or sustainable. After they learn how easy it is to activate Brain 3.0, how much better they can perform, connect, and create, and how much more alive they feel when they are in Brain 3.0, they readily embrace the paradigm shift.

The challenge that many encounter in making the shift stick over the long term is that it often feels like swimming against a current. The reality is that our society and institutions have built so many processes in Brain 2.0 that daily business routines automatically activate Brain 2.0 in employees, as well as customers, by dangling carrots for them to mindlessly chase; oftentimes, the carrots are out of realistic reach (or not equally accessible to everyone), and this inadvertently triggers Brain 1.0. Within these environments, a person needs the discipline to consistently notice when Brain 1.0 and Brain 2.0 have been triggered and then step back and create space for Brain 3.0 to activate, so he or she can look at the carrots with a sense of detachment and instead choose a more effective response in Brain 3.0.

Because human beings are social creatures, it is much easier to shift into

Brain 3.0 in a group than individually. Therefore, the long-term effects are greatest when teams work together and support each other to be in Brain 3.0. When teams become aware of their emotional states, check in with each other, and create new processes that help bring each other into Brain 3.0, they can start to break the behavioral cues and habits that tend to trigger Brain 2.0 and Brain 1.0 in themselves and in each other. Over time, the team starts to naturally trigger and activate Brain 3.0 in each other when they interact. As these Brain 3.0 interactions build and become reinforced, and Brain 3.0 emotional patterns get hardwired into the autopilot as habits and behavioral cues, everyone on the team benefits from a healthy, positive team culture that naturally lifts them up. (Please see part 2 for guidance on how to shift into Brain 3.0.)

The Present Day (2017)

About five years have passed since I ventured off in search of a way to live in greater alignment with my core values. After that mind- and heart-opening journey, when I returned to Philadelphia, a part of me wondered if my spiritual awakening might only be a passing phase. It was very possible that I could end up reverting to my previous conditioning when I reintegrated into society and recommenced working in business settings. Fortunately, understanding my emotional states using the Brain 1.0, 2.0, and 3.0 framework and actively cultivating Brain 3.0 have enabled me to redesign my life to be in greater alignment with my Higher Self.

Fostering Brain 3.0 has enabled me to create impact beyond what I thought was possible back in early 2012. Further, sharing what I learned with thousands of people, listening to their insights, and answering their questions have enabled me to deepen my own shift and further my own integration of body, heart, mind, and soul. In parallel, my thinking on how to create social impact has continued to evolve. Although the main focus of my work is secular (the workshops I run are primarily delivered in educational institutions and corporations), I find it intriguing that mystical experiences mysteriously continue to guide me forward.

At the very end of 2015, I knew I needed to finally go public with my struggles as a low-income, first-generation college student so I could lean into and overcome my fear of being stigmatized. In early 2016, I wrote about my experiences in my very first post on Medium.com, which was titled "Poor and

Traumatized at Harvard." To my surprise, the post immediately went viral. In a short period of time, it accumulated more than two hundred thousand views worldwide and was translated into other languages (without my permission). It also started a flood of messages through e-mail and social media. As I read the comments and messages from readers, it broke my heart to learn that the challenges I faced two decades ago are still experienced by low-income college students today. I felt that I had to do more to change the status quo and advocate on behalf of the students.

Conversations with students in Philadelphia who reached out to me prompted me to collaborate with them to start the Calm Clarity College Scholar Program for low-income, first-generation college students to attend Calm Clarity Weekend Retreats for free. Soon thereafter, I created a buy-one-give-one model for companies to help cover the costs of collaborating with universities and colleges to deliver training to low-income college students at a larger scale. Although students who participated in these events shared that the training had a profound impact on them, I saw that Calm Clarity training addressed only part of the challenges they faced. The students still lacked a strong social support system and access to mentors and role models to help them navigate the path forward. That was when the concept of building a social support network connecting disadvantaged college students to professional mentors began to emerge in my mind. Once again, the name would come to me in a meditative state: the Collective Success Network. Coincidentally, the domain name, collectivesuccess.org, was still available, so I immediately bought it. To attract collaborators, I simply talked to friends and posted a call to action on Facebook.

Even though I sensed that I was being guided to start this new nonprofit venture, doing so while writing this book, building Calm Clarity as a start-up social enterprise, and running training workshops seemed a bit overwhelming. I worried that I was taking on more than I could possibly handle. So I earnestly sought more guidance on the path forward and on how to complete this book in a way that would be of genuine benefit to readers. In early 2017, guidance came through two powerful mystical experiences.

The first happened at the end of January. I had a vision of Avalokiteshvara, the Bodhisattva of Compassion, who is also known as Chenrezig in Tibet and Guan Yin in China. As usual, the vision appeared to me in a hypnogogic state as I was waking up from sleep. It began with my seeing blue swirls that coalesced into a magnificent image of Avalokiteshvara with one thousand arms dynamically moving as if to signify to me that this great Being were somehow actively involved with my life's work and my life purpose. Then suddenly thousands of images of what looked like sacred iconography and artwork started flashing in front of my eyes, many of them featuring calligraphic script that I could not recognize but intuitively sensed could be ancient sacred teachings related to my soul's mission on this planet.

I opened my eyes in a state of bewilderment. Having a vision of Avalokiteshvara was a shock because I have never had any faith that Avalokiteshvara, or any bodhisattva, really exists. Unlike Siddhartha Gautama and Jesus Christ, bodhisattvas are not historical figures. Nonetheless, I couldn't help feeling a sense of reassurance from that specific image of Avalokiteshvara with one thousand arms. It felt like the one thousand arms were a symbolic message that I have backup and that more help is coming—that I wouldn't be building the Collective Success Network alone. Despite how baffling it was, the vision did give me the courage to continue to move forward and to do my best to trust that, in spite of my being overstretched for time and resources, the Collective Success Network would somehow manifest.

In hindsight, I could never have predicted how Collective Success was about to unfold—watching all the pieces come together has been like witnessing something miraculous arise out of nothing. In the next couple weeks, two highly respected leaders in the business and political circles in Philadelphia reached out to offer help and both agreed to serve on the founding board. By mid-February, the Facebook posts had drawn over fifty professionals (most of whom were complete strangers) to sign up, and about one-third of them volunteered to become actively involved in the steering committee to build the organization. Like me, many of the professionals had once struggled as first-generation college students and felt altruistically compelled to pay it forward to help students today benefit from the lessons they had to learn the hard way. By the end of March 2017, the Collective Success Network had become a nonprofit organization operating under the fiscal sponsorship of the Urban Affairs Coalition, and a group of impressive and inspiring leaders had come together to form the advisory board. By August 2017, more than 150 students had signed up to join, and more than 100 professionals had volunteered to support these students.

Although the vision clearly made an impression on me, I must confess that

I still find myself unable to make sense of Avalokiteshvara as a mythical figure. Instead, I have come to see Avalokiteshvara more as a symbol of the altruistic compassion that resides in every human being, the part of us that naturally wants to alleviate the suffering of fellow creatures. In getting to know the many wonderful volunteers who signed up to help, it is clear that the Collective Success Network serves as an outlet through which we can channel and express the compassion and goodwill within us. The motivation to serve had been lying dormant for years because we didn't have an appropriate platform to pay it forward.

The second mystical experience came soon after the first. In early February, I woke up with what seemed like fragments of a dream that felt like it might actually have been more than a dream. In it, I was in a space that was filled with light to receive advice from what seemed like a council of elders. None of us had bodies the way we have in the physical world. It was as if we were all part of the giant orb of light enveloping us, from which we dynamically emerged to communicate and into which we returned after speaking. With me in that orb of light were many of the sages whose teachings and writings I had come across in my journey, including MCKS, Paramahansa Yogananda, Lahiri Mahasaya, Christ and some of the apostles, as well as Buddhist sages and the Buddha. During the dialogues, someone would begin speaking, and then that speaker would smoothly morph into another person, as if to show me that I was really communing with one Collective Consciousness that contained everyone, including myself.

The part I remember vividly began with me being given the floor to speak, but I didn't know why I was there. I looked at the council and saw many revered spiritual teachers and saints known for performing great miracles. Then in a state of genuine confusion, I asked: "Why am I here? I don't know how to perform any miracles. I can't make the lame walk or the blind see. I can't walk on water. I don't have any powers. I'm not even religious. All I do is show people how to rewire their brains. What am I doing here?" Then the light answered me telepathically in energy, first as MCKS and then as Paramahansa Yogananda. Putting what they told me into words would go like this: "Miracles can be distracting, as we've seen again and again. Besides, miracles aren't important to your work. That's not what your work is about." So I asked with puzzled curiosity: "Then what is my work really about?"

In response, I heard the word "redemption" said out loud in clear American

English. I recognized the voice as the same voice I heard tell me to "come home" when I was in Kolkata. I couldn't see who said it. The voice appeared to come out of the orb of light directly. I was surprised by the answer, so I asked: "Can you explain what you mean by this word, 'redemption'?" And then the light answered telepathically: "Redemption means uniting with the Higher Self. It means returning to Source." I realized then, the energy speaking to me felt so very familiar—if the orb of light had an English name, the closest match would be Christ.

I woke up both inspired and disoriented. Was this only a dream? If so, why did it feel like it could have been real? Could this meeting have possibly taken place in a spiritual realm? I had no way to validate it one way or the other. The vision and the dream occurring so close together seemed to indicate they are interconnected aspects of a larger message. It felt like both were nudging me to own my mystical gifts. Since I started building the Calm Clarity Program, I had been more than happy to use science to show people how nurturing Brain 3.0 enables us to master our minds and become our best selves, but everything else that I could not ground in science, I had kept to myself, questioned, and doubted. In particular, I was concerned that talking about my mystical experiences would undermine my credibility as a rational, scientific thinker.

Ever since I first went to India in 2011, I have been baffled by the task of reconciling the Eastern mystical concept of Enlightenment with the scientific legacy of the Western Age of Enlightenment. Within me, there has been a constant argument between a skeptical scientist and a mystic seer. By default, given my many years of conditioning in a society that elevates science and reason, the scientist always got the upper hand. Since the mystical side of me was too mysterious to make sense of, I have metaphorically kept her hidden in the closet.

The vision and the dream provoked me to ask myself: Don't I also want to fully experience "redemption"? Is it possible that my self-imposed censorship has been limiting me? Instead of hiding the mystic inside me, wouldn't I feel more whole if I reclaimed all her gifts and talents? What would happen if I united and integrated the scientist and the mystic within me? Could bringing the two together lead to deeper understanding and insight?

Therefore, I have decided to take the risk of sharing the inner wisdom that has guided me through developing the Calm Clarity Program all these years: Source waits patiently for us to awaken and remember our true nature and where we come from. When we center ourselves and calm the whirlwind of the mind, we can tune in to Source, the metaphorical center of space and time, our true "home." Being there is like looking out into a multidimensional field of infinite potential from which all reality manifests. There, all time converges. All parallel universes converge. All sense of identity, differentiation, and "I"-ness collapses into oneness. Everything becomes nothing and nothing becomes everything.

Unfortunately, it's hard for a human mind to be at that center for longer than a moment. My human mind becomes overwhelmed and disoriented, unable to comprehend more than four dimensions, unable to comprehend how I could be both that ineffable creative center of force and this relatively feeble human being. Yet, in a manner that transcends space and time, all of us never leave that center, though a veil of ignorance keeps us from being aware of this truth.

When we live as souls out of the body, we do not experience Brain 1.0 and Brain 2.0. As souls, we are a form of consciousness composed of energies humans describe as Love, Light, and Wisdom. We are what humans would call "godly" and "divine." Relative to human perception, as souls we are unlimited. Yet when we incarnate, the embodied self is limited and constricted by the body—in particular, by Brain 1.0 and Brain 2.0—from fully expressing our innate spiritual qualities. By design, we have to develop and cultivate Brain 3.0 to tame and master these primal aspects of the human mind and body, so our soul, or Higher Self, can attain fuller expression in the physical realm. This is an important aspect of the grand eons-long creative spiritual experiment on this planet Earth. Now it involves 7 billion souls coming into human form—all evolving toward fuller expression of our true spiritual qualities while we are in a body, all interdependently interconnected, all collaborating toward a collective evolution.

Throughout human history, various mystics, sages, and shamans around the world essentially provided teachings, insights, and techniques to help people along this journey. Yet in many ways, spiritual teachings have been limited in their effectiveness because people couldn't help filtering the teachings through Brain 1.0 and Brain 2.0, coloring them with fear and desire and manipulating them to gain control and power. Even when the teachers were pure, there was no way to gauge whether mystical wisdom was being received accurately or being misinterpreted and misunderstood in people's minds. Gradually, as the teachings were passed down through successive generations and translated into new languages, the content became so distorted that the original essence got lost. Religious institutions became corrupted by people in Brain 2.0 seeking power and control rather than enabling genuine spiritual connection.

Ever since Copernicus, Galileo, and Newton shattered ancient misconceptions about the universe, scientific research has shed light on many of the mechanisms by which natural phenomena unfold, unapologetically liberating humanity from the shackles of ignorance, superstition, and religious dogma. We have mapped out our planet and solar system, as well as all the galaxies in our universe that our telescopes and satellites can detect. We have also made significant progress in understanding the physiological human organism by mapping out the human anatomy, the human genome, and, most recently, the human connectome. However, we must be careful about drawing the conclusion that there is nothing beyond our physical universe and bodies because scientific instruments and technologies are still limited in their ability to investigate spiritual matters. We can avoid swinging too far into the dogmatic views of scientific materialism by humbly acknowledging that we don't actually know for certain what is or isn't possible.

When I tune in, I see that humanity is on the verge of entering a new Age of Enlightenment arising from the convergence of the mystical wisdom of the Eastern Enlightenment tradition and the scientific empiricism of the Western Enlightenment tradition. As more and more advances emerge from scientific research, our Higher Selves are calling us to apply these insights to enhance our understanding of ourselves and free ourselves from inner obstacles. To gain inner freedom, we must also learn how to tune in to our inner wisdom to direct scientific knowledge and technology toward our greater spiritual evolution. I hope what follows can help people take a step in that direction.

Part 2

A Mind-Hacker's Guide to Shifting into Brain 3.0

We are already one. But we imagine that we are not. And what we have to recover is our original unity. What we have to be is what we are. —THOMAS MERTON, theologian

We evolved to see sacredness all around us and to join with others into teams and circle around sacred objects, people and ideas. . . . Most people long to overcome pettiness and become part of something larger. —JONATHAN HAIDT, psychologist[1]

BY TELLING MY STORY in part 1, I wanted to demonstrate that trauma doesn't have to be a life sentence. Pain can be transformed rather than transmitted. We always have a choice to empower ourselves to respond with wisdom rather than react mindlessly to the triggers that surround us. Yet we often forget that we have this power because cultural conditioning and trauma lead us to become fragmented and disconnected within. Anyone can regain this power by exercising and strengthening the "higher" neural pathways of Brain 3.0, which enable us to be more integrated. And as we become more integrated, we also enable our Higher Selves to be expressed and embodied through us. Then our Higher Selves can guide us to further transform and transmute our pain and suffering into wisdom and compassion and to remember and reclaim our true nature.

Here in part 2, I will build on the concepts shared in part 1 to help you make the shift into Brain 3.0 in a manner that is grounded in experience, science, and inner wisdom. This "mind-hacker's guide" is composed of two main sections. Chapter 7 provides an overview of what is involved in making the shift into Brain 3.0; an appreciation of the important functions of Brain 1.0, Brain 2.0, and Brain 3.0; and a more detailed explanation of the neural mechanisms underlying the shift. In chapter 8, the focus will be on helping you shift by providing a series of meditations and exercises that enable you to strengthen the neural networks of Brain 3.0.

In part 2, I will dive deeper into the scientific research that inspired the Calm Clarity Program to give you an understanding of how contemplative practices change the brain. In general, my goal is to summarize the science in a way that nonscientists find understandable, yet still provides enough details that readers who have a passion for brain science would find elucidating. If at any point the science becomes too heavy, feel free to skim these passages or go straight to the exercises.

Some people may find it helpful to do the exercises first and then read the explanations afterward to better understand what they experienced. Therefore, feel free to go through this mind-hacker's guide in the order that best suits the way you learn and assimilate insights.

7.

Overview of the Journey

> You cannot, in human experience, rush into the light. You have
> to go through the twilight into the broadening day before the
> noon comes and the full sun is upon the landscape.
>
> —Woodrow Wilson

> The "night sea journey" is the journey into the parts of ourselves
> that are split off, disavowed, unknown, unwanted, cast out, and
> exiled to the various subterranean worlds of consciousness. . . .
> The goal of this journey is to reunite us with ourselves. Such a
> homecoming can be surprisingly painful, even brutal. In order to
> undertake it, we must first agree to exile nothing.
>
> —Stephen Cope[1]

> The journey is the reward.   —Ancient Chinese proverb

Guidance for Going Inward

The universal challenge of being human is that we get easily hijacked by the
primal parts of the brain that unleash our Inner Godzilla and Inner Teen Wolf.
To unwind this cascade, we have to go inward to understand how Brain 1.0
and Brain 2.0 are conditioned by beliefs we hold about ourselves and how the
world works—beliefs that make us view each other as threats or competitors
when all of us are really kindred spirits. By centering ourselves and tuning in,
our Inner Sage can help us see through these illusory beliefs and gain more
and more clarity on how, in the larger scheme of things, we are all pilgrims on
the same quest to integrate with and embody our Higher Selves.

It may help to explain up front that within all of us is an instinctual "ego" that
enables us to navigate the world by drawing boundaries that separate "I" and
"mine" from "other" without realizing that these boundaries also separate us

from Source. As Walt Whitman once poetically declared, we are all "large and contain multitudes," but the ego's job is to pick and choose which aspects of our multidimensional self gets to be expressed. The ego is a part of the mind that weaves stories that create a sense of self and identity and builds mental models for how the world works. These stories are like scaffolding that enable the ego to create a sense of safety and security in a world full of uncertainty and risk, in a world where very little is actually within our direct control. However, at best, these stories can capture only a limited view of reality for a limited amount of time, and even when the stories no longer serve us, the ego continues to cling on to them the way a child attaches to a teddy bear or security blanket.

As the stories we repeatedly hear and tell ourselves about the world become a belief system, they get hardwired as self-triggering neural circuits in the brain and become part of our socially conditioned identity. Thus, when these stories are challenged, the ego feels threatened. As we act to protect our ego by preserving and defending these stories, we create self-sabotaging patterns: we push away evidence that contradicts these stories and even push away people we care deeply about when they raise doubts about these stories. If any aspect of our experience is dissonant with these stories, the ego pushes it out of conscious awareness into the "shadow," the part of the mind that is often called the subconscious. Thus, when we feed and defend our ego, we also strengthen our shadow. The more we live in denial, the bigger our shadow.

The shadow corresponds to the subcortical brain structures that make up our fast-thinking autopilot, including the key structures of Brain 1.0 and Brain 2.0.* Thus, it makes sense that our fears and impulses are largely irrational because they are "stored" underneath the cerebral cortex, which houses the human capacity for conscious reasoning. In fact, when we are in autopilot mode, scientists have noted there is very little activity in the frontal lobes (Brain 3.0). This is why we can do something on autopilot without having any conscious awareness or memory of having done it.

I think of the shadow as a dark, chaotic basement where we store the memories and experiences that our ego finds hard to deal with—experiences associated with shame, guilt, pain, anger, cruelty, weakness, and vulnerability—and all the aspects of ourselves that the ego doesn't want other people to know about

---

* "Subcortical" means underneath or inside the cerebral cortex, which is the outermost layer of the brain.

us, that the ego doesn't even want our conscious mind to know about ourselves. Some people see the shadow as negative or evil, but that is a very limited view. According to Jungian analyst and author Marion Woodman, "The shadow is anything we are sure we are not; it is part of us we do not know, sometimes do not want to know, most times do not want to know. We can hardly bear to look. Look. It may carry the best of the life we have not lived."[2]

The shadow also contains positive aspects of ourselves that we concluded weren't safe to express or feel in public. The things that get placed in the shadow for safekeeping and sometimes go forgotten include the joy and laughter of our inner child, the creativity of our inner artist, the wisdom of our Inner Sage, and the unconditional love and compassion of our Higher Self. If the shadow is where we have sealed away "the best of the life we have not lived" for protection, that means to live in a genuinely fulfilling manner, to truly unlock our full potential, we have to reclaim what we have put in there.

The challenge is that we have repressed these aspects of ourselves for so long that we have become used to living in disconnection and fragmentation and think this is normal. We have resisted and denied so many feelings and emotions because we don't like feeling vulnerable, exposed, weak, unsafe, uncomfortable, embarrassed, ashamed, or guilty. Because we somehow bought into an erroneous belief that denial and suppression make us stronger, we don't realize that they actually make us more fragile and deplete our energy. Whenever something triggers Brain 1.0 and sets off a freeze-flight-fight cascade in our body, many of us can't face these overwhelming feelings and sensations, so we seek to escape by numbing ourselves. We drink, smoke, eat, shop, watch TV, surf the Internet, etc. We keep our minds busy the whole day so we can distract ourselves from feeling uncomfortable emotions and sensations. We become like buildings with compromised structures that cannot withstand a minor earthquake.

As I shared in chapter 1, when people experience trauma, a common defense mechanism is to dissociate. Brain imaging studies have shown that dissociation correlates to the deactivation or impairment of key neural pathways across the brain, but especially in the frontal lobes. Dissociation may have initially been protective, but eventually it can make it very hard to function and experience life because encountering any situational or emotional cue that is associated with traumatic experiences can trigger a "reliving" of the traumatic episode, even when there is no risk or danger. Unfortunately, as I know from firsthand experience, dissociation generally becomes a negative spiral: the more we experience dissociation, the more impaired our brain functioning becomes, the more disconnected and anxious we feel, the more we try to repress these feelings, the more powerful our shadow becomes, the more frequently we dissociate.

To explain why we cannot heal by denying and repressing our sensations, feelings, and experiences, Bessel van der Kolk wrote: "As long as you keep secrets and suppress information, you are fundamentally at war with yourself. Hiding your core feelings takes an enormous amount of energy, it saps your motivation to pursue worthwhile goals, and it leaves you feeling bored and shut down. Meanwhile, stress hormones keep flooding your body, leading to headaches, muscle aches, problems with your bowels and sexual functions and irrational behaviors that may embarrass you and hurt the people around you." He also wrote, "The critical issue is allowing yourself to know what you know. That takes an enormous amount of courage."[3]

Allowing castaway aspects of ourselves to come to the surface of consciousness to be acknowledged, seen, and heard requires us to lean into fear. But the hard truth is, there is no other way to become integrated and whole. Without being whole, we will not be able to experience peace and harmony within. Without inner peace, we will not be able to manifest peace in our external world. Instead, we will continue to perpetuate the cycle of violence and trauma resulting in more suffering, heartbreak, and fragmentation. At some point, every one of us will get sick and tired of this cycle (like I did) and choose wholeness not because we are courageous but because we are fed up.

According to van der Kolk, "Neuroscience research shows that the only way we can change the way we feel is by becoming aware of our *inner* experience and learning to befriend what is going on inside ourselves."[4] Fortunately, there is now a wealth of resources and programs to help people do this. What the most effective programs have in common is that they involve enhancing self-compassion, metacognition, and interoception. With these skills, we can observe how the ego's defense mechanisms get triggered, learn to calm them down, and see underneath them to understand how these defense mechanisms are trying to protect us. Then we can find another means to create emotional safety and security. When we do this, we repair and strengthen the neural pathways connecting the structures of the conscious "higher" brain to the subcortical structures that correspond to the shadow. Thus, the more we exercise and strengthen Brain 3.0, the more we can shine the "light of consciousness" into our shadow.

It may also be helpful to explain here that higher consciousness envelopes and interpermeates lower consciousness. Therefore, it is not a coincidence that the neural pathways of Brain 3.0 envelope and interconnect into the neural pathways of Brain 2.0 and Brain 1.0. Spiritually, as we expand from individual to cosmic consciousness, it is like expanding our awareness as a container so that we grow to contain more and more multitudes. When we come into wholeness, we embody a spaciousness that holds the darkness as well as the light; we hold our suffering, vulnerability, and fragility as well as our compassion, joy, and transcendence in oneness. In Brain 1.0 and Brain 2.0, we are restricted to myopic "either/or" perspectives, whereas in Brain 3.0, we can thrive in the expanded consciousness of "And." Oneness is essentially a shift from "either/or" into "And."

What I have observed in my own experience is that when I resist nothing, the ego can no longer identify with any small part of the whole. In experiences of oneness, the ego simply becomes awestruck and bows to the sacredness of mystery beyond comprehension. As the ego humbly surrenders to oneness, the ego assumes a new role to serve as an instrument and channel for Spirit to flow through and become expressed in the physical world. Nevertheless, there is still an ongoing process of learning by trial and error. After the ego experiences expanded consciousness, it has to also learn from experience to not take this expansiveness for granted. Whenever my ego slips, forgets, and starts once again to resist, interfere, and attach to smaller parts of the whole, "I" can still get hijacked by Brain 1.0 and Brain 2.0. When this happens, my consciousness shrinks and becomes too small to contain the fullness of "And."

By observing this pattern, I became more willing to let go of attachments that pull me into Brain 1.0 and Brain 2.0 because whenever I recenter in equanimity in Brain 3.0, what I experience is often so uplifting, inspiring, and elevating that it cracks me open to experience more possibilities beyond my imagination. With enough experiences of "And," it becomes easier and more effortless to surrender my "either/or" limiting views of myself, other people, and the universe, and instead trust in the guidance and call of Spirit to embody more and more of my Higher Self and embrace oneness with the Collective Consciousness.

As you strengthen Brain 3.0, I hope you will see for yourself how Brain 3.0

functions as a neurobiological and spiritual bridge to our Higher Selves and the greater Collective Consciousness. When you intentionally and deliberately exercise and develop the pathways of Brain 3.0 that support you in aligning with and embodying your aspirations, your Higher Self will help you overcome your prior conditioning. By spending more of your life in Brain 3.0, you will naturally identify and shed self-limiting beliefs and outdated worldviews and bring more and more of your shadow into light. Then your consciousness will naturally expand so that you are able to embody more and more of your Higher Self.

Grounding Expectations

> *It is important that you do not consider awareness to be your "ally," called on to suppress the "enemies" that are your unruly thoughts. Do not turn your mind into a battlefield. . . . Opposition between good and bad is often compared to light and dark, but if we look at it in a different way, we will see that when light shines, darkness does not disappear. It doesn't leave; it merges with the light. It becomes the light.*
> —THICH NHAT HANH[5]

Before we dive in, I want to make clear that it is not realistic to expect yourself or anyone else to constantly be in Brain 3.0 all the time. The truth is that I still get triggered into Brain 1.0 and Brain 2.0 every single day. Each day has a way of throwing unanticipated challenges, emergencies, and mishaps that awaken my Inner Godzilla. Plus, I still live in a sea of temptations and am surrounded by advertising designed to trigger cravings to indulge in yummy foods, splurge on retail therapy, waste hours on social media, and get sucked into highly addictive television shows. When carrots are dangled in front of my eyes, the Inner Teen Wolf inside me still feels the impulse to chase them.

Brain 1.0 and Brain 2.0 are like two sides of one coin (their key structures are strongly interconnected in the brain). Our instinctive reaction to suffering and pain is to escape it as quickly as possible. So whenever we find ourselves in Brain 1.0, we tend to launch straight into Brain 2.0 to get immediate relief. Even anger, hatred, and revenge are really mechanisms that quickly launch people out of the discomfort of feeling fear, shame, and vulnerability in Brain 1.0, into a Brain 2.0–driven attack mode in which the reward is to deflect hurt and pain and defend one's ego. Although I appreciate the importance of antidepressants and anti-anxiety medication as someone who has benefited from them in the past, I see the fact that the global antidepressant and anti-anxiety market generates a staggering $20 billion in revenues a year as a testament to the degree to which human beings can't stand being in Brain 1.0. Whenever we wish for a magic pill or a quick fix to solve our problems, this is Brain 2.0 at work. The problem is that acting on Brain 2.0 impulses distracts us from being able to honestly observe, identify, and address the underlying patterns and mechanisms that keep us caught in a vicious cycle.

Shifting into Brain 3.0 doesn't mean repressing, suppressing, or struggling with our primal instincts. It means being aware when Brain 1.0 or Brain 2.0 gets triggered, and creating space inside so that we keep Brain 3.0 "online" rather than launch straight into mindless reactivity. We can do this by being the "observer." As I will explain in chapter 8, when we observe without inner judgment or criticism, we activate neural pathways for interoception and metacognition to tune in to the physiological stress cascade that gets set off inside the body, feel the emotions and impulses that get stirred up, observe the stories woven by the Inner Godzilla or Inner Teen Wolf unfolding inside the mind-track, and let these internal experiences and sensations arise and pass away without acting on them or resisting them. Shifting into Brain 3.0 involves having the ability to observe, acknowledge, and accept every aspect of the human experience with compassion and equanimity—without being at war within. Instead of pushing aspects of ourselves into the shadow, we use the light of conscious awareness to gently reclaim and reintegrate the shadow, transforming it into light. This is how shifting into Brain 3.0 enables us to live more fully and wholly.

Brain 3.0 Is Like a Battery That Needs Regular Recharging

Please keep in mind that Brain 3.0 is not an infinite resource. Just as a smartphone needs to be regularly recharged as it runs low on battery power, Brain 3.0 needs to be constantly rejuvenated and regenerated. One key difference between the brain and a smartphone is that when a smartphone is down to 20 percent battery life, you can still use all the apps on it. However, you have to be very close to 100 percent charged to enjoy the full functionality of Brain 3.0. When energy is scarce, the body will naturally prioritize giving energy to Brain 1.0 and then to Brain 2.0. It's only when you are full on fuel and well rested that Brain 3.0 can function at full capacity. That's why whenever you are hungry or tired, it's so much easier for your Inner Godzilla and Inner Teen Wolf to hijack your mind.[6]

Building Brain 3.0 Is Like Building Upper Body Strength

As I shared earlier in the introduction, one of the biggest breakthroughs in neuroscience was the discovery that the brain is "plastic." This means the brain has the remarkable ability to change its structure by building new neural pathways or pruning existing neural pathways according to how frequently the pathways are used. This also means we can direct neuroplasticity to create a "stronger" Brain 3.0 by intentionally and deliberately exercising the neural networks that enable us to thrive.

An analogy I find useful to help people understand "brain training" is to think of Brain 1.0 and Brain 2.0 like lower body strength and Brain 3.0 like upper body strength. Our legs are strong enough to carry our body weight all day long because they have done so since we first learned to walk. In contrast, we do not usually use our arms to carry our body weight. Without regularly building upper body strength, most people who are not athletes are unable to do more than twenty push-ups or more than one or two pull-ups. Similarly, for most of us, the neural pathways of Brain 1.0 and Brain 2.0 are very, very strong because they have been getting regularly triggered, activated, and used throughout our everyday lives, whereas the neural pathways of Brain 3.0 are relatively weak unless we intentionally exercise and activate them.

Another thing I want to point out using this analogy is that as we strengthen our upper body, our lower body doesn't get weaker, but rather our entire body gets stronger and more agile, so we can lift more weight and perform better at sports. Similarly, as Brain 3.0 gets stronger, it doesn't necessarily mean that Brain 1.0 and Brain 2.0 get weaker; it means that our entire brain gets rewired so that it is stronger, more integrated, and more agile, such that the entire brain performs better.

Changing Our Conditioning and Habits Is a Lifelong Process

Please keep in mind that unraveling a lifetime of past conditioning is a reiterative, progressive process that takes place over many, many years. There are no shortcuts or magic pills. There is no delete button for behavioral, emotional, or mental patterns that have been hardwired as habits into our autopilot. Even after lying dormant for many years, the neural pathways for strongly conditioned habits can easily reactivate whenever we find ourselves in circumstances and environments that are filled with cues associated with old habits. Since Brain 2.0 includes the neural structures that govern the habits and routines we perform without having to use our frontal lobes to consciously invest effort and concentration, Brain 2.0 automatically activates and takes over anytime we run through a habitual routine on autopilot and when our body is too tired or hungry to power Brain 3.0. Thus, the task of rewiring the brain involves a lifelong journey of proactively retraining habits (and their triggers or cues) and practicing self-care.

The good news is that we now have enough research on how habits are built and self-triggered to use it to our advantage. By intentionally practicing Brain 3.0 behaviors on a daily basis at certain times and/or in certain ways, we can create cues to build these practices into daily rituals we are able to do on autopilot. Using the same process that enabled us to train ourselves to brush our teeth in the morning and before going to bed, we can create situational cues to encode Brain 3.0 practices into a habitual Brain 2.0 routine. This means we can train Brain 2.0 to serve as an ally of Brain 3.0. Once that happens, we can leverage these cues to quickly catapult us into Brain 3.0. The use of cues and routines makes it very easy to prime and activate Brain 3.0 instantly and regularly. For instance, I have turned my morning cup of tea, afternoon walks, and riding the subway into cues to practice mindful drinking, mindful walking, and mindful breathing, respectively. (In the appendix, I include a template to help you plan a daily routine to activate Brain 3.0.)

Appreciating Emotions and Feelings

From an evolutionary point of view, emotions must be useful, or we would not have them. We must begin by seeing emotions as contributing to our ability to act intelligently, not as impediments to such action.

—GUY CLAXTON, psychologist[7]

Feelings are the expression of human flourishing or human distress, as they occur in mind and body. Feelings are not a mere decoration added on to the emotions, something one might keep or discard. Feelings can be and often are revelations of the state of life within the entire organism. . . . Life being a high-wire act, feelings are expressions of the struggle for balance.

—ANTONIO DAMASIO, neuroscientist[8]

By glorifying rationality and self-control, modern societies (especially in academic and business settings) seem unable to appreciate the value of "emotions" and "feelings," and seem to consider it a character defect that we have them at all. It's quite common to hear people say things like "that person is emotional" or "that person is soft" and know implicitly that they mean it as a negative. Even the concept of "emotional intelligence" is often brought up in the context of learning how to control runaway emotions so they don't control us. Unfortunately, many people "control" emotions and feelings by suppressing and denying them, and in doing so, become so unfamiliar with experiencing their emotions and feelings that they become afraid of them.

In contrast, the statements by Claxton and Damasio above point out the fact that emotions and feelings serve very important biological functions. By studying the neural correlates of affect (which includes emotions and feelings), neuroscientists have discovered that our capacity to effectively make "rational" decisions regarding our personal lives, relationships, financial investments, and future plans is actually built upon the neural networks that process emotions and feelings, and that without the capacity to feel and express emotions, we also lose our capacity to navigate our social world. I see the fact that the prefrontal cortex, once heralded as the seat of rationality and the critical brain structure that differentiates humans from other animals, is involved in the processing of emotions as a testament to the importance of emotions to being human.

Lisa Feldman Barrett, a leading researcher in the field of affective neuroscience, proposes in her book *How Emotions Are Made* that "emotions are not reactions to the world; they are your constructions of the world." This is based on recent discoveries demonstrating that a key function of the brain is to simulate, guess, and predict what is happening in the world ahead of time because this is much more efficient than processing every piece of information that floods in through your senses. She explains, "Your brain uses your past experiences to construct a hypothesis—the simulation—and compares it to the cacophony arriving from your senses. In this manner, simulation lets your brain impose meaning on the noise, selecting what's relevant and ignoring the rest."[9] In creating these simulations, the brain is continuously combining bits and pieces of the past and estimating how likely each bit applies to the current situation. She adds, "Prediction is such a fundamental property of the human brain that some scientists consider it the brain's primary mode of operation."

According to Barrett, our affective state (positive/pleasant, negative/unpleasant, or neutral) and the accompanying interoceptive sensations and feelings also emerge from our simulation and construction of reality as an immediate way to quickly make meaning of what is unfolding in our internal and external, or social, environments. Barret sees emotions as conceptual tools we learn from our society to encode combinations of affect, interoceptive sensations and feelings, and behaviors into memory. Every culture has its own unique set of prepackaged emotion concepts handed down from generation to generation that enables a group of people speaking the same language to quickly and efficiently read, interpret, and communicate complex and nuanced subjective internal states in a way that also prescribes how people should respond.[10]

In a recent *New York Times* article provocatively titled "We Aren't Built to Live in the Moment," Martin Seligman, the father of the field of positive psychology, explains that "the mind is mainly drawn to the future, not driven by the past." He adds, "We learn not by storing static records but by continually retouching memories and imagining future possibilities." Thus, the purpose of memory may be to enable us to "metabolize" past experiences by "extracting and recombining relevant information to fit novel situations." Seligman also explains, "Our emotions are less reactions to the present than guides to future behavior." He adds that depression is now understood as being caused by having "skewed visions of what lies ahead" and that depressed people tend "to imagine fewer positive scenarios while overestimating future risks."[11] In reading this article, I was struck by how well this current understanding aligns with my earlier experience of depression—I easily anticipated negative outcomes and could not naturally see positive possibilities.

A good portion of the brain's work in simulation, prediction, and emotion construction is carried out by the prefrontal cortex. As I shared in chapter 2, activity in the left and right prefrontal cortices have opposing impacts on affect. This is tied to how they simulate and predict the upside or the downside of any given situation. When there is stronger activity in the left prefrontal cortex, we tend to experience positive affect, anticipate the upside, feel motivated to pursue growth opportunities, and feel open to connecting with other people and building trust. This is what scientists call an "approach" orientation because in this state, we tend to move toward the object of attention. Combinations of positive affect and different bodily sensations can be interpreted and constructed as optimism, happiness, confidence, curiosity, attraction, drive, motivation, etc. On the other hand, when there is stronger activity in the right prefrontal cortex, we tend to experience negative affect and anticipate the downside, carefully look at risks and dangers, focus on problems, and defend ourselves from threats. This is what scientists call "avoidance" orientation because in this state, we tend to move away from the object of attention. Combinations of negative affect and different bodily sensations can be interpreted and constructed as pessimism, depressed mood, insecurity, caution, defensiveness, repulsion, reluctance, fear, doubt, etc.[12]

In line with what Seligman shares about seeing depression as having "a bleak view of the future," neuroscientists have found that depressed patients tend to have much higher activity in the right prefrontal cortex than in the left prefrontal cortex.[13] What this suggests is that in the brains of depressed patients, the right prefrontal cortex is very efficient and effective at simulating negative scenarios, but the left prefrontal cortex is comparatively inefficient and ineffective at simulating positive scenarios. As I shared in chapter 2, a key factor in my recovery involved experimenting with different ways to increase activity in my left prefrontal cortex.

Laura Carstensen, a researcher at Stanford University, found that as we age, we develop the capacity to experience the co-occurrence of positive and negative emotions, a capacity she calls emotional complexity. These mixed emotions often create a sense of poignancy, such as feeling both happy and sad to graduate from college. The findings suggest that the capacity for emotional complexity may render emotions less overwhelming and that increased emotional complexity may explain why older people are able to sidestep the emotional roller coaster that young people often feel.[14] While Carstensen's research studies did not involve brain imaging, one can infer that the co-occurrence of positive and negative emotions would be correlated with activity in both the left and the right prefrontal cortices, simulating both the positive and negative aspects of a situation at the same time. This is a contrast to the all-or-nothing simulations that tend to take place during adolescence.

David Rock has developed a useful framework that captures the types of situations that activate the avoidance system or the approach system, using the acronym SCARF, which stands for status, certainty, autonomy, related-ness (meaning a sense of belonging to a group), and fairness. According to Rock, a large body of research shows that whenever human beings experience a threat to or decrease in any of these five domains, Brain 1.0 gets activated and sets off the freeze-flight-fight cascade to defend ourselves against psycho-logical danger (such as a power struggle). On the flip side, human beings are also naturally motivated to approach opportunities to improve where we stand with regard to these five domains, what I call "SCARF rewards" for short.[15] SCARF rewards can activate Brain 2.0 and/or Brain 3.0, depending on our intentions and motivations for pursuing them. If the driving motivation is to benefit ourselves even at the expense of others, then Brain 2.0 is probably running the show and we are likely locked into a type of tunnel vision in which we see the situation as a "win-lose" competition. However, if the driving mo-tivation is to serve a greater purpose and to do no harm, then Brain 3.0 would also be activated, enabling us to see the situation from a bigger-picture per-spective and come up with creative win-win solutions. In this case, Brain 3.0 guides Brain 2.0 to achieve the best outcomes.

It's important to emphasize that both the "approach" system and the "avoidance" system are important aspects of Brain 3.0. Positive and negative affect are crucial to our ability to navigate the world and thrive. Positive affect guides us to move toward what's good for us, build social connections, explore novel experiences, and take risks to achieve our goals. Negative affect guides us to slow down, assess and take stock of what we have to lose, anticipate risks, avoid dangerous threats, and learn from mistakes so we don't repeat them. To demonstrate that we shouldn't be misled by the fact that one is labeled "posi-tive" and the other "negative," it's useful to think about how a battery or mag-net needs to have both a positive and a negative pole, and how without the negative end, batteries and magnets would be completely useless.

To further explain the importance of positive and negative affect, I want to go back to the car metaphor I introduced in chapter 2, in which I likened positive affect to a full gas tank and a functioning gas pedal, and negative affect to the braking system; I explained that in order to drive a car safely, we need to have both the gas system and the braking system in fully functioning condition. See figure 5.

To build further upon that metaphor, I explained that getting hijacked by

Brain 3.0 enables us to effectively navigate life

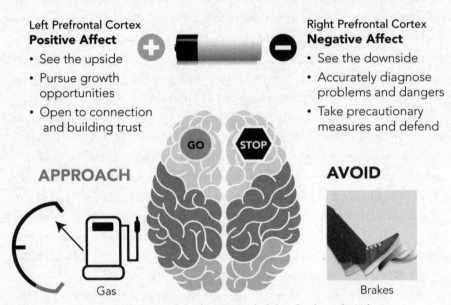

Left Prefrontal Cortex
Positive Affect
- See the upside
- Pursue growth opportunities
- Open to connection and building trust

Right Prefrontal Cortex
Negative Affect
- See the downside
- Accurately diagnose problems and dangers
- Take precautionary measures and defend

GO   STOP

APPROACH   AVOID

Gas   Brakes

Figure 5: The key structures of Brain 3.0, the left and right prefrontal cortices, have different functions in simulating the experience of affective states.

Brain 1.0 is like having the gas tank suddenly emptied, so there is no more energy to move forward toward our goals. Similarly, we can understand getting hijacked by Brain 2.0 as having the brakes temporarily disabled so that we are unable to control impulses or rein in reckless behavior. When the braking system becomes physiologically impaired (as in the case of addiction or brain damage), people lose the ability to stop themselves from acting on impulses, even if they know that the course of action will result in terrible consequences. See figure 6 for an illustration.

Thus, rather than see positive affect as good, and negative affect as bad, or

Challenge: Brain 3.0 gets hijacked by Brain 1.0 & 2.0

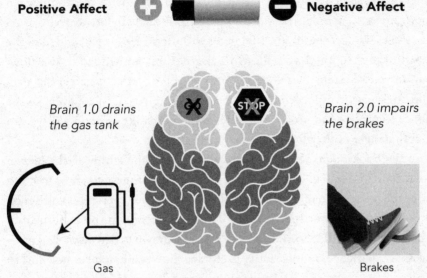

Left Prefrontal Cortex
Positive Affect

Right Prefrontal Cortex
Negative Affect

*Brain 1.0 drains
the gas tank*

*Brain 2.0 impairs
the brakes*

Gas

Brakes

Figure 6: In Brain 1.0, activity in key areas of the left prefrontal cortex is reduced; in Brain 2.0, activity in key areas of the right prefrontal cortex is reduced.

something to be avoided at all costs, we should appreciate our affective state and the emotions we use to characterize our affective state as a biological feedback mechanism that (1) conveys to us whether the simulations constructed by our brain are telling us to approach or avoid the object of our attention and (2) "reports" to us on the overall functioning of Brain 3.0 (whether our gas tank is empty or our braking system is impaired, or both), alerting us when Brain 3.0 needs to be regenerated by food, water, sleep, or other self-care activities such as meditation.

Appreciating Brain 1.0's Key Role in Self-Preservation

I want to take a moment to emphasize that without Brain 1.0, the human race would probably have become extinct long ago. Therefore, rather than see Brain 1.0 as a negative, we need to appreciate what Brain 1.0 does for us and how it is part of the universal human experience. Brain 1.0 plays a critical role in keeping us alive by ensuring that we learn from negative experiences and protect ourselves from danger. Brain 1.0 functions like a radar detector for threats, dangers, and risks. It makes us keep a safe distance from fire, be careful when using knives to cut food, and wait for a green traffic light before we walk across the street. The fact that Brain 1.0 (along with Brain 2.0) is embedded in the fast-thinking autopilot system, while Brain 3.0 is embedded in the slow-thinking pilot system, means that Brain 1.0 gets triggered by perceptual data from our senses in milliseconds, well before the prefrontal cortex even has the chance to receive and process the information. Thanks to Brain 1.0, we can react to danger even before we have time to think.

In the book *Buddha's Brain*, neuropsychologist and author Rick Hanson provides an in-depth understanding of how Brain 1.0 functions and why it's so strong. He writes, "Your brain is like Velcro for negative experiences and Teflon for positive ones—even though most of your experiences are probably neutral or positive." According to Hanson, the brain is "drawn to bad news" and gives negative experiences more priority in storage. As a result, it takes no effort to recall negative memories, but it takes active mental effort to recall positive ones. He also explains that "negative trumps positive," meaning that negative events have more emotional impact on us. To illustrate, he cites the following findings from research: "it's easy to acquire feelings of learned helplessness from a few failures, but hard to undo those feelings, even with many successes. People will do more to avoid a loss than to acquire a comparable gain [meaning sticks are stronger than carrots].... Bad information about a person carries more weight than good information, and in relationships, it takes about five positive interactions to overcome the effects of a single negative one."[16]

I find it helpful to think of Brain 1.0 as a big data server dedicated to tracking all of our fears and all of our negative experiences, especially all the dangerous and threatening situations we have ever experienced directly, witnessed indirectly, or learned secondhand from the media and other people, in order to help us learn from them and avoid getting into similarly dangerous situations in the future. Having a strong Brain 1.0 is not the problem. The problem is when Brain 3.0 is underdeveloped. When Brain 3.0 is strong, it can easily calm Brain 1.0 and help ensure that the Inner Godzilla doesn't take over unless we really need it to because we are in serious danger.

When Brain 3.0 is underdeveloped or impaired, we run the risk of living in

Brain 1.0. To help you visualize this scenario, imagine a seesaw where the Inner Godzilla is on one end and the Inner Sage is on the other, but the Inner Godzilla is more than twice the size of the Inner Sage. The Inner Sage is powerless to move the seesaw. Unfortunately, the more negative and threatening situations we have encountered in our youth, the stronger Brain 1.0 is and the less likely Brain 3.0 gets the opportunity to fully develop. This sets the stage for Brain 1.0 to dominate our lives. The more time we spend in a heightened state of alert, the less time we get to develop and strengthen Brain 3.0. With chronic hyper-vigilance, Brain 1.0 becomes extremely sensitive, letting even the slightest hint of possible danger trigger internal alarm bells. This means it will perceive threat even in generally safe circumstances. When a person is predominately in Brain 1.0, he or she is likely to fall into a vicious cycle of finding and creating more reasons to stay hyperalert in Brain 1.0. It is nearly impossible to relax and feel safe when one is caught in this cycle. To break out of it, people need to learn how to activate Brain 3.0 to "soothe and calm" Brain 1.0.

In line with Barrett's theory that emotions are constructed, one very simple way to calm intense emotions is to label them, which somehow also changes the simulation. "If you can name it, you can tame it" is one of the most widespread suggestions provided by leading psychologists such as Marc Brackett, director of the Yale Center for Emotional Intelligence.[17] The evidence for this comes from brain imaging studies conducted by Matthew Lieberman at UCLA, which found that when people named negative emotions, activity increased in the right ventrolateral prefrontal cortex (the key structure in Brain 3.0's braking system) and activity decreased in the amygdala (the key structure of Brain 1.0).[18] In an article about his research, Lieberman explains, "When you put feelings into words, you're activating this prefrontal region and seeing a reduced response in the amygdala. In the same way you hit the brake when you're driving when you see a yellow light, when you put feelings into words, you seem to be hitting the brakes on your emotional responses."[19]

In general, labeling your emotions is a way to turn "on" the observer function in Brain 3.0, which then creates space to shift out of Brain 1.0. To label an emotion, we have to use interoception to tune in to our bodily sensations so we can ascertain our emotional state, and metacognition to reflect on and name our emotional state. In my own personal experience, a clenched jaw, hands tightened in fists, tight shoulders, and a furrowed brow indicate that Brain 1.0 has been triggered; restlessness, impatience, a tense back, and a racing mind fixated on getting something let me know that I'm in Brain 2.0; a relaxed, open body posture, a spring in my step, and spontaneous smiling confirm that I am in Brain 3.0. Taking time to simply tune in to my physical sensations and notice when I'm not in Brain 3.0 is often all it takes to reactivate Brain 3.0 and prevent myself from becoming fully hijacked by my Inner Godzilla or Inner Teen Wolf.

By learning to keep the observer function "on" throughout the day (not just when I'm meditating), I've come to appreciate Brain 1.0 as a useful mirror that helps me notice, learn about, and understand my shadow. By looking into this mirror I have learned that, more often than not, I experience stress and negative emotions not because I am under actual threat of harm but because a story or simulation that my ego is entangled with is getting challenged. Thus whenever I get pulled into Brain 1.0, I try to calmly tune in and identify the hardwired beliefs and stories that my ego is trying to defend. By staying centered, my Inner Sage can guide me to look into my shadow and reclaim the part of me that my ego has been pushing into the shadow for self-protection. I will explain more about how to do this in chapter 8.

Appreciating Brain 2.0's Key Role in Motivation

As part of the fast-thinking autopilot system, Brain 2.0 is able to instantaneously process signals coming in from our environment in ways that bypass the slower conscious mind and plays a key role in conditioned learning through positive and negative reinforcement (such as carrots and sticks). It functions like a sponge for absorbing cultural conditioning and social mores. From an evolutionary perspective, Brain 2.0 also plays a critical role in ensuring the continuity of the human species, because without Brain 2.0, we would never feel any motivation. Brain 2.0 propels us to mate, to build a family, to take care of our physical, emotional, and social needs and those of our loved ones, and to acquire status in our community to expand our access to resources and cooperation. When Brain 2.0 is damaged, people lose their drive. They don't pursue even basic rewards like getting food when they are hungry. They don't bother putting in effort to achieve goals.[20] Therefore, Brain 2.0 should not be seen as a negative.

At the same time, one of the main reasons why Brain 3.0 is underdeveloped in many people is because it's very, very easy to spend most of our waking life in Brain 2.0 chasing immediate gratification and extrinsic rewards to feel happy. What changes when Brain 3.0 is sufficiently strong is that our Inner

Sage will instruct Brain 2.0 to selectively pursue goals and rewards that are aligned with our long-term goals, our core values, and our well-being. When Brain 2.0 is guided by Brain 3.0, we harness our motivation to move toward eudaimonia, the ancient Greek philosophers' concept of a much deeper sense of intrinsic happiness and fulfillment that is not dependent on external circumstances or sensory pleasures. Thus, having a strong Brain 2.0 is not a problem so long as we also have a very strong Brain 3.0 to guide it.

I often think of Brain 2.0 like a wild monkey that is easily distracted by anything exciting happening around it and that impulsively chases after any extrinsic reward (financial, physical, social, or emotional) that other people dangle in front of it to tempt it to do what they want, without ever questioning whether the reward is really worth chasing; the ability to step back and reflect is embedded in the prefrontal cortices, which get deactivated when we are fully swept up in Brain 2.0 mode. It is only when we are in Brain 3.0 that we have the neural capacity to consciously discern whether what we have to do to get these rewards is really the best use of our energy, and whether going after the reward moves us toward or diverts us away from our most important goals and aspirations. So long as Brain 3.0 is underdeveloped, impaired, or deactivated, there is no internal brake to stop the wild monkey from mindlessly chasing all the carrots dangled in front of us until we burn out.

Intrinsic motivation comes from having a strong Brain 3.0 guide Brain 2.0. This also underlies what researchers call having an "internal locus of control"—individuals with an internal locus of control believe they have the power to influence their environment and improve their personal circumstances, so they take ownership of their choices and contribution to any situation they find themselves in. In contrast, individuals with an "external locus of control" tend to see their fate as being shaped by uncontrollable external forces and tend to blame negative outcomes on their circumstances without acknowledging their own role in creating the situation or recognizing their agency to influence and change their situation.

Until Brain 3.0 is sufficiently developed to enable the capacity for an internal locus of control, people likely also won't have sufficient impulse control, a key network in the braking system that "resides" in Brain 3.0 and is impaired by Brain 2.0. When the braking system isn't fully developed and functioning, by default, Brain 2.0 has free rein to act out every impulse, which is why people feel like their life is out of their control. The fact that Brain 3.0 doesn't mature until our midtwenties may explain why teenagers, in general, tend to have an external locus of control, are very sensitive to peer pressure, and experience high levels of anxiety. These tendencies are all associated with living primarily in Brain 2.0.

Recall in chapter 3, I explained that Brain 2.0 corresponds to the dopamine pathways in the brain and that electrically stimulating these pathways can induce an overwhelming craving that makes people feel agitated, tense, restless, and anxious until they satisfy the craving. What this reveals is that having carrots stimulate Brain 2.0 is like creating an itch that drives you nuts unless you scratch it. When Brain 2.0 is triggered, you believe that getting the carrot will make you happy, but what having the carrot really provides is the alleviation of your agitation about not having the carrot. Once you get the carrot, you get only a temporary period of relief before the cycle starts over again.

In Brain 2.0, we tend to mindlessly chase and accumulate symbols of social validation and status to achieve a sense of security, stability, and control. The problem is that the ego tends to confuse our achievements and possessions with our identity and self-worth. When our egos are built up based on what we have or where we stand in a social hierarchy, we lose touch with who we really are deep inside. Sometimes, the progressive deadening we feel inside can itself lead to a wake-up call that forces us to acknowledge that the chase for extrinsic rewards is a never-ending cycle that does not lead to the truly fulfilling type of happiness associated with eudaimonia. When I reached this point in my early thirties, I began to wonder: is this really all there is to life? That was when I started plotting my escape from the carrot-chasing maze I had lost myself in. Only after the spell of the chase was broken did I finally start to wonder about what life could be like outside the maze.

For me, the path out of the maze involved acknowledging and accepting that the urges and impulses biologically preprogrammed in Brain 2.0 are part of the universal experience of being human. As I learned to create space in Brain 3.0 to observe these urges and impulses as they arise within me with a sense of curiosity and equanimity—without judging them as good or bad and resisting or suppressing them—I learned to not identify with them. When I created space to gently hold my urges and impulses, I found that as they arise, the energy within them can get released by simply experiencing them as sensations and feelings within my body. I learned that after these sensations and feelings are brought into conscious awareness, often the urges and impulses dissipate on their own without needing to be acted out. Then I realized that these urges and impulses are simply messages from within coming into conscious awareness. By developing space to hear these messages, I've also come to appreciate Brain 2.0 as a mirror revealing my unconscious beliefs around things I need to do or gain or earn to feel happy, worthy, or secure. Now, whenever Brain 2.0 is triggered, I do my best to create space by activating Brain 3.0 to tune in to my body as a vehicle of inner wisdom rather than allow my ego to exploit my body as a carrot-chasing machine.

Interestingly, the main structures of Brain 2.0, the basal ganglia, seem to also function as a storage vehicle for all our habits. When scientists look at an expert performing a skill, they see very little activity in the frontal lobes and a hot spot of activity in the basal ganglia, which led to the conclusion that these structures must play some role in coordinating nonconscious habitual motor movements. (It also led to the conclusion that the ability to beautifully perform a skill without exerting much conscious effort is what distinguishes masters from amateurs in any field.) It turns out any skill or activity that we repeatedly perform over and over eventually becomes hardwired as a habit into the basal ganglia, such that we no longer need to exert any conscious effort to carry it out.[21] Therefore, we can intentionally rewire Brain 2.0 by practicing and drilling life-enriching Brain 3.0 activities and behaviors until they become embedded into our autopilot, so it no longer takes much conscious effort to activate and rev up Brain 3.0.

Understanding the role of the basal ganglia in the reward system helped me realize that Brain 2.0 could be transformed into an ally because aligning Brain 2.0 with Brain 3.0 by mindfully crafting cues, routines, and thoughtfully chosen rewards can accelerate the shift into Brain 3.0. Various forms of meditation and contemplative practices like prayer from around the world are essentially exercises handed down from ancient sages that enable us to encode Brain 3.0–enhancing activities and behaviors into the basal ganglia, thus rewiring Brain 2.0. This also explains why meditation needs to be practiced on a regular basis to sustain an enduring transformation. To my relief, I did find that as I experimented with this approach to align Brain 2.0 with Brain 3.0, Brain 2.0 naturally stopped behaving like a "wild monkey" chasing all the rewards dangled in my environment. Instead, it "learned" to selectively focus on the rewards that aligned with my higher aspirations and values.

There is now preliminary evidence in neuroscience that may shed light on how these changes in my brain unfolded. A 2015 study conducted at Harvard Medical School by Tim Gard and Sara Lazar, a pioneer in using brain imaging research to shed light on meditation, investigated differences in whole brain resting state connectivity among experienced yoga practitioners, experienced meditators, and matched controls. One of the key differences they found was that long-term yoga practitioners and meditators had much greater widespread connectivity between an area of the basal ganglia associated with flexible goal-directed learning and most of the cerebral cortex, including frontal brain regions. Together with findings from a 2012 study showing that flexible goal-directed behavior was predicted by white matter connectivity between this area and the ventromedial prefrontal cortex (a key structure of Brain 3.0 that will be discussed in the next part), these insights suggest that the increase in mindful responding and decrease in mindless reactivity that I, and many other yoga practitioners and meditators, experience may be explained by increased connectivity between the prefrontal cortices (Brain 3.0) and the basal ganglia (Brain 2.0).[22]

Appreciating Brain 3.0 as the Vehicle of Wisdom and Grace

Albert Einstein's famous statement that "no problem can be solved from the same level of consciousness that created it" succinctly captures the importance of Brain 3.0 in enabling humanity to solve the problems we create when we get locked in Brain 1.0 and Brain 2.0. The key structure of Brain 3.0, the prefrontal cortex, is the nexus where emotion, reason, and intuition converge. All of the abilities we need to see the bigger picture; to integrate and synthesize knowledge from multiple domains; to form a system of morality, ethics, and etiquette to guide social interactions; to study and understand the root causes of complex social issues; and to resolve social tensions and conflicts are enabled by the neural structures underlying Brain 3.0.[23]

According to Guy Claxton, author of *Intelligence in the Flesh*, the notion of the brain as the central command center of the body is a myth. Claxton proposes instead that the main function of the brain is to host conversations. He explains, "Brains evolved to help increasingly complicated, mobile bodies deal with problems of coordination and communication that they could not solve on their own. The brain is the central information exchange of the body."[24] Thus, intelligence is not concentrated only in the brain, but rather is distributed throughout the body and "flows" into key hubs (dense clusters of nerves) where the information is synthesized to determine a "local" course of action. Many of these distributed hubs do all their work below conscious awareness.

Eventually, information from all the distributed hubs in the body flow into centralized, specialized hubs in the brain to be synthesized into global maps. These hubs in the brain combine in various ways to create neural networks underlying key functions. The prefrontal cortex, the key structure of Brain 3.0, serves as the brain's primary "conference center" (a bit like a neural "United Nations"), where information from all the hubs and networks are integrated and the various maps are combined to construct a dynamic model simulating our internal world, our external world, and the relationship between the two.

One of the most important networks to our well-being and ability to function biologically is the interoceptive network, which processes signals coming in from the body. The interoceptive network itself is part of another very important network called the "salience" network, which enables us to selectively direct attention to stimuli we perceive as being "salient" (meaning most in need of our attention at any given time). Together, the interoceptive and salience networks underlie our capacity for emotion, wisdom, and intuition.

The most straightforward way to appreciate the function of a particular area of the brain is to look at what happens to people in whom that area is damaged. In *Descartes' Error*, neuroscientist Antonio Damasio describes the cases of patients who had severe damage to the ventral and medial parts of the prefrontal cortex (the area of the brain from behind the eyebrows to the center of the forehead). At first, the doctors concluded that the patients were fine because the damage did not cause any movement, sensory perception, attention, memory, language, or arithmetic impairments. Even their IQ scores were largely unchanged. Nonetheless, these patients were not capable of holding a job (they struggled to make decisions and organize processes) or managing their own finances (they had a tendency to become bankrupt from making poor investment decisions).[25]

Eventually, Damasio noticed that these patients were emotionally flat as they talked with extraordinary detachment about the tragedies they had experienced. Upon further investigation, he realized these patients didn't experience pleasure or emotional pain and exhibited a lack of empathy and sense of caring for others because they no longer had the ability to feel or express emotions or to read emotions expressed by other people. According to their families and loved ones, their personalities were so dramatically changed that they were no longer the person they used to be.

Over time, it became clear that their preserved intellectual and analytical faculties were worthless without the guiding wisdom once provided by the ventromedial prefrontal cortex. The curious observation that patients with damage in this area hopelessly make the same bad decisions over and over again, even though they consistently lead to negative outcomes, led Damasio and his colleagues to hypothesize and confirm through tests that this area plays an important role in enabling people to learn from mistakes by emotion-ally tagging bad decisions in "embodied memory" so they are not repeated. Interestingly, the fact that these patients also tend to ignore social conventions and etiquette without any sense of embarrassment or shame also led Damasio and his colleagues to hypothesize and confirm that the ventromedial prefron-tal region plays an important role in detecting the emotional significance of stimuli that trigger socially mediated emotions such as sympathy, embarrass-ment, or guilt.[26]

The reason Damasio's accounts of these patients stood out to me is because the Buddhist and yoga traditions I have studied aim to proactively cultivate equanimity by training the mind to become less attached to pleasure and less averse to suffering—but the results are strikingly different. Advanced medi-tators and yogis I have come across or whose writings I have studied, such as the Dalai Lama, Lama Zopa Rinpoche, Pema Chödrön, Thich Nhat Hanh, S. N. Goenka, and Paramahansa Yogananda, all share an elevated joyful vitality that is grounded in emotional depth and a contagiously genuine compassion, empathy, and altruistic concern for the welfare of other people. Furthermore, these people are role models of wisdom who bring light to our world—the opposite of being rudderless. Millions of people around the world turn to them for inspiration and read their books to learn from their insights.

This contrast led me to ponder how equanimity in the Buddhist and yogic sense is distinct from the emotional flatness and absence of affect experienced by patients with damaged ventromedial prefrontal cortices. As a brain geek, I naturally looked for clues in neuroscience. I found answers by understanding the relationship between the ventromedial prefrontal cortices (VMPFC) and two very important areas of the brain that make embodied wisdom, emotions, empathy, and intuition possible.

The VMPFC connects directly with the insular cortex (also called the insula) and the anterior cingulate cortex (ACC), key structures that are also vital to experiencing emotions. Perhaps not coincidentally, these three interconnected structures are also strengthened by meditation and spiritual practices.[27] In 2005, Sara Lazar published a groundbreaking brain imaging study that confirmed that long-term meditators have greater cortical thickness in the right anterior insula and in the prefrontal cortices than nonmeditators. In addition, independent studies by Norman Farb and Véronique Taylor showed that connectivity within the medial prefrontal cortices, and between the medial prefrontal cortices and the insular cortex as well as other parts of the brain, changed as a result of sustained meditation practice. Altogether, these findings suggest that long-term meditation may dramatically slow the age-related degeneration of the prefrontal cortex and may also rewire neural connections within the prefrontal cortices in ways that increase moment-to-moment awareness and reduce self-centered rumination and worry.[28] (We will talk more about these studies in chapter 8.)

The insular cortex in each hemisphere processes interoceptive signals from all over the body and builds an internal representation of the body (a body map). The left and right insular cortices are located at the bottom of the lateral sulcus of each hemisphere.* The differential impact of strokes on the right insular cortex versus the left insular cortex reveals that the right insular cortex serves as the primary central interoceptive hub, where all signals from the body come together to be synthesized into a global body map and an integrated embodied sense of self. Only when the right insula is damaged do stroke patients suffer a condition called anosognosia, borrowing a Greek word that means "the inability to acknowledge disease in oneself." When these patients are asked how they are, even though the left side of their body is completely paralyzed, they sincerely respond that they are fine because they are cognitively unaware of their paralysis. Neurologists attribute this condition

---

* The lateral sulcus is the very deep fissure that separates the frontal lobe and the parietal lobe from the temporal lobe on each side of the brain.

to the fact that their right insula can no longer process and synthesize intero-ceptive signals to form an up-to-date body map.[29]

The ACC is the front portion of the cingulate cortex, a structure that wraps around the corpus callosum, the area where the two hemispheres of the brain join together. The cingulate cortex is thus ideally situated to gather and redis-tribute information from and to a wide range of brain regions. The anterior portion of the cingulate cortex sits between the medial frontal lobes and the limbic system (which "houses" Brain 1.0 and Brain 2.0) and seems to play a role in the integration of sensory, emotional, and cognitive information. Brain imaging studies have found that the ACC is involved in emotional self-control, focused problem solving, error recognition, and adaptively responding to changing external and internal conditions.[30] The ACC is particularly active when experiencing pain (physical and emotional) and when perceiving other people's pain and suffering, and thus may play a crucial role in empathy, com-passion, and social awareness.[31] Malfunctioning of the ACC is associated with compromised communication skills and impaired ability to accurately sense what emotions other people are feeling and expressing.[32]

The insula and ACC are the key structures of the previously mentioned salience network, which is believed to help bring conscious attention to salient information—such as novel or unanticipated events, errors, conflicts, and emotions (ours and other people's)—and is also involved in task switching (the ability to immediately drop one task and start another). In line with this understanding of the salience network, Wendy Hasenkamp at Emory Univer-sity found that during focused attention meditation, when the meditator fo-cuses on breathing, in the moment that a meditator becomes aware that he or she got distracted by mind wandering, there is significant activity in the an-terior insula and the anterior cingulate cortex. This noticing then allows the meditator to shift from mind wandering back to observing the breath.[33]

According to Andrew Newberg, author of *How Enlightenment Changes Your Brain*, "the areas that generate intuition are located in the insula and anterior cingulate cortex, and they contribute to our ability to understand the world in more global and comprehensive ways."[34] The reason why the insula and ACC play an important role in intuition may stem from the fact that they are the only parts of the brain that contain rare spindle-shaped nerve cells called von Economo neurons (VENs), named after the scientist who discovered them in 1929.[35] These rare neurons have only been found so far in humans, great apes, elephants, and some whales and dolphins—species notable for having large brain sizes and being very social.[36] VENs are unusual because they are thin and elongated and have only one dendrite (a branchlike projection from a neuron that receives signals from neighboring neurons) whereas more common types of neurons have pyramidal or star-shaped bodies with multiple dendrites.[37]

Intriguingly, VENs are four times bigger than most other brain cells, and in the nervous system, size tends to correlate with speed. Therefore, John Allman, the leading neuroscientist studying VENs, hypothesized that VENs expedite communication from the insula and ACC to the rest of the brain and that VENs may form a specialized system that serves as a high-speed communication platform to rapidly receive, process, and transmit critically important information across great distances (in the context of the brain). Thus, the specific location of VENs in the ACC and insula (the brain's "social hot spots") led Allman to propose that the VEN system enables people to make accurate, split-second, intuitive judgments in the reading of complex social situations—an ability that probably conferred an evolutionary advantage to our ancestors.[38]

I would like to propose that the evidence that meditation, and spiritual practices in general, stimulates and strengthens the medial regions of the prefrontal cortex, the insular cortex, and the ACC (and the VENs within the insular cortex and ACC) can help explain how long-term meditators (and spiritual practitioners in general) experience a sense of "grace," that is, of being guided by and assisted by a higher being. It is probably not a coincidence that the ventromedial prefrontal region happens to be directly behind what the yogis call the ajna chakra (a.k.a. the "third-eye" chakra). As I shared in chapter 5, this is the chakra that was remarkably made known to me as "real" through a vision of an upside-down indigo triangle creating a vortex that led to my seeing the opening of an ethereal iridescent portal. Though I cannot know this for certain, I have come to understand the opening of that portal as a symbol for gaining access to higher realms of spiritual consciousness. Therefore, it is possible, though very hard to scientifically validate at the present, that the VMPFC's crucial role in the experience of wisdom may have something to do with its connection to the ajna chakra, and the ajna chakra serving as a "portal" to commune with higher consciousness.

Speaking from my personal experience (a field experiment where $n = 1$),

I have definitely observed that as my interoception became enhanced through meditation and yoga practices, I have also been able to access more and more embodied wisdom. My own ongoing transformation leads me to propose that it is possible that over time, spiritual practices lead to a reorganization of the salience network such that it no longer "marks" as salient social stimuli related to status, popularity, power, wealth, and influence (the carrots that activate Brain 2.0). Instead, interoceptive awareness reorients to tune in to stimuli of a higher spiritual nature, such that the salience network gains the ability to discern and mark as salient any messages coming from higher realms of consciousness. This "reprogramming" of the salience network could possibly be the neural correlate for pratyhara, the fifth limb of Ashtanga yoga, which literally means the withdrawal of the senses from the external world. Somehow as these areas of the brain rewire and attune to the Higher Self, we receive a positive biological feedback signal in the form of waves of bliss emanating from the heart that overcome the interoceptive network, creating the yogic experience of "sat-chit-ananda" ("eternal truth, consciousness, unconditional joy").

To return to the earlier question around the contrasting manifestation of equanimity between spiritual sages and patients with damage to the VMPFC, in my view, genuine equanimity most likely arises from the reorganization of the salience network, as I described above. This leads me to propose that developing embodied wisdom, intuition, and grace is very much tied to the integration of body, heart, mind, and soul in Brain 3.0 and that it is neurologically facilitated by a mysterious, self-organized rewiring of the neural pathways interconnecting the VMPFC, insula, and ACC (and the VEN system within the insula and ACC) to serve as a neurobiological bridge, or "transmission system," for the Higher Self. When this happens and we tap into the perspective of the Higher Self, we experience something like the overview effect, the transformative shift in awareness described by astronauts who look back at the planet Earth from outer space.[39] Once we experience this shift in consciousness, whenever we look at the rewards and prizes we were previously so attached to acquiring and hoarding, they appear to us like toys we obsessed over as children and subsequently lost interest in as we matured.

Calm Clarity's Key Mechanisms

I want to take time now to walk through the key mechanisms that will hopefully help you understand how to navigate your own progressive shift into Brain 3.0. The diagram I created below illustrates the key mechanisms by which the Calm Clarity Program enables people to rewire the embodied brain to experience greater well-being, resilience, and integration in Brain 3.0 (see figure 7).* I also refer to this diagram as a road map for enhancing the "emotional immune system" because these mechanisms also reduce our vulnerability to emotional contagion. With a strong emotional immune system, we can be exposed to triggers, yet remain grounded and centered.

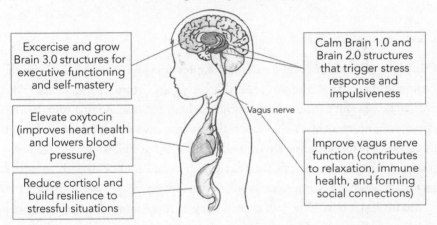

How to strengthen the "emotional immune system"

Calm Clarity's Key Mechanisms

Excercise and grow Brain 3.0 structures for executive functioning and self-mastery

Calm Brain 1.0 and Brain 2.0 structures that trigger stress response and impulsiveness

Elevate oxytocin (improves heart health and lowers blood pressure)

Vagus nerve

Improve vagus nerve function (contributes to relaxation, immune health, and forming social connections)

Reduce cortisol and build resilience to stressful situations

Figure 7: These are the key mechanisms by which Calm Clarity techniques promote well-being.

Brain 1.0, Brain 2.0, and Brain 3.0 can be seen as embodied emotional states in that these three patterns of brain activation extend beyond the brain,

---

* The "embodied brain" refers to the fact that the brain and body inextricably intertwine to generate the experience of having a mind and thus also includes the nervous system, which connects the brain and body.

into the autonomic nervous system and into the rest of the body. There are physiological differences throughout the entire body when we are in Brain 1.0, 2.0, or 3.0. Being in Brain 1.0 and Brain 2.0 is strongly tied to the increased activity of the sympathetic nervous system (also known as the freeze-flight-fight system), which redirects energy toward our legs and arms to prepare us to take action. As I shared earlier, Brain 1.0 and Brain 2.0 are highly inter-twined and function like two sides of the same coin. In Brain 2.0, people typically have high levels of dopamine and serotonin, but if they don't get the "carrot" they want, levels of dopamine and serotonin quickly plunge, putting them in Brain 1.0. In contrast, being in Brain 3.0 is strongly tied to the para-sympathetic nervous system (also known as the rest-and-digest system).

Since oxytocin and the vagus nerve play a central role in switching the autonomic nervous system from sympathetic mode (stressed and aroused) to parasympathetic mode (calm and at ease), they also serve as key levers in helping us shift out of Brain 1.0 or Brain 2.0 into Brain 3.0. Therefore, I'd like to take a moment to share useful information about them.

Oxytocin and the Tend-and-Befriend Response

Oxytocin (as I mentioned earlier in chapter 6) is nicknamed the "cuddle hormone" because it promotes social bonding and gestures of affection. In-creasing oxytocin can set off a healing biochemical cascade that reduces the circulation of stress hormones and inflammatory cytokines (molecules used by cells in the immune system to send signals), dilates blood vessels (thereby reducing blood pressure), facilitates the regeneration of heart cells to repair any micro damage, stimulates vagus nerve activity, and restores blood flow to the frontal lobes.[40] The fact that the oxytocin system mitigates (and pos-sibly even offsets) the harmful effects that high levels of stress can have on our physiological well-being is seen by numerous scientists in the field of positive psychology as evidence that human beings are wired for social con-nection.

Shelley Taylor's research team at UCLA has found that stress doesn't al-ways have to elicit the well-known freeze-flight-fight cascade, which escalates arousal by releasing stress hormones that flood the body and brain. Stress can also elicit another behavioral pattern, especially among women, that Taylor calls the "tend-and-befriend response" because it involves seeking and providing social support. The tend-and-befriend response unfolds because stress can also trigger the release of oxytocin, which prompts us to connect with our social support network.[41] Unlike the freeze-flight-fight cascade that reinforces the activation of Brain 1.0, tending-and-befriending activates biological feedback mechanisms that "turn on" Brain 3.0 and take us out of Brain 1.0. Taylor's research on the tend-and-befriend response has led scientists to propose that this instinctive mechanism evolved to give mothers the courage to risk danger in order to protect offspring and other loved ones by building strong social alliances and pooling resources to create safety nets.[42]

Kelly McGonigal, author of *The Upside of Stress*, concludes from reviewing the research on oxytocin that "oxytocin isn't just about social connection. It's also a chemical of courage. Oxytocin dampens the fear response in your brain, suppressing the instinct to freeze or flee. This hormone doesn't just make you want a hug. It also makes you brave." McGonigal adds, "Unlike the fight-or-flight response, which is primarily about self-survival, a tend-and-befriend response motivates you to protect the people and communities you care about. And, importantly, it gives you the courage to do so."[43]

In looking at the research on stress, McGonigal was surprised to learn that when people turn on the tend-and-befriend response, high levels of stress are not associated with negative health problems. This is because the body releases oxytocin and other substances, like DHEA, that provide protective effects by transforming what could otherwise be experienced as toxic stress into "eustress" (good stress).[44] McGonigal then realized that by seeing a greater meaning and social purpose for undergoing adversity, people can transform stress into a challenge response that turns on beneficial biological mechanisms that enable the body to harness the energy unleashed by the stress cascade to improve performance and concentration under pressure.[45] These findings led her to conclude that "caring for others triggers the biology of courage and creates hope."[46]

McGonigal also believes that anyone can intentionally activate the bravery of the tend-and-befriend state under stress by choosing to connect with and help others: "Whether you are overwhelmed by your own stress or the suffering of others, the way to find hope is to connect, not to escape." McGonigal adds, "In any situation where you feel powerless, doing something to support others can help you sustain your motivation and optimism."[47]

As I will explain in chapter 8, the biological mechanisms of the tend-and-befriend response make the compassion meditation a powerful means for activating Brain 3.0 and building resilience.

The Vagus Nerve and Vagal Superstars

The vagus nerve ("vagus" means "wandering" in Latin) is the largest cranial nerve in the body and has the very important job of connecting all our internal organs with the brain. It starts at the brain stem, and from there wanders through the neck downward into the torso, where it innervates (meaning it supplies nerves to) all our vital organs. One of its branches connects to the ears, where it helps us pick out the sound of a loved one calling for us in a loud noisy room. It innervates muscles in the neck that help us nod our heads and orient our gaze toward other people and vocalize our emotions. It connects to the heart and lungs and helps coordinate the interaction between breathing and heart rate. It connects to the liver and enteric nervous system and helps modulate digestive processes. The vagus nerve also helps mediate the immune system by innervating the thymus and spleen. As I shared earlier, about 80 percent of the nerve fibers of the vagus nerve are afferent, which means that the vagus nerve serves as a mostly one-way information superhighway for organs to send interoceptive messages to the brain.[48]

Because the vagus nerve is so vital to well-being and health, scientists created an index called vagal tone to measure the health of the vagus nerve. In general, scientists have found that higher vagal tone is associated with greater empathy, prosocial orientation, feelings of connectedness, more experiences of love and transcendence, a stronger immune system, and reduced inflammation. In contrast, low vagal tone is associated with high stress vulnerability, negative moods, loneliness, a weaker immune system, and heart attacks.[49]

Dacher Keltner, founding director of the Greater Good Science Center at the University of California, Berkeley, believes that the vagus nerve, which he describes as one of the "great mind-body nexuses" in the human nervous system, is evidence that human beings are "wired for compassion." His research team measured the activity of the vagus nerve as research subjects looked at photos of suffering and distress. They found that the more compassion the participants felt, the stronger the vagus nerve response. Furthermore, his team identified a subset of participants with unusually high vagal tone as "vagal superstars" and found that "these folks have more positive emotion on a daily basis, stronger relationships with peers, better social support networks." In addition, they found that "fifth graders who have a stronger vagal profile are the kids who intervene when a kid is being bullied. They're more likely to cooperate, and will donate recess time to tutor a kid who needs help on homework."[50]

Barbara Fredrickson, one of the pioneering researchers in the field of positive psychology, describes the important role of the vagus nerve in the neurophysiology of love in her book *Love 2.0*. She writes: "Your vagus nerve is a biological asset that supports and coordinates your experiences of love. . . . Your vagus nerve stimulates tiny facial muscles that better enable you to make eye contact and synchronize your facial expressions with another person. It even adjusts the minuscule muscles in your middle ear so you can better track the other person's voice against any background noise."[51] She hypothesizes that higher vagal tone increases the odds that people will make deep and meaningful connections.

The impact of sympathetic arousal on the vagus nerve explains why when we are stressed, we often lose our empathy, find it harder to feel connected with other people, and even feel disconnected within. Whenever Brain 1.0 and/or Brain 2.0 gets triggered, the arousal of the sympathetic nervous system automatically "deactivates" the vagus nerve and "pauses" parasympathetic functions to free up energy for our arms and legs to run or fight. So that means under stress, we lose our primary channel of interoceptive communication between our vital organs and our brain.

Given these mechanisms, resilience can be defined as the ability to quickly "reactivate" the vagus nerve and the parasympathetic nervous system and keep Brain 3.0 "online." We can therefore infer that higher resilience must correspond to having a healthy vagus nerve that is "strong" enough to modulate a freeze-flight-fight cascade. Fortunately, for those of us who may not be vagal superstars, we can use deep breathing to increase vagus nerve activity, calm a racing heart, and shift our bodies and brains from sympathetic mode (Brain 1.0 and Brain 2.0) to parasympathetic mode (Brain 3.0). (See chapter 8 for guidelines on deep breathing on page 232.)

"Outside-In" Approach

To rewire the brain, I believe the easiest and most effective sequence is to work outside in by first strengthening Brain 3.0, then reconditioning Brain 2.0 through retraining our habits, and then reconditioning Brain 1.0 by gently examining and releasing our fears, traumas, and self-limiting beliefs. I recommend this sequence because physiologically, Brain 3.0 is the easiest to access as it corresponds to the newest structures of the human brain to evolve, which enable us to "consciously" experience life. Brain 2.0 is the part of the autopilot that stores our habits, so it can be rewired by consciously, deliberately, and diligently cultivating habits and routines. Brain 1.0 is the most challenging to change because it corresponds to the deepest, hardest-to-access areas of our subconscious mind where we "store" our fears, shames, and traumas. The truth is, it's nearly impossible to rewire Brain 1.0 until we have sufficient "power" in Brain 3.0 to hold space to compassionately experience, honor, and release the negative emotions and memories that are stored in Brain 1.0 without giving in to urges to escape into Brain 2.0.

The key is to continuously exercise, prime, and strengthen the neural pathways connecting Brain 3.0 into Brain 2.0 and into Brain 1.0. When these pathways are stronger, Brain 3.0 can guide and calm Brain 2.0 and Brain 1.0 so they are less likely to hijack us. These integrating pathways also spiritually enable the Higher Self to calm and soothe the Inner Teen Wolf and Inner Godzilla and to shine light into the shadow. To deepen this transformation, we have to progressively build up Brain 3.0 to have more and more conscious access to Brain 2.0 and Brain 1.0. In the next chapter, I will introduce techniques to build these neural pathways.

8.

Nurturing and Strengthening Brain 3.0

As a single footstep will not make a path on the earth, so a single thought will not make a pathway in the mind. To make a deep physical path, we walk again and again. To make a deep mental path, we must think over and over the kind of thoughts we wish to dominate our lives.
—Henry David Thoreau

Similar to the design of the Calm Clarity workshop program, this chapter incorporates four key components: introducing, explaining, and interweaving scientific and contemplative concepts; reflection worksheet exercises to tie the concepts to everyday life; contemplative practices to develop an embodied understanding of the concepts; and reflection questions after the practices and exercises. I highly recommend that you take time to write your answers to these questions in order to further assimilate and encode your insights and learnings into your long-term memory. I also provide additional comments following the exercises and practices to help you understand how I experience and use them, and how they can deepen your understanding as you practice them over time.

This chapter contains six sections:

- "A Quick Intro to Meditation" explains the Calm Clarity approach to meditation, including guidelines on posture and deep breathing that will apply to all the meditations presented subsequently.
- "Cultivate Compassion" introduces the Calm Clarity Compassion Meditation, beginning with background on the practice and how it works.

- "Know Thyself" helps you to better understand yourself by making conscious how you unconsciously shift between Brain 1.0, Brain 2.0, and Brain 3.0.
- "Mind Your Autopilot" helps cultivate mindfulness to further understand your autopilot and retrain it as you strengthen your capacity for metacognition, interoception, and proprioception.
- "The Voice in Your Head" enables you to apply what you learned from the previous sections to observe your mind-track and transform your Inner Critic into an ally.
- "Embody Your Higher Self" enables you to experience the overview effect by connecting to your Higher Self and harnessing energies and wisdom from your Higher Self to self-actualize.

To re-create the feel of a group workshop, you may find it beneficial to do these exercises with a partner or group of friends with whom you can share and discuss learnings and insights. If you feel you may benefit from experiencing a more formal and structured in-person group workshop, you may want to consider joining one of the open-registration Calm Clarity Weekend Retreats, which we offer several times a year (see calmclarity.org for more information).

Lastly, please keep in mind that the instructions to the exercises are not meant to be rigid. I invite you to experiment and find your own way of achieving the intent of these exercises in a manner that is most effective for you.

A Quick Intro to Meditation

My mouth shall speak of wisdom; and the meditation of my heart shall be of understanding.   —PSALM 49:3, King James Bible

To preface this topic, I feel it is important to clarify that I do not consider meditation to be the be-all and end-all solution to life's challenges and the problems facing our world. Meditation by itself cannot solve hunger, poverty, a broken educational system, or global warming or bring about an end to war and violence. However, it is meditation's impact on enhancing Brain 3.0 that makes it so worthwhile. Without developing Brain 3.0, we cannot have creative breakthroughs, harness the power of collaboration, or make wise decisions to benefit the greater good. Given the mounting scientific evidence confirming the power of meditation to strengthen Brain 3.0, meditation deserves to be an activity we all undertake to develop our potential and to accelerate humanity's progressive evolution toward self-actualization.

For me, the word "meditation" refers to a large basket of contemplative practices that modulate the embodied brain by exercising, priming, and strengthening specific neural pathways and networks. The English word "meditate" simply means "to engage in contemplation or reflection."[1] In explaining the etymological evolution of the word, Jon Kabat-Zinn (the creator of the mindfulness-based stress reduction program) shares that "meditation" and "medicine" are derived from the same Latin word, "mederi,"which means "to heal," and that "mederi" can itself be traced to the Proto-Indian root "med," which means "to measure" and could also mean "right inward measure," a concept that refers to the characteristics that make people who they are or give things their specific properties.[2] Thus, Kabat-Zinn sees meditation as a means to perceive the right inward measure of one's own being, and medicine as a means to restore right inward measure when it is disturbed by injury and ill health.[3]

In my wanderings to learn how to meditate, I came across quite a few people who dogmatically insisted that only their own tradition constitutes real meditation, so no other practice can correctly be called "meditation." I found that this narrow view did not sit well with me because no single group, culture, or tradition can make a credible claim to own the word "meditation." Just about every spiritual tradition around the world has its own form of structured contemplation and reflection, such as ritualized prayer (which is sometimes done with the aid of a rosary or a mala), communal singing, chanting and dancing, periods of retreat for quiet introspection, and longer-term spiritual formation programs.

Although, the word "meditation" today is frequently associated with the Eastern spiritual traditions of Buddhism and yoga, this wasn't always the case. For instance, the words "meditate" and "meditation" appear throughout the King James Bible, which was completed in 1611. Later that century, René Descartes, widely revered as the father of modern Western philosophy and credited with laying down the rational philosophical foundation that catalyzed the Age of Reason (or Age of Enlightenment), titled his 1641 masterpiece *Meditationes de prima philosophia, in qua Dei existentia et animæ immortalitas demonstratur* (translation: *Meditations on First Philosophy, in which the existence of God and the immortality of the soul are demonstrated*). Therefore, we should recognize that "meditation" is also strongly rooted in a Western context.

Furthermore, there really is no reason to restrict the definition of "meditation" to an activity we do sitting still with our eyes closed and legs crossed on the floor. Once we learn how to activate the neural circuits of Brain 3.0 through various forms of meditation practices, the key to self-actualization is to practice keeping these neural circuits "primed" throughout the entire day. We can enter a meditative state while walking, eating, cooking, swimming, doing chores, and just about any activity just by being in the present moment and being aware of our thoughts, feelings, emotions, and sensations. In this manner, all of life becomes meditation. While I very much value and appreciate learning traditional approaches to meditation, I am more of a tinkerer than a purist when it comes to my own practice. I can't help but use insights from science to experiment with tweaking traditional forms of meditation to enhance their physiological benefits and make them easier for beginners to understand and practice.

The meditations I share in the Calm Clarity Mindful Leadership Program are recognizable to people who are familiar with the traditional versions they are based on, but are also creatively enriched by science. I'm proud to share that people who have meditated for decades have told me that doing the Calm Clarity meditations provided them with new insights on how to practice and teach meditation.

Next, I want to take a moment to walk you through guidelines on posture and deep breathing, which will apply to all the meditations in this book.

Meditation Posture

Every meditation tradition more or less has its own guidelines on meditation posture, but so far, there is no scientific rationale to support the idea that there is only one correct posture. The most important thing is to find a comfortable and ergonomic position that doesn't create any biomechanical strain on the body. For instance, I don't take to sitting cross-legged for more than a few minutes because inevitably, my legs fall asleep and then the pins-and-needles discomfort becomes an agonizing distraction.

I generally recommend sitting in a chair with your legs parallel and feet flat on the floor, as this provides a grounding effect. Try to keep your spine from your neck to your sacrum (where your spine connects to your hips) straight and aligned, so you aren't leaning forward, right or left. Ideally, when you find an ergonomic sitting position, you won't feel the need to use a backrest. However, if you are tired, it is fine to use the backrest so long as you make sure not to slouch. A straight spine helps to increase the activity of the vagus nerve and also helps to open the chest because this creates more room for the lungs to expand when you breathe in. Gently pull the shoulder blades back as much as possible without creating any strain. Hands can rest on the lap where they are comfortable, but be mindful that they are not so far forward that the arms pull the body to lean forward. It does not really matter if the hands face up or down or are intertwined. There is no need to hold any mudras (ritualistic hand positions or gestures).

The chin should be parallel to the floor and tucked slightly in. Try not to bend your head downward or upward, as that puts pressure on the spine. I also suggest people curl the tongue upward and place the tip of the tongue on the palate. One of my meditation teachers shared that this reduces drooling and helps keep the mind alert. Since I do have a tendency to drool, I tried it and found what he said to be true.

The eyes can be open or closed depending on your state of mind and the level of challenge you want. When the eyes are open, the brain tends to generate beta waves, which are associated with active concentration and busy, anxious thinking. When the eyes are closed, the brain tends to generate alpha waves, which are associated with relaxation. However, the challenge is that when people are sleepy, closing their eyes increases the likelihood of falling asleep. To meditate with eyes open, the idea is to have the eyes partially open and softly focused on a single point on the ground about one to two feet in front of you. Since meditating with eyes open is much more challenging than with eyes closed, I recommend if you are a beginner to start with your eyes closed, and if you get sleepy, simply open your eyes for a short period of time to become more alert.

Deep Breathing

All of the Calm Clarity meditations begin with deep breathing to activate the vagus nerve. The idea is to fully expand and empty the lungs to create a bellows effect that signals to the autonomic nervous system that it is time to relax and switch gears from sympathetic mode into parasympathetic mode. There are many approaches to deep breathing, so there is no "right" or "wrong" way to do it.

The approach I prefer is to have people inhale slowly for a count of six seconds; hold the breath for a count of three seconds, which lets the air expand the rib cage more fully; then exhale slowly for a count of six seconds; and hold the breath for a count of three seconds, which creates a vacuum in the lungs. The shorthand for this particular breathing cycle is "6-3-6-3," and it results in about three breathing cycles per minute (each breathing cycle is a full inhale and exhale). I chose this sequence based on a "Goldilocks" rule of thumb that the length of time for inhaling, exhaling, and holding the breath should not be so long that it creates strain and stress, nor so short that it fails to produce a noticeable physiological effect. Also, the 6-3-6-3 cycle pattern is easy to remember, which is very important from a practical point of view.

Keep in mind that the muscles that control inhaling and exhaling (the diaphragm and the intercostal muscles of the rib cage) are not under direct conscious control in the same manner as the muscles in our arms and legs. We indirectly make them contract and relax through conscious breathing. The key is to breathe very slowly and steadily so that as the diaphragm naturally contracts downward, the intercostal muscles are also able to stretch the rib cage to its maximum capacity. From my own experiments, I have found that quick deep breaths do not allow the intercostal muscles to fully expand the rib cage. *If you are curious, I suggest you place your hands on the front sides of your lower rib cage and try different breathing rates to see for yourself the degree to which your rib cage expands and contracts.*

Before each meditation, I generally guide people to do three of these 6-3-6-3 breathing cycles, which is usually sufficient to feel the autonomic nervous system switch gears. This deep breathing exercise can also be done as a stand-alone practice for several minutes whenever you find yourself overwhelmed by a high level of stress as a way to calm your sympathetic nervous system and come back to center.

Cultivate Compassion

A human being is a part of a whole, called by us "Universe," a part limited in time and space. He experiences himself, his thoughts and feelings as something separated from the rest—a kind of optical delusion of his consciousness. This delusion is a kind of prison for us, restricting us to our personal desires and to affection for a few persons nearest to us. Our task must be to free ourselves from this prison by widening our circle of compassion to embrace all living creatures and the whole of nature in its beauty. —ALBERT EINSTEIN[4]

Our human compassion binds us the one to the other—not in pity or patronizingly, but as human beings who have learnt how to turn our common suffering into hope for the future. —NELSON MANDELA

Compassion is the keen awareness of the interdependence of all things.
—THOMAS MERTON, theologian

To provide some background, compassion meditation (also known as loving-kindness meditation) is rooted in a 2,500-year-old metta tradition that was originally taught by Siddhartha Gautama, the historic Buddha. Recall from chapter 6 that "metta," a Pali word, can be translated as loving-kindness, benevolence, friendliness, and compassionate goodwill. Siddhartha Gautama taught that metta can be cultivated by actively wishing that all beings be happy and free from suffering. The traditional Metta Bhavana (meaning "cultivation of loving-kindness") practice consists of methodically sending benevolent wishes first to oneself, then to our loved ones, to a person we feel neutral toward (such as a stranger), to a person with whom we have a difficult relationship, and finally to all living beings. The wishes are stated like blessings, such as "May all people be happy."

In designing the Calm Clarity Mindful Leadership Program, I chose to introduce the compassion meditation as the very first meditation practice because it is a relatively easy meditation for beginners to learn, and thereby have a positive initial experience with meditation, and because the underlying biological mechanisms it activates are very compelling. Earlier in chapter 6, I shared that scientists such as David Kearney and Barbara Fredrickson, who have studied metta meditation practices, found that it is associated with a reduction in symptoms of PTSD and depression, increased positive emotions, enhanced social connection, and improved vagal tone.[5] Furthermore, preliminary brain scanning studies revealed that this practice over the long term may also be associated with increased activity in the left prefrontal cortex. These intriguing changes stood out to me because many people who experience chronic stress (meaning their sympathetic nervous system has been highly aroused in freeze-flight-fight mode for a prolonged period of time) have low vagal tone and low activity in the left prefrontal cortex, so could very much benefit from using the compassion meditation to rewire their brain and nervous system.

Richard Davidson's research team at the University of Wisconsin was one of the first to conduct brain imaging studies of Tibetan Buddhist monks practicing meditation compared to a control group of volunteers (college students with no exposure to meditation). In 2004, his team reported that when the monks practiced metta meditation to generate a state of "unconditional loving-kindness and compassion," they found very high brain activity in the left prefrontal cortex (a region associated with positive affect and stronger immune functioning) and unusually powerful and coordinated gamma waves, a phenomena called gamma synchrony. According to the *Washington Post* reporter who covered the story, "The intense gamma waves found in the monks have also been associated with knitting together disparate brain circuits, and so are connected to higher mental activity and heightened awareness."[6]

In *Train Your Mind, Change Your Brain*, Sharon Begley explains that gamma waves "appear when the brain brings together different sensory features of an object, such as look, feel, sound, and other attributes that lead the brain to its *'aha!'* moment" of recognizing what that object is.[7] No one had ever seen gamma wave activity like this before, so the research team published a journal article specifically to report this remarkable finding. They wrote: "The high-amplitude gamma activity found in some of these practitioners is, to our knowledge, the highest reported in the literature in a nonpathological context."[8]

As the monks in the study had engaged in tens of thousands of hours of meditation over their lifetimes, Davidson was naturally cautious about extrapolating these findings to a broader population.[9] I found encouragement in a 2010 study by a team of researchers at Oxford University investigating the impact of a short session of either metta meditation or mindful breathing meditation on people recovering from depression, which found stronger relative left prefrontal activation after both types of meditation. The simple experiment involved having the volunteers follow a guided recording of a fifteen-minute meditation with no other instruction and without any way for the volunteer to ask a meditation instructor for clarification on the technique. The researchers ended the paper by writing: "In conclusion, the current findings suggest that brief meditation techniques can bring about significant changes in brain activity underlying affective state."[10]

Based on the available research to date, I believe practicing compassion, or metta, meditation can lead to decreases in symptoms of PTSD and depression because it activates key mechanisms underlying the tend-and-befriend response to rebalance the autonomic nervous system and boost parasympathetic functions by stimulating the vagus nerve and increasing levels of oxytocin, which helps reduce stress hormones like cortisol. The increased feelings of positive emotions, connectedness, and well-being are also most likely tied to increased activity in the left prefrontal cortex, along with the increase in vagal tone and oxytocin.

By taking these mechanisms into consideration, I then designed the Calm Clarity version of the compassion meditation to boost these physiological benefits during a short period of practice. After years of guiding various audiences to activate Brain 3.0 by doing the Calm Clarity Compassion Meditation, I can share that each time, I have witnessed a dramatic before-and-after difference as people's bodies and brains physiologically shift from Brain 1.0 and Brain 2.0 into Brain 3.0. In general, participants are usually struck by how calm and centered they feel afterward. Several people who explained that they had been so stressed by work that they had not been able to sit still and relax for a very long time shared that they didn't expect this meditation to work, yet it actually made them the most relaxed they had felt in years.

Calm Clarity Compassion Meditation

The Calm Clarity Compassion Meditation builds from the traditional Metta Bhavana practice, which involves repeating a series of altruistic wishes in a genuine manner, and combines new elements to enhance the physiological mechanisms. It begins with a deep breathing exercise to activate the vagus nerve and stimulate the parasympathetic nervous system. It also incorporates visualization, smiling, and a gratitude exercise to further strengthen the activation of the vagus nerve and release oxytocin (thus setting off biochemical cascades that lower blood pressure and induce relaxation) and further prime and strengthen Brain 3.0 neural pathways. I ask people to verbally say the wishes out loud as they smile to enhance what is called the ideomotor effect (which I will explain in the next section), so that the ideas captured by the wishes and the accompanying muscle movements deepen the physiological and biochemical cascades. As people say the wishes, the key is to intentionally self-generate the feeling of unconditional love and compassion that an ideal mother feels for her child toward the person or people the wishes are for. Saying the wishes in a cursory or mindless manner will not produce the same degree of physiological changes.

Instructions for the Calm Clarity Compassion Meditation

For your convenience, a guided recording of the Calm Clarity Compassion Meditation can be found on the Calm Clarity website at calmclarity.org.

1. POSTURE

Assume a comfortable meditation posture in which your spine from the top of your neck to your sacrum is straight and aligned, and your chest is open. Place the tip of your tongue on your palate.

2. DEEP BREATHING

Begin with slow, deep breathing. Start by exhaling to empty the lungs.

Inhale slowly to a count of six. Hold for a count of three. Exhale to a count to six. Hold for a count of three.

Do two more cycles and then normalize your breathing.

3. PRIMING POSITIVITY

Think of a happy memory, a time when you felt joyful, carefree, and safe. Smile as you remember and relive this moment. See if you can bring your body into this state of happiness now.

4. WISHES FOR OURSELVES

Smile as you say these lines out loud. Saying them aloud increases the physiological effects of these wishes on your body:

"May I be Happy. May I be Healthy. May I be Safe. May I be Peaceful. May I be Prosperous. May I live in Harmony with others."

Now visualize a picture of yourself as a child in front of you and send love to yourself. Picture yourself happy, healthy, and thriving. If you like, give your inner child a virtual hug.

5. Wishes for family

Now visualize your entire family in front of you and lovingly share these wishes with them as you smile:

"May my family be Happy. May my family be Healthy. May my family be Safe. May my family be Peaceful. May my family be Prosperous. May my family live in Harmony."

Picture your family in front of you and send love to everyone. Picture your family happy, healthy, successful, and thriving. Picture everyone united in love and harmony.

6. Wishes for friends and loved ones

Now visualize your friends and loved ones in front of you and smile as you share these wishes with them as a group:

"May you be Happy. May you be Healthy. May you be Safe. May you be Peaceful. May you be Prosperous. May you live in Harmony with others."

Now send love to your friends. Picture everyone happy, healthy, successful, and thriving.

7. Wishes for a difficult relationship

Now think of one person with whom you have a difficult relationship. Perhaps this person is going through a challenging situation that triggered them into Brain 1.0 or Brain 2.0. Maybe they can benefit from kindness and goodwill, so let us also wish them well. Picture this person in front of you and say to them with sincere compassion:

"May you be Happy. May you be Healthy. May you be Safe. May you be Peaceful. May you be Prosperous. May you live in Harmony with others."

Visualize this person getting the help and support they need to overcome their challenges and come into Brain 3.0. Imagine them living in harmony with everyone around them, including you.

8. Wishes for all people

Now think of all 7 billion people on this planet. While we don't know most of them, every person is probably similar in that we all want to be happy and to be free from suffering. So picture the planet Earth and all people on it and say with heartfelt goodwill:

"May all people be Happy. May all people be Healthy. May all people be Safe. May all people be Peaceful. May all people be Prosperous. May all people live in Harmony."

Feel the love and kindness in your heart expand. Picture the planet Earth and send love and peace to everyone throughout the world. Imagine everyone happy, healthy, joyful, hugging each other and helping one another. Imagine everyone working together to solve the problems of our world. Imagine everyone living in peace and harmony.

9. Visualization to activate the vagus nerve

Now visualize a lightbulb above your head, very much like what we see in cartoons when someone has a bright idea.

Think of this as the light of inspiration inside you. As you breathe in, direct the love from your heart up to this light; imagine the energy of your love fueling this light, making it shine brighter and brighter until it becomes brilliant and radiant.

As you breathe love into this light, you activate the vagus nerve and strengthen the connection between your heart and brain.

Now take a few minutes to focus your awareness on this light. Keep breathing love into this light, making it brighter and brighter. As the light gets brighter, let it fill you with peace, calmness, clarity, and inspiration. (Silence for about two minutes or as long as you wish.)

10. Setting intentions

As you end this meditation, set a few intentions. Ask yourself to hold on to this state of calmness and clarity no matter what happens today. Form an intention to be mindful and aware of what you think, feel, say, and do.

As you go about your day, whenever you see an act of kindness, appreciate it and celebrate it. When kindness is shown to you, can you make a point to express gratitude? When you see an opportunity to show kindness, can you make the effort to do so?

11. Gratitude practice

While you are in this state of peace, think of the wonderful things that bring you happiness and joy and allow them to bring a smile to your face. As you smile, think of five things that you are thankful for.

When you are ready, open your eyes and write down the five things.

Post-Meditation Reflection

To further hardwire this learning experience, please take time to write your answers to the following questions:

1. How was this experience? What insights, if any, arose?
2. What stood out most to you?
3. How can you incorporate the compassion meditation into your everyday life?

Additional Comments on the Compassion Meditation

This compassion meditation is my primary go-to practice to keep me grounded in Brain 3.0 and connected to my Higher Self when I have challenging interactions. I have to confess: at times, this meditation is all that stands in the way between me and a full-blown amygdala hijack. When Brain 1.0 gets triggered and I feel my Inner Godzilla stir inside, saying the compassion meditation wishes for myself and the people I am getting upset with keeps me from going into full freeze-flight-fight mode. I can feel the physiological cascade it sets off, keeping my heart from closing and keeping the blood flowing to Brain 3.0, so I can continue to feel compassion and empathy for the people whose behavior may be aggravating my Inner Godzilla. So long as I keep my heart open, I can tune in to guidance from my Inner Sage on how to respond with wisdom.

In the Tibetan Buddhist tradition, the spirit of metta is captured in a short prayer called the Four Immeasurables, which is often chanted at the beginning of teachings and gatherings. The Four Immeasurables prayer is often translated into English as follows:

May all sentient beings have happiness and the causes of happiness;
May all sentient beings be free from suffering and the causes of suffering;

May all sentient beings never be separated from the happiness that is free from suffering;
May all sentient beings live in equanimity, free from attachment and aversion.

In my own practice, I often also recite the Four Immeasurables because this form of Metta Bhavana reminds me to stay connected with my Higher Self, the source of the "happiness that is free from suffering."

As I shared in chapter 6, the spiritual connection that arises from my doing the compassion meditation often brings on mystical experiences. In generating the feeling of metta and sending it out into the world, it's like my consciousness merges with the energy of unconditional love in the Collective Consciousness. I often feel this energy as the Divine Mother, holding all of humanity as a baby in her loving and nurturing arms. In addition, I also feel surrounded by and infused with "agape," the Greek word for a very pure form of unconditional divine love that was used by Jesus Christ in the original Greek New Testament when he instructed people to "love the Lord thy God with all thy heart, and with all thy soul, and with all thy mind" and to "love thy neighbour as thyself."[11] When I feel agape, I sense that it is a sign a great presence, which many New Age mystics refer to as the Christ Consciousness, is connecting with me. In turn, I do my best to use my heart as a conductor to transmit and share these energies of agape and metta with the people around me and with the world. Over time, the more I let these energies flow through me, the deeper my experiential and visceral understanding of the meaning of agape and metta and how the True Self is essentially composed of these energies.

The power of the compassion meditation is beautifully captured by Albert Einstein in the epigraph at the beginning of this section in which he states: "Our task must be to free ourselves from this prison by widening our circle of compassion to embrace all living creatures and the whole of nature in its beauty." At a higher level, we all arise from one Source, so the True Self represents the unity of life. The esoteric truth that underlies the compassion meditation is that regardless of whom we say the wishes for, we are really saying the wishes for our True Self. Sharing divine unconditional love with all people enables us to break through the delusion of separation, and as that veil erodes, cosmic consciousness shines through.

Know Thyself

There are three things extremely hard: steel, a diamond, and to know one's self.
—BENJAMIN FRANKLIN

He who knows others is intelligent; he who knows himself is enlightened.
—LAO-TZU

No one is free who has not obtained the empire of himself. No one is free who cannot command himself.
—PYTHAGORAS

"Know thyself"—this mysterious instruction has been given by sages in different forms across many cultures throughout history to seekers of wisdom because it is one of the most challenging undertakings that any human being can embark on. Fortunately, recent scientific research has shed light on why it is actually very difficult to know oneself. Thanks to our autopilot, we have huge blind spots that make it nearly impossible for us to fully see and understand ourselves because the conscious pilot, by default, is able to observe only what we are consciously aware of. To fully know ourselves, we have to turn to science to systematically and methodically study and understand our autopilots so we can uncover our blind spots.

As Daniel Kahneman explains in *Thinking, Fast and Slow* (which I touched upon earlier in chapter 5), the fields of behavioral economics and behavioral neuroscience emerged from efforts to make sense of intriguing findings that there is a near universal discrepancy between how people see and describe themselves as rational agents and how they actually behave in remarkably irrational ways in everyday life, such that when the observations are reflected back to them, even they are taken by surprise. Thankfully, these fields have provided a wealth of observations and insights that enable us to better understand the nearly universal patterns for how the autopilot (System 1) seems to behave under certain conditions. Kahneman even cites a book by Timothy Wilson with the evocative title *Strangers to Ourselves* to explain that the stranger is System 1, which "may be in control of much of what you do, although you rarely have a glimpse of it."[12]

In that book, Wilson, a professor of psychology at the University of Virginia, argues: we are strangers to ourselves because we have developed a story about ourselves that is out of touch with how the adaptive unconscious (the autopilot, or Kahneman's System 1) actually functions. Beneath conscious awareness, the autopilot continuously performs a set of pervasive, sophisticated processes that construct and adjust mental maps of the world around us, set implicit goals, and initiate actions toward them. Therefore, to truly know yourself, you have to study your autopilot. To get more objective data, you need to pay attention to what you actually do and to what other people observe about you.[13] What you learn may come as a surprise, but chances are good that the insights you (and other people) discover about your autopilot (which includes your shadow) may apply to the universal human condition.

Gaining self-knowledge requires unbiased self-observation and self-examination. To do this, you have to be willing to test and challenge the self-image built up by your ego. You have to be willing to challenge and to let go of stories and mental models you have unconsciously created about how the world works and how you are supposed to navigate it. If you rigidly cling to a narrow self-image and to your mind's projections onto the world, the simulation engine of your mind will distort how you see yourself and the world around you to confirm your views and expectations. Until you are willing to observe and examine that distortion process, your mind will not let you see who you really are, nor will you be able to interact with the world as it really is. Furthermore, whatever is suppressed or denied will get stored in your shadow, usurp enormous amounts of energy, and subconsciously hinder you from fully integrating and connecting to your Higher Self.

How we perceive ourselves and the world around us is determined in part by the neural circuits that are activated, that is, "primed," in our brain. This is documented in studies of priming, a nonconscious form of suggestion and influence. According to *Psychology Today*, priming "refers to activating particular representations or associations in memory just before carrying out an action or task. For example, a person who sees the word 'yellow' will be slightly faster to recognize the word 'banana.' This happens because yellow and banana are closely associated in memory."[14] Priming can also make us become more acutely aware of a certain type of object, concept, emotion, or experience. For example, a person who just bought a new car may start to notice other cars of the same make and model more frequently; a person learning about punctuation will take note of punctuation marks more when reading;

a person who is angry at someone will quickly perceive more reasons to justify that anger; and a person who experiences discrimination becomes more sensitive to noticing subtle acts of bias and prejudice.

According to Kahneman, priming works through "associative activation," a process in which ideas that have been evoked trigger many other related ideas in a spreading cascade of activity in the brain, such that each element is connected to and supports and strengthens the others, yielding a self-reinforcing pattern of cognitive, emotional, and physical responses. Once activated, neural networks are primed to fire strongly associated neural circuits much more readily. Kahneman further explains, "Only a few of the activated ideas will register in consciousness; most of the work of associative thinking is silent, hidden from our conscious selves."[15]

Figure 8 below illustrates how associative activation works. I depicted the elements in a circular ring to show that the cascade can happen in any direction. Actually, each of the elements can trigger any one of the others, in no particular order. For instance, the term "ice cream" can evoke an image of an ice-cream cone in your mind; make you think of your favorite flavors of ice cream; and bring back fond childhood memories of enjoying ice cream, which make you happy and put a smile on your face. Conversely, making certain muscle movements, facial expressions, and postures can also start a cascade. For instance, raising your hand up in the air can bring back memories of being in a classroom and trying to get a teacher's attention, and raising both hands up in the air and pumping them can make you feel excitement. Recalling specific memories can also bring flooding back to life the emotions and body sensations associated with that memory.

The unconscious physiological effects of priming are so remarkable that scientists created a term, the "ideomotor effect" (often used interchangeably with "ideomotor response"), to describe when a thought, idea, or image triggers an automatic muscular reaction or behavior outside of the awareness of the person. For instance, anyone who has sucked on a lemon or lime will find that their mouth instantaneously waters whenever they see a photograph of a lemon or lime. It also works in the reverse. When people carry out a muscle movement, it can also prime them to act, feel, and think in line with actions, feelings, and thoughts previously associated with the motion.

For instance, when people unconsciously mimic the muscle movements of a smile by holding a pencil in their mouth using their teeth, they find jokes

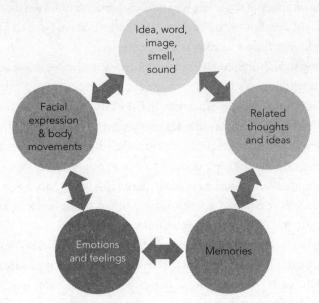

Figure 8: Associative Activation Cascade

funnier; and when people unconsciously mimic the muscle movements of a frown or grimace by holding a pencil in their mouth without using their teeth (by pressing their lips tightly around it), they have a stronger emotional response to upsetting images.[16] Similarly, a recent study found that people rate the funniness of cartoons higher when repeatedly articulating the vowel sound "i" (which mimics the muscle movements of a smile and positively charged words such as "like," "alive," and "hi") than when articulating the vowel sound "o" (which mimics the muscle movements of a frown and negatively charged words such as "alone," "no," and "woe").[17] As you have seen from doing the Calm Clarity Compassion Meditation, Calm Clarity techniques often incorporate smiling to prime positive affect, which boosts activity in the left prefrontal cortex.

The growing evidence around the body's role in cognition has led many scientists, such as Daniel Kahneman, to conclude: "Cognition is embodied. You think with your body, not only with your brain." Oftentimes, the body actually trumps the brain in terms of speed. For instance, take the issue of frustrating doors that have a handle that you normally pull but that have a sign that says "push." As soon as people place their hand on the handle, the muscle memory automatically triggers a reflex for pulling. By the time it takes the pilot (System 2) to read and process the sign to push the door, the autopilot has already learned from the fact that the door didn't open from pulling that it is either broken or needs to be pushed.[18]

Through priming, our emotions can subconsciously influence the way our brain filters sensory stimuli and processes information. When the network for a specific emotion is activated, this primes pathways that have become associated with this network to fire, thereby triggering more memories and ideas associated with that emotion to come into consciousness, and rendering our attention more sensitive to information that reinforces our emotional state. This, in turn, colors our perception, interpretation, and internal storytelling to intensify our emotional state. Scientists use the term "affective recall" to refer to how our emotions influence our memory, and "affective bias" to refer to how our emotions influence what we pay attention to and how we interpret information.

Thus, what we pay attention to and how we interact with the world can vary greatly depending on whether we are in Brain 1.0, 2.0, or 3.0. For instance, when I am in Brain 1.0, I am more easily disappointed and overwhelmed. If I am around people who are upset, I tend to get upset along with them and even escalate the situation emotionally. In addition, I tend to be more paranoid that the system is out to get me and become hypervigilant toward signals of aggression and oppression. I tend to get pessimistic about the future and become much more sensitive to signs that things are going wrong, such that I read too much into minor setbacks, making them into catastrophes.

When I get triggered into Brain 2.0 by something I want to have, I turn everything I see around me into evidence or justification for why I should go for it. For instance, if I see someone wearing a stylish new dress I like, by the tenth time I see someone wearing that style, I have already concluded: if everyone has that look, don't I need it too?! Next thing I know, it's in my closet. (Thanks to Brain 2.0, there are quite a few things in my closet that I have yet to wear even once.)

When I am in Brain 3.0, I feel very solid, centered, grounded, and connected, such that bad news doesn't easily shake me. I am able to see setbacks as one door closing so another door can open, and remain optimistic that life will work out fine. I don't experience emotional contagion when I am around people who are upset. Instead, I am able to feel compassion for what they are experiencing and do what I can to help. In addition, I generally feel content with what I have and only go shopping when there is a clear functional need to buy an item.

Throughout each day, every person moves between Brain 1.0, Brain 2.0, and Brain 3.0 depending on what events happen, the people they interact with, and what triggers they come across. The shift into Brain 3.0 begins with *your* becoming aware of how *you (and your autopilot)* move between Brain 1.0, Brain 2.0, and Brain 3.0. Therefore, it is very important now to take time to understand your own triggers for each of these three emotional states.

Knowing Your Brain States Worksheets

This exercise gives you the opportunity to become aware of and identify triggers that bring you into Brain 1.0, Brain 2.0, and Brain 3.0 and how your emotional state affects the way you feel, think, and act. On the following pages are three worksheets to help you reflect on your specific triggers. For each trigger, a list of questions will guide you to reflect on what unfolds internally when you are triggered in this manner. Completing this exercise should make it easier for you to assess your emotional state by tuning in to observe the physical and emotional signals in your body and the stories unfolding in your mind. An example has been provided in each table to help illustrate how to fill them in.

Post-Exercise Reflection

To further hardwire this learning experience, please take time to write your answers to the following questions:

1. What did you learn about yourself from identifying your triggers?
2. What stood out most to you?
3. How can you do more of the activities that bring you into Brain 3.0?

Additional Comments on Knowing Your Brain States

These reflection exercises enable you to activate Brain 3.0 to slow down and unveil how your autopilot simulates the world and constructs emotions in line with how it interprets what is happening. As you become conscious of your triggers for Brain 1.0 and Brain 2.0, the triggers start to lose their power over

Brain 1.0: Self-Preservation

Triggers that activate Brain 1.0 (my Inner Godzilla)	Trigger example: Being stuck in rush-hour traffic	Trigger 1	Trigger 2	Trigger 3	Trigger 4
Physical sensations: what happens in my body?	I clench my hands; my arms and shoulders get tense; I have knots in my stomach	....	....	....	....
What emotions do I feel?	Irritated and impatient, anxious	....	....	....	....
What thoughts and stories play out in my mind?	I think people who cut me off are jerks; I imagine worst-case scenarios about being late	....	....	....	....
How do I behave differently from normal?	I switch lanes, honk a lot, and run more yellow and red lights than usual	....	....	....	....
How do I treat people?	I get mad at slow pedestrians, yell at other drivers, and don't let cars pass in front of me	....	....	....	....

Brain 2.0: Reward and Acquisition

Triggers that activate Brain 2.0 (my Inner Teen Wolf)	Trigger example: An all-you-can-eat buffet	Trigger 1	Trigger 2	Trigger 3	Trigger 4
Physical sensations: what happens in my body?	My mouth waters and stomach rumbles	....			....
What emotions do I feel?	Anticipation and craving	....			....
What thoughts and stories play out in my mind?	I want to get my money's worth and binge on my favorite foods	....			....
How do I behave differently from normal?	I eat two to three times more than normal, like there is no tomorrow	....			....
How do I treat people?	I compete to grab my fill of the best foods before they run out	....			....

Brain 3.0: Self-Mastery and Well-Being

Triggers that activate Brain 3.0 (my Inner Sage)	Trigger example: Witnessing someone do an act of kindness	Trigger 1	Trigger 2	Trigger 3	Trigger 4
Physical sensations: what happens in my body?	My heart feels warm and fuzzy, my chest opens, my eyes water	....	....	....	....
What emotions do I feel?	Elevation, inspiration, admiration, connectedness	....	....	....	....
What thoughts and stories play out in my mind?	I admire that person and want to be more like him or her	....	....	....	....
How do I behave differently from normal?	I smile and don't sweat the small stuff	....	....	....	....
How do I treat people?	I become nicer, friendlier, and warmer, even to strangers	....	....	....	....

your autopilot. Eventually, with increased awareness, you will be able use Brain 3.0 to observe Brain 1.0 get triggered and be able to slow down the freeze-flight-fight cascade so you are not completely swept into Inner Godzilla mode. You will also be able to use Brain 3.0 to observe Brain 2.0 get triggered and be able to create space to feel cravings and impulses arise and pass away without being taken over by your Inner Teen Wolf's urge for immediate gratification. Furthermore, paying attention to what triggers bring you into Brain 3.0 will enable you to intentionally spend more time doing and savoring life-giving activities and priming yourself to feel, think, and act like your Inner Sage.

Now that you are aware of your triggers, the next step is to cultivate mindfulness and to learn how to create space to curiously observe, feel, and release the energies unleashed by triggers for Brain 1.0 and Brain 2.0 as raw sensations, feelings, and impulses instead of suppressing, resisting, or denying them.

Mind Your Autopilot

Virtually all of our suffering comes from our mindlessness.

—ELLEN LANGER, psychologist

Between stimulus and response there is a space. In that space is our power to choose our response. In our response lies our growth and our freedom.

—ATTRIBUTED TO VIKTOR FRANKL, psychiatrist and Holocaust survivor[19]

Carefully watch your thoughts, for they become your words. Manage and watch your words, for they will become your actions. Consider and judge your actions, for they have become your habits. Acknowledge and watch your habits, for they shall become your values. Understand and embrace your values, for they become your destiny.   —ATTRIBUTED TO MAHATMA GANDHI

Here is where I want to make clear that we should not see the autopilot as a negative. Rather, the autopilot should be celebrated and appreciated as a miraculous feat of evolution. The autopilot is a sophisticated, energy-efficient mechanism for storing skills and habits. To illustrate, when we first learn a skill, it takes a lot of conscious effort and concentration to perform it. It feels awkward and uncomfortable because the neural pathways to carry out the skill are being fired and wired together for the first time. Then as we repeatedly practice the skill, it takes less effort and concentration to carry out the actions because these neural pathways are getting stronger. Eventually, when we become very good at the skill, we can perform it with no effort at all because the neural pathways to perform the skill have gotten hardwired into the autopilot (into the basal ganglia). This frees up energy and attentional bandwidth in our prefrontal cortex to focus on more pressing matters. Thus, the autopilot is what makes it possible for us to do more than one thing at a time, such as walking and talking, driving and listening to a podcast, and taking notes while interviewing someone.

Nevertheless, it is essential that we mind our autopilots for two reasons. First, when we spend most of our lives on autopilot, we can end up going through the motions of our daily routines in a mindless, robotic state in which we miss out on enjoying and savoring life. Second, while there are a lot of great skills and habits that we intentionally develop and store in our autopilot, it also holds plenty of not-so-great habits and routines that we mindlessly absorb from our environment and experiences. As long as what we do on autopilot remains below conscious awareness in our shadow, we cannot do anything to change it.

Earlier in chapter 5, when I introduced key ideas from Daniel Kahneman's book *Thinking, Fast and Slow*, I explained that Brain 1.0 and Brain 2.0 are part of the fast-thinking autopilot (System 1), while Brain 3.0 is linked to the slow-thinking pilot (System 2). I also shared that in order to guide and retrain the fast-thinking autopilot, we have to strengthen the "muscles" of the slow-thinking pilot, which are attention and awareness. Mindfulness develops and strengthens the neural networks that enable us to focus attention and open awareness. With mindfulness we can observe our autopilot and identify habits and patterns that do not serve us, and hardwire new habits and patterns that better serve us. The more we use the neural networks that support mindfulness, the stronger they become. Mindfulness practiced over the long term can eventually change the neural structure of the brain such that present-moment awareness becomes hardwired into a natural default setting.

Calm Clarity's presentation of mindfulness builds on integrating scientific findings with my experiences during the Vipassana retreat in Dharamsala and my ongoing practice, and with insights gained from reading the Mahasatipatthana Sutta ("The Great Discourse on the Establishing of Mindfulness"), in which Siddhartha Gautama expounds on the concept of Samma Sati, which is the seventh component of the Noble Eightfold Path.[20] Samma Sati is often translated as "Right Mindfulness," but this term does not capture the full depth and weight of the meaning, which is more like "conscious remembering, application, and discernment of higher truth and wisdom."

To provide a synopsis, the Mahasatipatthana Sutta contains Siddhartha Gautama's instruction on how to develop the ability to unpack perception, so we can attend to the raw sensory input that comes in through our sense organs and distinguish these sensations from our mind's reaction to and interpretation of them. This ability then enables us to calm the whirlwind of the mind and develop equanimity from craving and aversion, thus gaining freedom from self-generated suffering.

The sutta begins with Siddhartha Gautama instructing people to pay attention to the body in and of itself by observing bodily phenomena. He instructs people to first observe themselves breathing, then walking, standing, sitting, and lying down, and then in all their daily movements, actions, and activities. Then he instructs them to meditate on the various components that make up the body, and on the fact that bodies are by nature impermanent and destined to decay.

Next, he adds more granular detail on how to observe one's inner experience. He instructs people to pay attention to and discern sensations and feelings in and of themselves, and how they instinctively categorize the sensations and feelings as pleasant, painful, or neutral (neither pleasant nor painful). He then instructs people to pay attention to and discern the mind in and of itself, noticing when the mind is affected by desire, aversion, and delusion, and when the mind is restricted, scattered, expansive, transcendent or elevated, and concentrated.

Next, he instructs people to pay attention to and discern mental phenomena. He begins by touching upon five hindrances to spiritual growth: craving, ill will (rejection/aversion), sloth (sleepiness/sluggishness), restlessness (distractedness), and wavering doubt—not suprisingly, these represent different aspects of being in Brain 1.0 or Brain 2.0. He simply tells people to honestly recognize and acknowledge when these factors are present within them (rather than suppress and deny that these states have arisen) and to also acknowledge when they are not present. Similarly, he instructs people to honestly recognize when seven factors of awakening are present within themselves: mindfulness (sati), investigation of phenomena (dhammavicaya), energy and determination (viriya), joy and rapture (piti), inner peace (passadhi), concentration (samadhi), and equanimity (uppekha).

In addition, he instructs people to pay attention to and unpack subjective experience by breaking down the experience of life into five component parts, which in Buddhist terminology are called the "Five Skandhas," often translated as the "Five Aggregates." Interestingly this ancient framework actually closely aligns with the modern scientific understanding of perception, which I will use to explain the five items here. The Five Aggregates are: (1) rupa, which means form or matter (the body and its component parts); (2) vedana, which means sensation and feeling (the interaction of stimulus with a sense organ) and the immediate affective classification as pleasant, unpleasant, or neutral; (3) samjna, which means perception and recognition (the processing of sensory information in the brain); (4) samskara, which means the preconditioned patterns (how neural networks have fired and wired together) and priming that drive cognition (thinking and mental activity) through associative activation; and (5) vijnana, which means sentience and consciousness (the subjective experience of a stream of consciousness).[21]

After explaining the Five Aggregates, he then dives deeper into sensory perception, instructing people to observe and discern the relationship between the sense organs, the object being perceived, and the sensations that arise when the object is perceived and recognized. Then they can observe how a "fetter" of aversion or of craving and attachment arises from the mind's reaction to the sensation, rather than the object itself. Only by observing and understanding this process can a person release and "abandon" the fetter. (In my experience, my fetters take the form of a story embedded in Brain 2.0 about how I *need* something to happen in a very specific way to be happy and that if I don't get what I *need*, I have a reason to be upset.)

In the final sections of the sutta, Siddhartha Gautama explains how to apply Samma Sati to meditate on the Four Noble Truths and gain a deeper direct understanding of these teachings. He instructs people to observe how distress and suffering arise from triggers that bring on craving, and from yearning for pleasant physical sensations, emotions, feelings, expectations, and desires that cannot be realistically fulfilled. Then he instructs people to observe how cravings are initially formed from pleasurable experiences. By doing this, we see it is possible for cravings to be abandoned at the point when they arise. Then he explains that we can cultivate the ability to release our attachment to fulfilling cravings by following the Noble Eightfold Path, which, as I shared earlier in chapter 5, is essentially a road map for cultivating habits and routines that strengthen Brain 3.0 and bring one into alignment and integration with one's Higher Self (or Buddha Nature, the term preferred by Buddhists).

This teaching becomes even more powerful when overlaid with the latest understanding from neuroscience. Lisa Feldman Barrett explains, "the brain is not a simple machine reacting to stimuli in the outside world. It's structured as billions of prediction loops creating intrinsic brain activity. Visual predictions, auditory predictions, gustatory (taste) predictions, somatosensory (touch) predictions, olfactory (smell) predictions, and motor predictions travel throughout the brain, influencing and constraining one another. These predictions are held in check by sensory input from the outside world, which your brain may prioritize or ignore." According to Barrett, we can experience reality only through our brain's simulation of reality because this ongoing prediction process influences our perceptions—our simulations actually change how our sensory cortices process sensory information.

To illustrate, she explains that in nature, a rainbow is a continuous spectrum of light without any stripes; however, humans tend to perceive a rainbow as having distinct stripes of color, and those stripes of color are influenced by our cultural conditioning. For example, Americans are conditioned to see the rainbow as having six colors (red, orange, yellow, green, blue, violet), whereas Russians are conditioned to see the rainbow as having seven colors because they distinguish between light blue and dark blue, and in turn, actually perceive these colors in a rainbow.[22]

The brain is continuously constructing and updating our mental model of the world through prediction and correcting prediction error. As the brain simulates reality, "meaningful" information gets tagged with previously associated affective signals (pleasant = approach; unpleasant or painful = avoid) in such a way that we can quickly interpret and react to the information. Anything that the prediction got wrong gets marked as "salient," so that our attention homes in on that information in order to learn and more accurately simulate this scenario in the future. Barrett explains that the brain also has another alternative: to be "stubborn" and filter the sensory input to make it consistent with the original prediction. This means that there is clearly a risk that our construction of the world may fall out of sync with what is actually unfolding. If we get attached to a certain worldview, we can force the sensory data to support and reinforce our beliefs, opinions, and stories, rather than invest energy in updating our construction of the world. Thus, Barrett cautions, "your brain is wired for delusion."

She further explains that predictions "not only create your perception and action but also explain the meaning of your sensations."[23]

The cultivation of Samma Sati provides a means to observe and deconstruct how the mind simulates and constructs reality, and in so doing, we become aware of inaccurate simulations that no longer serve us, which take the form of beliefs, opinions, stories, and worldviews that are really habitual mental and behavioral patterns. We learn to unpack and unwind the simulation process so that we stop forcing sensory data to reinforce our beliefs, opinions, stories, and worldviews, and thus become able to discern when our simulations are not aligned with reality.

In the Vipassana, or insight, meditation tradition as it is taught today, the conventional instruction approach involves telling people to sit quietly, usually cross-legged on the floor, and focus their attention entirely on their breathing. Inevitably, the mind will wander, so meditators have to vigilantly keep guiding their attention back to the breath. Almost everyone gets extremely frustrated because no matter how hard they try to focus on the breath, they find that their mind jumps all over the place. This is what is called the "monkey mind." Unfortunately, the conventional instruction often gives people a misguided notion that mind wandering is a problem, so people fall into the trap of getting caught up in wrestling matches with their monkey mind that they cannot win. Then struggling with the monkey mind becomes a distraction that prevents them from experiencing the essence of Samma Sati. At this point, it is normal for people to conclude that because they could not follow the instructions, meditation is a pointless exercise for them to continue (like I did when I was younger).

When I finally decided to learn meditation, as a brain geek, I found it interesting that most meditation teachers don't explain up front that the mind is supposed to wander. In fact, researchers have found that the typical mind wanders within twelve seconds, and they hypothesize that mind wandering evolved as a way to give our executive functioning neural pathways a break to refuel and to allow new insights to surface in consciousness.[24] As mentioned in chapter 5, the areas of the brain that activate during mind wandering correspond to the default mode network (DMN), so named because scientists discovered that whenever subjects in MRI scanning machines take a break in between tasks during a brain imaging study, this neural network seems to consistently turn "on" almost like a default setting. Intriguingly, according to Scott Barry

Kaufman, director of the Imagination Institute at the University of Pennsylvania, the DMN is also what scientists have identified as the "imagination" network.[25] This means that in insight meditation (a.k.a. Vipassana), the very thing people struggle with—mind wandering—could ironically be the underlying mechanism that brings about the chief aim of the practice: insight.

The discovery of the DMN about two decades ago revealed that even when people are not doing any activity, the brain is still continuously consuming energy. Therefore, scientists wanted to know: what function(s) does the DMN accomplish as it burns all this energy? Researchers soon realized that the DMN is active during mind wandering, and that people's minds don't wander randomly, but rather, they usually wander to thoughts about themselves and other people—what psychologists call "social cognition."[26] Today, scientists have found that the DMN plays a primary function in social cognition, self-conception (the process of building a sense of self and answering the question, "who am I?"), planning, imagination, daydreaming, and rumination.

According to Barrett, the DMN is a key part of the brain's simulation engine, helping to synthesize past experiences into simulations that anticipate the future; a key region that is activated during mind wandering is the ventromedial prefrontal cortex, which, as I shared in chapter 7, is involved in reading and processing emotions in other people and ourselves, and in the experience of wisdom, intuition, and grace.[27] In line with this, Martin Seligman explains that what happens when the mind wanders is that "it's simulating future possibilities. That's how you can respond so quickly to unexpected developments. What may feel like a primitive intuition, a gut feeling, is made possible by those previous simulations."[28] The way evolutionary biologists see it: what evolution has chosen as our default mode must be a critically important activity. Therefore, now that the evidence shows that our brains evolved to process and synthesize information that enables us to navigate our social world, anticipate the future, and imagine new possibilities as its default mode, it simply doesn't make sense to see mind wandering categorically as negative.[29]

Nevertheless, there are pitfalls in the way that the DMN tends to operate. According to Kahneman, one of the built-in shortcomings of the fast-thinking autopilot is that it has a tendency to create coherence from partial information by filling in missing information through weaving stories, which often involves making unjustified assumptions and jumping quickly to mistaken conclusions.[30] This internal storytelling continuously fuels a tornado of mental activity into which we get lost inside our heads. Often the voice inside our head projects a fantastic narrative that is far removed from what is actually unfolding right in front of us in the present moment. By getting swept away in our internal ramblings, we become ungrounded from our bodies and spin out into the past or the future. Because of these mechanisms, we spend much of our lives disconnected from the present moment.

Given this knowledge of the DMN, I reasoned that it was futile to fight against how our brains are evolved to function. Instead, Samma Sati must stem from becoming familiar with how our minds actually function and unpacking the process. According to Yongey Mingyur Rinpoche, the fact that the Tibetan word for meditation, "gom," means "to become familiar with" supports this notion of meditation as a process for becoming familiar with one's mind and the way the mind constructs reality.[31]

Today, neuroscience can greatly enhance our understanding of the human mind by shedding light on how the brain works through the associative activation of neural networks. To understand our own brain activity, it helps to see that thoughts correspond to neurons firing (though this does not mean that consciousness can be reduced to neuronal activity, because we don't yet understand how neuronal activity creates thoughts—and there is still the possibility that the process unfolds the other way around, that thoughts somehow cause specific neurons to fire and wire together and specialize to take on particular functions). Then we can understand the monkey mind as continuous chain reactions of neurons firing as the DMN runs simulations. Activity in the DMN is constantly fueled by incoming sensory input interacting with thoughts already in our mind. This in turn triggers neural pathways for associated memories and ideas to fire and feed into the DMN's simulations. When our mind wanders, our brain is simply doing what it was designed to do: use spare neural capacity to run simulations that help us make sense of our experiences, learn and prepare for what might happen next, and decide how to interact with other people.

In my perspective, Samma Sati involves developing three interrelated skills: metacognition, interoception, and proprioception, all of which I touched upon in chapter 5. To provide a quick review, the word "metacognition" etymologically means thinking about thinking. It involves knowledge, awareness, and understanding of one's own thought processes and the ability to steer and regulate one's thinking. Metacognition is strengthened by developing the capacity of the pilot (the slow-thinking System 2) to observe and retrain the autopilot (the fast-thinking System 1). A critical component of metacognition is interoception, the ability to sense and appraise internal bodily signals. When we consciously tune in to our feelings and emotions, we are using the interoceptive system to read and respond to messages from our body. When we move, another critical component of metacognition is proprioception, the ability to sense the relative position of all the parts of the body, as well as the strength and effort we are using to coordinate specific movements.

Neuroscience has now provided preliminary confirmation that long-term meditators have structural differences in brain areas associated with metacognition and interoception (MRI machines do not lend themselves well to studying proprioception). In 2005, Sara Lazar at Harvard Medical School conducted a groundbreaking study comparing the brain structure of long-term insight meditation practitioners with a control group of nonmeditators matched for gender, age, education, and race. She found significantly greater thickness in areas of the prefrontal cortex called Brodmann area 9 and 10 (the frontmost region of the brain), which play a key role in metacognition, executive functioning, and social cognition.* Furthermore, these areas of the prefrontal cortex tend to thin as people age, but astonishingly, the long-term meditators did not show any age-related decreases, while the control group did, which suggested that meditation may slow age-related brain degeneration. Not surprisingly, the study found the greatest between-group difference in cortical thickness in the right insula, a key brain structure involved in processing interoceptive signals and generating body maps. Lazar also found greater thickness in a region of the somatosensory cortex, another structure that plays an important role in interoception and proprioception.[32]

In 2007, a study titled "Attending to the Present: Mindfulness Meditation Reveals Distinct Neural Modes of Self-Reference" by Norman Farb and his colleagues at the University of Toronto shed new light on the neural mechanisms involved with mindfulness practices. The team designed the study to explore the proposed theory that there are two neurally distinct modes of self-reference: a self-centered narrative focus (NF) mode, which could also be called "me-talk," and an experiential focus (EF) mode, in which we pay

---

* Brodmann area 9 and 10 enable us to construct and revise mental maps, read people, effectively navigate our social worlds, and make moral and emotional decisions.

attention to the present moment and have enhanced sensory awareness. The researchers hypothesized that because mindfulness practices must correlate to EF mode, as it involves tuning in to both interoception (inner bodily senses) and exteroception (externally oriented senses of sight, sound, touch, smell, and taste), they might be able to distinguish which neural networks are involved in EF mode versus NF mode by imaging the brains of people practicing mindfulness. Therefore, they recruited people who had signed up for an eight-week mindfulness-based stress reduction (MBSR) course offered at a nearby hospital and randomly assigned them to either a pre-MBSR-training wait-list control group or a post-MBSR-training experimental group. The researchers had both groups of participants adopt a narrative focus and an experiential focus while they monitored the participants' brain activity in an fMRI scanner.

In line with their hypothesis, they found a seesaw effect between the brain networks for NF and EF, such that engaging in NF mode increases activity in the default mode network (DMN) and reduces activity in the sensory awareness system (which includes the insula and right lateral prefrontal cortex), and engaging in EF mode reduces activity in the DMN and increases activity in the sensory awareness system. In the control group, because the participants had a harder time sustaining EF mode as their minds wandered, the reduction of activity in the DMN in EF mode was relatively restricted. In the post-MBSR group, the seesaw effect was much greater, confirming that the post-MBSR group was more aware which mode they were in, could more easily shift between NF and EF modes, and could more reliably sustain EF mode.

Interestingly, the MBSR group also showed distinct changes in brain structure before and after the training. Before MBSR training, functional connectivity analysis showed a strong coupling between the right insula and a key structure of the DMN, suggesting a default connectivity whereby interoceptive signals trigger the activation of the DMN. After the MBSR training, the right insula had become decoupled from the DMN and instead showed an increased coupling with the dorsolateral prefrontal cortices, a key part of the brain's braking system involved with controlling attentional resources. The researchers determined that this cortical reorganization was consistent with the observed shift away from self-centered "me-talk" toward a nonnarrative, present-moment sensory awareness.[33]

A growing body of neuroscientific evidence indicates that it is possible that the quieting of the inner chatter experienced by meditators is tied to the rewiring of the DMN, which possibly makes it easier to shift the DMN into Brain 3.0. A 2011 study led by Judson Brewer at Yale University found that "the main nodes of the default-mode network (medial prefrontal and posterior cingulate cortices) were relatively deactivated in experienced meditators" and that experienced meditators had stronger functional connectivity between the posterior cingulate cortex (a key structure in the DMN that activates during mind wandering) and the dorsal anterior cingulate (a key part of the salience network involved in conflict monitoring), and between the posterior cingulate cortex and the dorsolateral prefrontal cortices (involved with controlling attentional resources), both at baseline and during meditation. The researchers wrote, "the consistency of connectivity across both meditation and baseline periods suggests that meditation practice may transform the resting-state experience into one that resembles a meditative state, and as such, is a more present-centered default mode."[34]

In line with these findings, a 2012 study conducted by Wendy Hasenkamp at Emory University also found that experienced meditators exhibited increased connectivity within attentional networks, as well as between attentional regions and medial frontal regions (key structures of the DMN).[35] According to Hasenkamp, these brain changes "might explain how it feels easier to 'drop' thoughts as you become more experienced in meditation—and thus better able to focus. Thoughts become less sticky because your brain gets re-wired to be better at recognizing and disengaging from mind-wandering."[36] How this translates into everyday life is that these changes to the brain make it much easier for us to catch ourselves ruminating and worrying in Brain 1.0, or becoming distracted by rewards (that don't actually align with our well-being and priorities) in Brain 2.0, and then disengage from these thoughts to shift gears into Brain 3.0.

A 2013 study conducted by a team at the University of Montreal compared the functional connectivity of DMN regions between experienced meditators and novice meditators in a resting state. They found that the experienced meditators had weaker functional connectivity between DMN regions involved in self-referential processing and emotional judgments (the dorsomedial PFC and the ventromedial PFC). They also found increased connectivity between the dorsomedial PFC and right inferior parietal lobule (IPL), which scientists hypothesize enables us to construct a high-level spatial representational system and plays a primary role in maintaining focus on a task in the face of distraction, detecting salient new events, and flexibly switching tasks as external demands change.[37] Therefore, these connectivity changes likely support the reduction in self-referential thinking (such as rumination) and the enhanced global awareness and present-moment awareness observed in experienced meditators in a restful state.[38]

Since proprioception usually involves body movement, it is at present rather challenging to study the neural correlates of proprioceptive ability because MRI scanners require subjects to be very, very still. Scientists believe proprioceptive awareness is embedded in the body representation system (BRS), which comprises two networks: (1) the sensorimotor control network (which includes the motor and somatosensory cortical regions, basal ganglia, thalamus, and cerebellum) that contributes to the formation of body representations and fast corrections of movement and (2) the frontoparietal network (which includes portions of the lateral prefrontal cortex and posterior parietal cortex) that integrates exteroceptive environmental information (mainly from sight, touch, and sound) with bodily signals (such as balance and positioning) to construct a mental representation of the body in an environmental context.[39] (Virtual reality technologies work by manipulating our sense of proprioception.)

Preliminary insights into the impact of mindful movement on brain structure come from a 2013 brain imaging study conducted by the National Center for Complementary and Alternative Medicine (NCCAM) at the National Institutes of Health, which found that yoga practitioners could keep their hands in a freezing cold water bath more than twice as long as the control group, and that cold pain tolerance correlated positively with increased gray matter volume (GMV) in the left and right insula. They also found that the yogis had larger brain volume in areas also associated with the body representation system, such as the somatosensory cortex, supplementary motor area, the superior and inferior parietal lobules (also involved in directing attention), and the visual cortex (involved in visualization). Interestingly, the number of years of yoga practice correlated with greater GMV in the left insula, which scientists hypothesize may be associated with parasympathetic activity, positive affect, and affiliative emotions. These findings suggest that increased pain tolerance in experienced yoga practitioners may be a consequence of adaptive insular changes, mediated by increased parasympathetic activity and interoceptive processing.[40]

Thus, these initial findings from neuroscience support that mindfulness meditations and mindful movement practices provide a very effective means to develop metacognition, interoception, and proprioception by carving out time for us to just observe the raw sensory input and sensations that come into our sensory nervous system; notice what they trigger in terms of emotions, feelings, memories, thoughts, and other sensations; and notice what stories arise in our minds like a domino effect. By doing these practices, we learn how our autopilot works by attentively observing how perception and subjective experience unfold in the embodied mind. By observing that sensations, feelings, and thoughts continuously arise and pass away (as neurons fire and quiet), we lose our attachment or aversion to these phenomena. By learning to detach from the stories woven by the voice inside our heads, we eventually develop equanimity and inner freedom. As the tornado of activity inside the mind calms because we stop fueling it, this allows us to be in tune with the present moment and experience a higher consciousness.

The dramatic changes that I have personally experienced and the studies I have referenced have led me to conclude that Samma Sati arises from rewiring the brain so that moment-to-moment sensory awareness replaces narrative "me-talk" (such as rumination and worry) as a default state. As someone who once struggled with depression and anxiety, I used to have a very strong tendency to get swept up in epic emotional dramas inside my head because my default state was to take everything that happened so seriously and so personally. My mind would then end up ruminating obsessively on things that were going wrong in my own life and things that were going wrong in the world. Because I was reacting to stories in my past, I could not actually be in the present moment with what was unfolding, which rendered me rather ineffective at handling the challenging circumstances I was in. Because my mind was stuck re-creating the past (and at times, reenacting trauma), it was often impossible for me to take in new information and respond with wisdom to what was actually unfolding before me.

Since doing the Vipassana retreat in 2012 and practicing meditation ever since, I no longer get stuck in torrents of mind-racing rumination and catastrophic simulations of the future. I have found that tuning in to my interoceptive and proprioceptive systems naturally grounds me in my own body. This is important because it is by being grounded in my body that I can experience the present moment. With metacognition, I'm able to recognize sensations as sensations, feelings as feelings, thoughts as thoughts, and stories as stories, and as a result, I'm no longer lost in a whirlwind of mental activity.

Together, metacognition, interoception, and proprioception enable me to activate Brain 3.0 and create space for Samma Sati to emerge as calmness and clarity.

When I used to hear people talk about "beginner's mind" to describe a Zen mindset, I had only processed it intellectually to mean that experts can be bogged down by their expertise, such that their knowledge and ingrained ways of thinking impede curiosity, openness, and creative problem-solving. This phrase only started to really make embodied sense after I learned how to reduce the inner chatter of my mind by tuning in to sensory awareness. Once my mind was free of distracting inner chatter and could actually be aware of and take in the present moment, I realized that beginner's mind is more like experiencing life through the eyes of a child free of preconceptions and filled with curiosity and wonder. It is like having the mind become a clean white canvas on which the brushstrokes of my inner wisdom can clearly be perceived and appreciated. Furthermore, quieting the inner chatter unleashes an internal wellspring of joy, energy, and creativity.

In designing mindfulness practices for the Calm Clarity Mindful Leadership Program, I intentionally "hacked" traditional mindfulness practices to give people a taste of Samma Sati. I've seen time and time again that after I guide participants to directly experience what freedom from their inner stories feels like, and how it enables them to respond more effectively to what's in front of them, they are naturally inspired to continue the practices on their own. This has confirmed for me that when people experience what it is like to quiet the default mode network and create space for inner wisdom to arise, they experience a permanent shift in the way they see themselves and each moment of life.

In the following pages, I provide instructions on the Calm Clarity version of three common mindfulness practices: focused attention and open awareness meditation, mindful walking, and mindful eating. I recommend you try these practices in the sequence provided, as each exercise will build on the previous ones.

Calm Clarity Focused Attention and Open Awareness Meditation

The Calm Clarity Focused Attention and Open Awareness Meditation is a condensed Vipassana practice that is designed to enhance metacognition, executive functioning, focused attention, and sensory awareness in a short period of time. It begins by having you focus your attention on your breath; then shifting your focus to different senses one by one, such as hearing, touch, and interoception; and then opening up your awareness to all sensory input coming in at once. This is a helpful exercise for people with short attention spans to strengthen their neural capacity for focus so they can better concentrate without getting sidelined by distractions.

Feel free to try this meditation with your eyes closed and with your eyes open. Most people start meditation as an isolated sitting activity. Eventually, as these neural pathways get stronger, you'll be able to bring more metacognitive and interoceptive awareness into all your daily activities, which means you can continue practicing mindfulness while standing, walking, eating, washing the dishes, doing the laundry, buying groceries, writing a memo, listening to someone, participating in a meeting, etc.

For your convenience, a guided recording of this meditation can be found at calmclarity.org.

Instructions for the Calm Clarity Focused Attention and Open Awareness Meditation

1. POSTURE

Assume a comfortable meditation posture in which your spine from the top of your neck to your sacrum is straight and aligned, and your chest is open. Place the tip of your tongue on your palate.

2. DEEP BREATHING

Begin with slow, deep breathing. Start by exhaling to empty the lungs.

Inhale slowly to a count of six. Hold for a count of three. Exhale to a count of six. Hold for a count of three.

Do two more cycles and then normalize your breathing.

3. TUNING IN

Gently sense how your body feels in your seat. Take notice of how gravity pulls your weight to the ground. Try to settle into a posture in which it's effortless to keep your body and neck straight and aligned. Allow your muscles to relax. See if you can release the tension in your body.

4. FOCUSED ATTENTION ON BREATHING

Gently focus your awareness on your breathing while your body relaxes. Feel the physical sensations of breathing—like the air coming into and out of your nostrils, how your breath blows on the area above your mouth, and the gentle rise and fall of your rib cage and your belly as your lungs expand and contract. Now choose one of these sensations to focus on. This will be your anchor. Whenever your mind wanders, come back to this anchor.

Thoughts, emotions, feelings, or sensations may arise as you focus on your breath. When they do, try to observe them without engaging, resisting, or judging them. See if you can calmly and peacefully acknowledge and accept what unfolds in your mind without any attachment or struggle. Then return your attention to your breath.

Whenever you notice yourself distracted, gracefully return your attention to your breathing. There is no need to berate yourself for having a wandering mind. It is part of being human. The important thing is to recognize that your mind has wandered and to gently redirect your attention back to the breath.

To be present with your breath, it helps to appreciate each breath. Thinking of each breath as a precious gift enables you to see each moment of life as a present.

Please observe in silence for several minutes. Whenever your mind wanders, simply use your anchor to come back to the present exercise.

5. FOCUSED ATTENTION ON HEARING

Now gently focus your attention on your sense of hearing. As you do this, observe how your mind perceives the sounds in the environment and any story that may arise. Is it possible to attend only to the raw sensation of sound? Please observe in silence for several minutes. Whenever your mind wanders, simply use your anchor to come back to the present exercise.

6. FOCUSED ATTENTION ON TOUCH

Now gently focus your attention on your sense of touch, the physical sensations on your skin, such as temperature, tension, weight, pressure, vibrations, or pulsations. As you do this, observe how your mind perceives these sensations and any story that may arise. Is it possible to attend only to the raw sensation of touch? Please observe in silence for several minutes.

Whenever your mind wanders, simply use your anchor to come back to the present exercise.

7. FOCUSED ATTENTION ON INNER SENSATIONS

Now gently focus your attention on your inner sensations. What does it feel like to be in your body? What does it feel like to be sitting? How does your posture feel? Are you slouching? Is your body leaning right or left? Are you balanced? Is there tension inside your body?

Observe how your mind perceives these inner sensations and any story that may arise. Is it possible to attend only to the raw interoceptive sensations without your mind adding a story? Please observe in silence for several minutes. Whenever your mind wanders, simply use your anchor to come back to the present exercise.

8. OPEN AWARENESS

Now open your awareness and notice the various forms of sensory input coming in through all your senses. Also observe how your mind perceives these sensations and any story that may arise. Is it possible to attend only to the raw sensations? Please observe in silence for several minutes. Whenever your mind wanders, simply use your anchor to come back to the present exercise.

9. SETTING INTENTIONS AND CLOSING

As you come to the end of this meditation, set an intention to continue being mindful of the present moment throughout your day by being aware of your thoughts, feelings, emotions, and physical sensations. Use your anchor to stay connected with the observer inside you.

Give yourself a big smile. Gently move your fingers, hands, toes, and feet. Feel free to give yourself a nice stretch. When you are ready, open your eyes.

Post-Meditation Reflection

To further hardwire this learning experience, please take time to write your answers to the following questions:

1. How was this experience? What insights, if any, arose?
2. What stood out most to you?
3. How can you incorporate this focused attention and open awareness meditation into your everyday life?

Mindful Walking

The majority of our routinized daily movement activities, such as walking and typing, are hardwired into the fast-thinking autopilot (in particular, the basal ganglia, a key structure of Brain 2.0), such that they don't require any conscious thinking or effort to carry out. Mindful walking, along with more complex mindful movement practices such as yoga and tai chi, enhance proprioception and interoception by intentionally bringing bodily sensations back into conscious awareness. In mindful walking, we slow the process down and break it apart until it feels like we are learning how to walk all over again.

Human beings aren't born with walking stored in our autopilot, a fact that can be observed by watching toddlers learn how to walk for the first time. When they start, they have no idea how to use their muscles to support and balance their body weight as they move one leg after the other, so they stumble and fall frequently. Each step requires intense concentration and focus. But once they get the hang of it and walk frequently, the movements become automatic, such that walking no longer requires active concentration. Within only a few months, walking becomes a nonconscious activity that they can do on autopilot, which frees up their conscious mind to think about other matters as they walk.

Instructions for Mindful Walking

In mindful walking, you will try to consciously focus your full attention on the sensations of walking, tuning in and observing how your body feels as you walk. Try to walk in silence, because engaging in conversation (or any other form of multitasking) will prompt you to walk in autopilot mode, such that you stop paying attention to your sense of proprioception and interoception.

In the beginning, it is best to choose a quiet, safe, secluded path, away from other people and moving vehicles. Some people choose to pace back and forth along a walkway of about ten feet, so they can pay attention to the walking process without a distracting goal of getting anywhere.

Since we evolved to walk in autopilot mode so we can multitask, whenever the mind wanders, we tend to slip out of mindful walking back into walking in autopilot mode. As soon as you notice your mind wandering, gently bring your full attention back to the mindful walking practice.

When you begin, it helps to move slowly so you can feel all the micro movements and sensations involved in taking each step. When people start mindful walking, they tend to focus their attention mainly on their feet. After a while, people notice how their entire body is involved in walking (muscles in the back, core, and neck are also flexing when people walk).

Observe how you take each step and all the muscular and skeletal movements involved in walking:

- How do your balance and weight shift?
- With each movement, which parts of your body bear your weight?
- As you move, which muscles contract and which relax?
- Where does your foot strike? When does the other foot come up?
- How is your gait? (Neutral, pigeon-toed, or duck-footed?)
- What are your arms doing?
- Where do your eyes look? (Hint: your eyes don't need to look at your feet to walk mindfully.)
- How is your posture?

Tune in to sense and feel what it is like to be in your body.

- How does being grounded in your body enhance your sensory awareness?
- How does mindful walking affect your sense of presence and connectedness?
- What emotions do you feel?

Try varying your walking rate.

- How does speed affect your ability to walk mindfully?
- With continued practice, can you walk mindfully at your normal speed?

Post-Mindful-Walking Reflection

To further hardwire this learning experience, please take time to write your answers to the following questions:

1. How was this experience compared to how you normally walk? What insights, if any, arose?
2. What stood out most to you?
3. How can you incorporate mindful walking, or mindful movement in general, into your everyday life?

Mindful Eating

Eating is another activity that we tend to assign to our autopilot in order to free up mind space to multitask. This seems like a productivity enhancer, but it comes at a great cost: we become insensitive to signals from our body telling us when we have had enough food, or enough sugar, starch, fat, and salt. We also miss signals from our bodies to eat the nutrients that it needs more of. Over time, eating on autopilot tends to result in overeating and a diet skewed toward foods that are low in nutrient content, such as fast food.

As with mindful walking, mindful eating requires us to slow the process down and break it apart so that we relearn to eat with beginner's mind. Mindful eating also enhances exteroception, interoception, and proprioception. As you follow these instructions, you will frequently find that whenever your mind wanders, your autopilot automatically kicks in to carry out your ingrained eating habits. It will take active, conscious effort to break your eating habits in order to follow the instructions.

Instructions for Mindful Eating

Try to eat in silence because engaging in conversation (or any other form of multitasking) will prompt you to eat in autopilot mode. Whenever your mind wanders, gently bring your full attention back to the mindful eating practice.

When you begin, it helps to move slowly so you can feel all the micro movements and sensations involved in each step of eating.

Engage your exteroceptive senses: sight, smell, touch, taste, and sound.

- Notice what your food looks like with your eyes.
- Notice how it smells with your nose.
- Depending on what you are eating, if it makes sense, touch the food and notice its texture in your hands.
- Notice the initial flavor and the sound as you bite into it.
- Notice its texture in your mouth and how the texture changes as you chew.
- Notice how different flavors get released as you chew.

Observe each step of the process. Notice what muscle movements are required for eating:

- How do your hands bring food to your mouth?
- How do your body and head move as you eat?
- How does your jaw work to bite and chew this particular food? How does the way you bite and chew vary with different types of food?
- How does your tongue work to taste food and move it around in your mouth?
- When do you swallow?
- How do the muscles of your throat contract to swallow the food?

Give yourself plenty of time to savor every bite. Wait until you finish each mouthful before reaching for the next portion.

Tune in to sense and feel what it is like to be in your body. Pay attention to all the flavors that are released and all the sensations in your body as you eat (and drink).

- How does being grounded in your body enhance your sensory awareness?
- How does mindful eating affect your sense of presence and connectedness?
- Can you feel yourself becoming full? Does your body tell you when you've had enough sugar, starch, fat, or salt?
- What emotions do you feel?

Try varying your eating rate.

- How does speed affect your ability to eat mindfully?
- With continued practice, can you eat mindfully at your normal eating rate?

Post-Mindful-Eating Reflection

To further hardwire this learning experience, please take time to write your answers to the following questions:

1. How was this experience compared to how you normally eat? What insights, if any, arose?
2. What stood out most to you?
3. How can you incorporate mindful eating into your everyday life?

Additional Comments on Mindfulness Practices

When you first begin to practice mindfulness and you are finally paying attention to how you switch between being in pilot mode and autopilot mode, you will be astounded by the frequency with which you slip back into autopilot mode. Just remember, you are not alone. All of us end up in autopilot mode whenever our mind wanders and we get unconsciously swept up in habitual patterns for thinking, feeling, and moving. When you notice that you are no longer consciously paying attention to your breath, your walking, or your eating (or any other activity you are applying mindfulness to), there is no reason to beat yourself up. Your brain is simply conserving energy by activating the DMN and switching into autopilot mode.

Please note that when you are tired or hungry and thus have too little energy to fuel Brain 3.0, fighting this built-in tendency to conserve energy is like swimming against a strong current. It is very hard to mindfully observe and retrain your autopilot when Brain 3.0 is at reduced functioning. Thus, practicing self-care may be a better use of your time and remaining energy because you need to refuel and regenerate Brain 3.0. Whenever you are tired or stressed and need to boost your vagus nerve and rev up your parasympathetic system, I would recommend doing the compassion meditation before doing any mindfulness practices.

People often ask me: what is the ideal amount of time and frequency to practice meditation? The truth is that there is no one-size-fits-all answer, as it depends on your personal situation, the structure of your brain (your connectome), what your intentions are, and how you define meditation. If you define meditation narrowly as a practice you do while sitting down with your eyes closed in a quiet space, this idea will limit how often you exercise the corresponding neural networks. Instead, if you expand your definition of meditation as a way of living with greater awareness and presence, then you'll realize that your entire life can become meditation.

I often tell people that meditation is to mindfulness what practice is to basketball, or rehearsal is to musical performance. This is very consistent with the explanation offered by Bhante Henepola Gunaratana (one of the most respected teachers of mindfulness in the Theravadan tradition living in the United States), who says that "the meditation you do on the cushion is your homework. The rest of life is your fieldwork. To practice mindfulness, you need both."[41] You need to activate and exercise the neural pathways for your sensory awareness system on a regular basis if you want to rewire your brain to quiet your monkey mind and improve your capacity to be grounded in present-moment awareness, even in the midst of stress. The easiest way to do that is to incorporate a variety of mindfulness practices into your daily routine, so that you create space to tune in regularly. (See the appendix for a template to help you plan a daily routine to activate Brain 3.0.)

As you develop your capacity for mindfulness, you may begin to experience how Samma Sati seems to enable your Higher Self to grow deeper and deeper roots into your cells. As that happens, tuning in to your body gives you a means to tune in to your soul. Ultimately, Samma Sati is not about what you do, but rather how you do what you do. Samma Sati means being present to what is unfolding, rather than trying to force something else to happen. Instead of resisting the flow of life, with Samma Sati, you can appreciate the gifts of each unfolding moment and tap into your inner wisdom to guide and harness the energies being released to move toward greater self-actualization. In the next section, I will talk about how to deepen the experience of Samma Sati by making the mind the object of meditation.

The Voice in Your Head

Your brain is wired for delusion: through continual prediction, you experience a world of your own creation that is held in check by the sensory world.
        —LISA FELDMAN BARRETT, neuroscientist and psychologist

Where there is perception, there is deception.   —THICH NHAT HANH

Mindfulness gives you time. Time gives you choices. Choices, skillfully made, lead to freedom. You don't have to be swept away by your feeling. You can respond with wisdom and kindness rather than habit and reactivity.
        —BHANTE HENEPOLA GUNARATANA[42]

Now that you have experienced how tuning in to bodily sensations naturally quiets the monkey mind by rewiring the default mode network, you can understand how mindfulness practices strengthen the interoceptive, proprioceptive, and sensory networks that ground you in the present moment. As these networks strengthen, you can now make your monkey mind the object of your mindfulness practice. Mindfully paying attention to the voice in your head will help you understand that thoughts arise and pass away as neural circuits activate and quiet.

As I shared in chapter 5, "mind-track" is the word I use to describe the stream-of-consciousness self-talk that unfolds inside our minds. We all experience the mind-track as an inner dialogue made up of thoughts that pop up into consciousness in a seemingly random manner as the people we interact with and the things we see, hear, touch, smell, and taste trigger neural pathways we've associated with them to activate. The mind-track comes about as a chain reaction of neural pathways firing and continuously triggering more neural pathways to fire. As you observe your mind-track, you will begin to cultivate the ability to see how your mind constructs simulations of the world and how these simulations are based on your preconditioned beliefs about the world. As you start to unpack these simulations, you will have a deeper understanding and experience of Samma Sati.

During the Vipassana meditation retreat, as I learned to activate Brain 3.0 to tune in and listen deeply to my mind-track as a messenger of insight, I noticed that my mind-track sounded very different depending on whether I was in Brain 1.0, 2.0, or 3.0. When I'm in Brain 3.0, my Inner Sage coaches and guides me on the path forward by helping me step back, come out of tunnel vision, and see the bigger picture. When my meditations bring me into Brain 3.0, sometimes I feel like my Higher Self uses my mind-track to help me see interconnections and interdependencies that show me how I am part of something much bigger than myself and to guide me to see and break out of conditioned patterns that no longer serve me.

In contrast, when I get triggered into Brain 1.0 and Brain 2.0, feeling stressed out tends to amplify an "Inner Critic" that points out the negative, beats up on myself, and criticizes other people. Unfortunately, listening to and believing this Inner Critic leads to my getting disconnected from my Higher Self. This negative inner voice is often also called the "Inner Saboteur" because it causes people to behave in self-sabotaging and self-destructive ways and serve as their own worst enemy. After reading David Rock's book *Your Brain at Work*, I realized that the volume of my Inner Critic is usually correlated to the degree of threat I feel in one of the five domains in his SCARF model: status, certainty, autonomy, relatedness, and fairness.[43] Interestingly, all five of these domains are related to our social world—how we see ourselves in relation to society or the people around us.

During the Vipassana meditation retreat, I also learned that by using mindfulness to activate Brain 3.0, I can use my "observer mode" to attend to my Inner Critic without resisting or struggling with it. By observing it over time, I began to appreciate that the Inner Critic is nothing more than a defense mechanism created by my ego to help me avoid repeating negative experiences I had in the past by replaying negative things people have said to me. One day it dawned on me that my Inner Critic is often just a broken record repeating warnings and reprimands my parents said throughout my childhood. Then I realized that, like my parents, my Inner Critic is trying to help protect me from feeling disappointed, hurt, or betrayed and from experiencing failure, embarrassment, shame, rejection, discrimination, and stigmatization by preemptively making me feel and anticipate those negative outcomes so I will avoid them.

These insights helped me to create space in Brain 3.0 to reflect on how to make peace with my Inner Critic (because fighting with it or ignoring it doesn't actually work). Eventually, I developed a counterintuitive approach that finally worked for me: turn the Inner Critic into an ally. What I do now is whenever I hear my Inner Critic, I thank it for trying to help and protect me. Then I take time to reflect on how this current situation is different from the past. I also take time to see how the Inner Critic's advice is not the best way to handle the situation by walking myself through what would happen if I followed its advice. Usually, what I see is that the course of action could make the situation worse or negatively affect other people and myself. Then I meditate on how the Inner Critic is simply a mirror reminding me of what I am afraid of so that I can become aware of and let go of the self-limiting belief or beliefs underlying the fear. As I meditate, eventually my mind-track will show me what self-limiting beliefs and experiences gave rise to the messages from my Inner Critic, and how these beliefs have driven and continue to subconsciously influence my behavior and emotional reactions.

In case it is helpful, I have included figure 9 to illustrate how our environment and experiences condition us to construct a system of beliefs about the world (our worldview), which in turn shapes our mindset, personality, and trajectory, and form unconscious life scripts that we mindlessly follow without questioning if there are other possibilities.

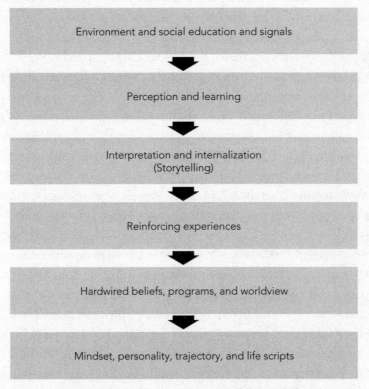

Figure 9: How conditioning shapes our beliefs and mindset

Even after we identify a self-limiting belief, the next part, shedding it, is not easy, especially if we are very attached to it and identify with it. Superficial approaches, such as doing positive affirmations ("I am strong and brave"; "I am beautiful"; "I am lovable"; etc.), don't work for me. The lack of authenticity makes these approaches feel to me like a state of denial or wishful thinking. As a lifelong defensive pessimist and skeptic, I have a very strong right prefrontal cortex (a.k.a. my bullshit detector) so it requires a more robust approach. What works for me is to use the strength of my analytical mind to unpack the self-limiting belief and replace it with a revised belief that is more aligned with the person I am today and my current worldview.

To do this, I meditate on what led me to construct this limiting belief in the first place, what experiences reinforced it, and how at one point that belief may have served me. I then think about how the self-limiting belief is now obsolete because it no longer matches my current situation. This then opens space in my mind to consider how to revise this belief or worldview to be more in line with the person I am today and the strengths I bring to the table. I know I have found the "right update" for the belief when I feel a shift in my energy toward eudaimonic joy. This happens when the revised belief is able to strongly activate my left prefrontal cortex to match the power of my right prefrontal cortex.

It can take time to assimilate a new belief into our worldview and our brain's simulations of reality. Oftentimes, when I get stuck on how to move forward, I find I need to intentionally turn "on" my Inner Sage for guidance to think about the situation from a higher perspective. To do this, I often picture a role model I look up to, such as Oprah Winfrey, Martin Luther King, Gandhi, Socrates, Siddhartha Gautama, Lao-Tzu, or Christ, and ask what he or she would do in my situation. Other times, I imagine a loved one in a similar situation and reflect on what advice I would give that person. When I'm really stuck, I try to purposefully activate the overview effect by turning toward my Higher Self to help me understand what is unfolding from a higher perspective. Then I tune in to my body as a biofeedback mechanism to sense how looking at the situation from these other perspectives makes me feel. If it lifts my spirits and opens my mind, then it is moving me in the right direction. Then I visualize what I would do if I listened to the advice of my Inner Sage, and how that would affect me and other people involved in the situation. Usually that visualization exercise naturally guides me to the course of action that leads to a positive outcome.

To help you build a healthy distance from the voice in your head and make peace with your Inner Critic, this section includes two exercises. The first exercise is a mindful breathing meditation to help you step back and observe your mind-track without getting lost in the tangle of thoughts, feelings, emotions, and sensations inside your head. By mindfully paying attention to your mind-track, you learn to see that you are not your thoughts, feelings, emotions, and sensations, and that there is no need to take them as truth or to personally identify with them. You see how your mind-track is made of associative cascades of neuronal activity. By watching the associative activation process, you will see patterns for how you've been conditioned to associate certain thoughts, feelings, emotions, ideas, images, memories, and actions together. By watching your mind-track, you can gather insights about your habitual mental patterns. Once you see these patterns, if you realize that any of your habits do not serve you, you can use mindfulness to change your habits by replacing them with new habits that better serve you.

The second exercise is a self-reflection worksheet that captures the process I use to turn my Inner Critic into an ally. The first half guides you to tune in to how the Inner Critic is trying to help you and then identify the self-limiting belief that is driving it. Then you can finally see and feel its negative impact on you (how it weighs you down and makes you feel anxious, isolated, and fragmented) and recognize that this belief was created from an outdated worldview. The second half guides you to further activate Brain 3.0 and get help from your Inner Sage to build new neural pathways and replace the self-limiting belief with a more inspiring and energizing belief that leads to integration and wholeness.

Calm Clarity Mindful Breathing Meditation

This meditation is designed to help you observe and understand your mind-track and detach from the voice in your head, which is often an Inner Critic. The goal is to improve the ability of the pilot to observe the autopilot without being swept away by habitual ways of perceiving, thinking, and feeling. It is very similar to the first part of the focused attention and open awareness meditation, but this time, the mind-track becomes part of the meditation. The key is to observe the mind-track with curiosity so you can learn about your individual conditioning. Like opening up the hood of a car and seeing how all the parts are assembled together, you will gain insight on how thoughts, ideas, beliefs, memories, feelings, emotions, and sensations have fired and wired together in your brain. This practice will further enhance metacognition, equanimity, and beginner's mind.

For your convenience, a guided recording of this meditation can be found at calmclarity.org.

Instructions for the Calm Clarity Mindful Breathing Meditation

1. POSTURE

Assume a comfortable meditation posture in which your spine from the top of your neck to your sacrum is straight and aligned and your chest is open. Place the tip of your tongue on your palate.

2. DEEP BREATHING

Begin with slow, deep breathing. Start by exhaling to empty the lungs.

Inhale slowly to a count of six. Hold for a count of three. Exhale to a count of six. Hold for a count of three.

Do two more cycles and then normalize your breathing.

3. TUNING IN

Gently sense how your body feels in your seat. Take notice of how gravity pulls your weight to the ground. Try to settle into a posture in which it's effortless to keep your body and neck straight and aligned. Allow your muscles to relax. See if you can release the tension in your body.

4. ATTENDING TO THE BREATH

Gently focus your awareness on your breathing while your body relaxes. Feel the physical sensations of breathing—like air coming into and out of your nostrils, how your breath blows on the area above your mouth, and the gentle rise and fall of your rib cage and your belly as your lungs expand and contract. Now choose one of these sensations to focus on. This will be your anchor.

To be present with your breath, it helps to appreciate each breath. Thinking of each breath as a precious gift enables you to see each moment of life as a present.

5. OBSERVING THE MIND-TRACK

By simply being the observer, you may notice thoughts, feelings, emotions, and physical sensations arise and pass away as neural circuits activate and then quiet.

As you focus on your breath, when thoughts, emotions, feelings, or sensations arise, see if you can observe them without engaging or resisting them, without judging them as good or bad. Then you can see they are not you. You are not your thoughts, feelings, emotions, or sensations. You don't have to get tangled up in them, identify with them, or wrestle with them. You can choose to let them arise as neural circuits fire, let them be, and let them pass away as the neural circuits naturally quiet.

Whenever your attention wanders, simply take note of what it has wandered to. See if you can calmly and peacefully acknowledge and accept what unfolds in your mind with detached curiosity, like a scientist. Then gently return your attention to your breath.

Can you connect what unfolds in your mind to how neurons have fired and wired together in your brain? As your monkey mind jumps around and makes associations, can you see patterns? What can you learn about yourself?

Continue to observe in silence for several minutes. Whenever you notice yourself distracted, simply use your anchor to come back to your breath and ground yourself in the present moment.

(Continue for anywhere from one minute to ten minutes or longer if you wish.)

6. SETTING INTENTIONS AND CLOSING

As you come to the end of this exercise, set an intention to continue being mindful of the present moment throughout your day by being aware of your thoughts, feelings, emotions, and physical sensations. Use your anchor to stay connected with the observer inside you.

Give yourself a big smile. Gently move your fingers, hands, toes, and feet. Feel free to give yourself a nice stretch. When you are ready, open your eyes.

Post-Meditation Reflection

To further hardwire this learning experience, please take time to write your answers to the following questions:

1. How was this experience? What insights, if any, arose?
2. What stood out most to you?
3. How can you practice mindful breathing and watch your mind with curiosity in your everyday life?

The Voice in Your Head Reflection Exercise

The following exercise captures how I turn my Inner Critic into an ally in helping me identify and change self-limiting beliefs. For example, I have a very strong fear of failure, such that every time I have to get out of my comfort zone, my Inner Critic comes up with thousands of ways that things can go wrong and warns me that I am not up to the task, that I am going to fail, and that people will think my work is awful or see that I'm incompetent. These messages put me on the verge of a panic attack—my heart races and my stomach tightens. If I were to listen to and believe these messages, I would run and hide in my room and never come out. I would never try anything new and would miss out on a lot of wonderful experiences. However, then I remember that my Inner Critic is only trying to protect me from experiencing shame and embarrassment because somehow I internalized the limiting beliefs that failing is catastrophic and that I must avoid it at any cost because failing would doom me as a failure for life. Underneath this fear of failure is a self-limiting belief that failure leads to social rejection, which makes me preemptively reject opportunities where there is potential for failure.

By looking at these beliefs, I realize they are not in line with reality. Failure and success are two sides of the same coin. For instance, venture capitalists build failure into their investment process, knowing that only one or two out of every ten start-ups they invest in will provide the return on investment they are looking to achieve. Most successful people say that they had to fail first to learn the lessons that enabled them to succeed. The most successful risk takers have learned to embrace rather than reject failure as an integral part of a learning curve.

Seeing failure from different perspectives then enables me to step back further and see my situation from a bigger-picture perspective. I then see a pattern that the underlying cause of my anxiety is a discomfort with uncertainty. My desire to control the outcome and to have certainty that I will succeed drives my anxiety because that level of control and certainty is not realistically possible. My Inner Sage then helps me to see that I have to befriend uncertainty by focusing not only on what can go wrong, but also on what can go well. Then I start to see that there are many more potential positive outcomes than negative outcomes that could come out of the situation, more possibilities than I can possibly anticipate or imagine. As a result, I start to feel in my heart that I want to go ahead with the project because I can now see the positive impact it can make on the larger community and the opportunities it will give me to grow and learn. I can then choose to move forward with a more open mind and heart and a stronger sense of resolution.

As you can see, you need to first develop some degree of mindfulness (in particular, strong neural networks for metacognition and interoception) to even notice that your Inner Critic is pulling you into Brain 1.0 and unleashing a freeze-flight-fight cascade. If you are in a state of high stress, I would recommend doing a few cycles of deep breathing to calm and recenter yourself, and if there is time, consider doing the compassion meditation or the mindful breathing meditation to prime Brain 3.0 before you do this reflection exercise.

The Voice in Your Head: Tune In to the Inner Critic and to the Inner Sage

I: Tune in to the Inner Critic	1. What is your Inner Critic telling you? What sensations and emotions does it bring up in your body?
	2. If you listen to and believe your Inner Critic, what would you do?
	3. What would be the impact on yourself and other people?

I: Tune in to the Inner Critic	4. How is your Inner Critic trying to help you? What is it trying to protect you from?
	5. What beliefs make you feel that this type of help or protection is needed?
II: Tune in to the Inner Sage	6. Are these beliefs out-of-date? Are they unrealistic? How would you revise them in a way that activates Brain 3.0?
	7. Ask your Inner Sage to help you see the situation from a higher perspective. What sensations and emotions does this perspective bring up in your body?
	8. If you listen to your Inner Sage, what would you do?
	9. What would be the impact on yourself and other people?
	10. How do you choose to move forward?

Post-Exercise Reflection

To further hardwire this learning experience, please take time to write your answers to the following questions:

1. How was this experience? What insights, if any, arose?
2. What stood out most to you?
3. How can you observe the voice in your head and turn it into an ally in your everyday life?

Additional Comments on the Mind-Track

For people like me, who have had many traumatic and painful experiences growing up, the default volume setting of the Inner Critic can be so loud that it drowns out everything else, including the Inner Sage. As we practice mindfulness, we develop present-moment sensory awareness by grounding ourselves in our bodies, tuning in to raw sensory information, and noticing thoughts and emotions without adding embellishing storylines to them. The effect is that of turning down the volume setting for the Inner Critic—it never really goes away, but it no longer dominates our field of attention. Then, in this quiet inner space, we may start to pick up the signal of a "higher" and more compassionate voice—that is the Inner Sage or the Higher Self emerging into conscious awareness. With this emergence, new insights and connections we've never seen before bubble to the surface of consciousness.

So long as the noise inside our minds is loud and cacophonous, we may never notice that our Higher Self is trying to guide us because the signal-to-noise ratio is too low. Although the Higher Self is always with us, it's hard to recognize or distinguish its "voice" when there is so much internal chatter distracting us. In the next section, I share a set of reflection exercises to further enhance the signal-to-noise ratio and strengthen your connection with your Higher Self.

Embody Your Higher Self

The greater danger for most of us lies not in setting our aim too high and falling short; but in setting our aim too low, and achieving our mark.
—ATTRIBUTED TO MICHELANGELO

You are not here merely to make a living. You are here in order to enable the world to live more amply, with greater vision, with a finer spirit of hope and achievement. You are here to enrich the world, and you impoverish yourself if you forget the errand. —WOODROW WILSON

Your time is limited, so don't waste it living someone else's life. Don't be trapped by dogma—which is living with the results of other people's thinking. Don't let the noise of others' opinions drown out your own inner voice. And most important, have the courage to follow your heart and intuition. They somehow already know what you truly want to become. Everything else is secondary.

—STEVE JOBS

As your cultivation of mindfulness deepens, you will start to have more profound experiences of Samma Sati: you may begin to "remember" your true nature and how aspects of consciousness cannot be contained in a human body. You may be reminded of an unconditional intrinsic joy that does not depend on external circumstances. You may have flashes of higher consciousness that break the spell of your attachments to things of this physical world. It will be like waking up from a dream, wondering how you were so caught up in such "limited" and "petty" ways of experiencing life on this planet. It will be like countless thousands of veils of ignorance being gradually lifted one by one, each one shaking you to your core. As Samma Sati deepens, you gradually realize that it is naturally leading you to the next phase of the Noble Eightfold Path, Samma Samadhi: aligning and further integrating your embodied self with your Higher Self.*

The Higher Self is like a spiritual sun that is constantly shining, but which we cannot see because the sky of the mind is obscured by many clouds. The thing is, those clouds blocking the sun are illusory simulations created by our own minds when we are in Brain 1.0 and Brain 2.0. When we shift into Brain 3.0 and stop creating these clouds, the sky of the mind naturally clears. Only then can we see just how bright the spiritual sun is and how vast the mind can be when our awareness is not shrouded.

Furthermore, these changes are observable to my eyes: I see a radiant ethereal star like a mini spiritual sun above people's heads when they are spiritually connected. The more spiritually connected people are, the more light they radiate all over as they become infused with life-giving spiritual energy from

---

* Buddhists prefer to use the term "Buddha Nature" over words like "Higher Self" or "True Self" to emphasize that the sense of a permanent absolute self at any level of consciousness is an illusion because all notions of selfhood arise from interdependence through interactions and relationships with other forces and the self is continuously changing and impermanent (everything that is born dies and decays). Furthermore, at a "higher level of consciousness," all sense of individuated selfhood dissolves into oneness. Nevertheless, for the purpose of communicating with embodied minds, concepts like "Higher Self" are very useful for emphasizing that we are more than our bodies (and brains). I must also clarify that the "Higher Self" is continuously evolving, interdependent on the Collective Consciousness, and impermanent in its own way. Paradoxically, the True Self is Self-less.

the Higher Self. It is like a brilliant pillar of light is enveloping them. In my own personal experience, when I am strongly connected with my Higher Self, I can see the air and space around me radiate like luminous mist. This happens frequently when I facilitate Calm Clarity training. It is a wondrous phenomenon that will likely continue to dazzle and awe me for the rest of my life.

When we cultivate Brain 3.0 by practicing Samma Sati, we observe and deconstruct the process by which the mind creates simulations and, as shared earlier, the neuroscientific evidence shows that doing this somehow rewires our default mode network to shift its efforts away from building simulations based on self-centered narratives to paying more attention to raw sensory data coming in from our senses (the external senses and inner senses). I believe this rewiring of the DMN initiates the process of clearing the obscurations from our minds. As the mind becomes clearer, the signal-to-noise ratio of the Higher Self becomes much stronger. Then the DMN becomes a vehicle for insights and guidance from the Higher Self to bubble into conscious awareness.

As the mind's awareness becomes more spacious, we can begin to discern the presence of the Higher Self through wisdom, grace, and unexplainable synchronicities. The stronger Brain 3.0 becomes, the more intermittent experiences we have of transcendence, the more we understand that who we really are is the Higher Self. When I "sense" the energies of the Higher Self, what I feel is unconditional benevolence and agape (the ancient Greek concept of divine love in the New Testament). Because the Higher Self is a microcosm of the True Self/Source, through aligning with and embodying our Higher Self, we open a portal to reconnect with Source.

I believe Jesus Christ may have been speaking as a manifestation of this portal when he said: "I am the way and the truth and the life. No one comes to the Father except through me. If you really know me, you will know my Father as well. From now on, you do know him and have seen him."[44] To me, the Holy Trinity has a symbolic meaning: the Father corresponds to the True Self/Source, the Son corresponds to the Higher Self, and the Holy Spirit is the aspect of Source that manifests as the physical realm, provides the life force that sustains and animates the universe, and fosters humanity to manifest and express our innate divine qualities. As I have come to know my Higher Self, I have also come to experience and understand that the Higher Self is like a "cell" within the True Self and that all our Higher Selves are interconnected like the neurons of a great, ineffable cosmic being. Shifting from Brain 1.0 and

Brain 2.0 into Brain 3.0 enables humanity as a collective to manifest and express our benevolence and interconnectedness.

Since I first became conscious of the presence of my Higher Self in 2012, I have slowly come to realize that I had already been acquainted with my Higher Self throughout my life through my experiences of inspiration and aspiration. Instances of inspiration are really moments of spiritual connection and a "downloading of energy" from the Higher Self. My aspirations are essentially qualities of my Higher Self that I am being called to embody or an intention of my Higher Self that I am being called to fulfill. I have also learned that the easiest and simplest way to connect with my Higher Self is to think and act like my Higher Self. This naturally activates neural pathways in Brain 3.0, which are able to calm Brain 1.0 and Brain 2.0. The more I act on and express these higher qualities, the more aligned, uplifted, and energized I feel. Thus, I have come to realize that my shift into Brain 3.0 can be best described as the journey of embodying more and more of my Higher Self—so that I become the best version of me, so that I model the ideals I want humanity to exemplify, and so that I actively create the legacy I want to leave behind when the life span of my embodied self comes to an end.

Over the last two decades, researchers in the field of positive psychology have found that when people express benevolence and altruism, they actually experience greater physical, emotional, and mental well-being. These studies have also led researchers to begin to uncover the biological mechanisms to explain why. For instance, numerous studies have proven that the "helper's high" is real: altruistic actions and intentions are associated with increased activity in the brain's reward circuitry, increased levels of oxytocin, and reduced levels of stress hormones, thus promoting resilience and enhancing connectedness.[45] Furthermore, these mechanisms support increased longevity. A University of Michigan study tracking more than four hundred elderly married couples over five years found that helping and supporting others cut the risk of dying by half.[46] Another longitudinal study, from the University of Miami, found that among retirees, those who volunteered had a significantly higher degree of life satisfaction, a stronger will to live, and fewer symptoms of depression and anxiety.[47]

What the science is pointing toward is a higher truth that acting like our Higher Selves can set off biochemical cascades that "reward" our bodies with greater health and well-being and naturally align Brain 2.0 with Brain 3.0 in the service of the greater good. As research in positive psychology continues, I believe that scientists will gather more and more evidence that human beings have built-in biological feedback processes that help us to flourish and experience a deeper, more meaningful sense of happiness when we express our Higher Selves and support one another to shift out of Brain 1.0 and Brain 2.0 into Brain 3.0. Our bodies (including our brains) are wired such that the more we embody our Higher Selves, the more spiritually harmonized and integrated we become, the greater our well-being and the deeper our experience of eudaimonia. These positive feedback loops encourage us to invest energy and time in developing and cultivating Brain 3.0 so that our Higher Selves can more deeply integrate into the core of our embodied selves.

As the embodied self becomes more and more integrated with the Higher Self, we will also have a clearer understanding of our life purpose and be able to harness energy from the Higher Self to fulfill all that we are being called to do. As this metamorphosis unfolds, every moment of life becomes endowed with meaning, wisdom, and grace.

What follows here are a series of self-reflection exercises to help you kick-start the process of becoming better acquainted with your Higher Self.

Tune in to Your Higher Self Exercise

In order to embody your Higher Self, it is very important to begin to get to know the qualities of your Higher Self. The following five-part exercise is designed to help you do that and see your life from a much higher perspective. Completing this exercise will hopefully also make it easier for you to tune in to your inner wisdom by amplifying the voice of your Inner Sage/Higher Self. It may be helpful to activate and prime Brain 3.0 by doing the compassion meditation before starting these exercises.

Part 1. People Who Inspire You

Name five people who inspire you and reflect further on the things that they do that inspire you. What does this reveal to you about the qualities you aspire to have? *Note: Some participants find it easier to skip the first column and just list the actions people do in general that inspire them in the middle column. If that's your preference, please feel free to do so.*

Person who inspires me	What this person does/did that inspires me	Qualities I aspire toward
1.		
2.		
3.		
4.		
5.		

Part 2. Core Values

List five things that people do that provoke a sense of moral outrage within you. What is it about these behaviors that upsets you? What does it tell you about your values?

Action or behavior that provokes a sense of moral outrage	Why this upsets me	Core values that matter to me
1.		
2.		
3.		
4.		
5.		

Part 3. Understanding Your Aspirations

This next set of questions will help you create a clearer picture of the qualities of your aspirational Higher Self.

a.	What patterns and themes can you see emerging from the two previous lists? What do they reveal about your values and aspirations?
b.	What experiences have given you a sense of fulfillment? What ideas, activities, and interactions energize you? How do they tie to your values and aspirations?
c.	If your present self were to meet your future self in ten to twenty years' time, what qualities would you want to see in your future self?
d.	What would make you feel most proud of the person you will become?

Part 4. Visualization Meditation

A portrait of your aspirational Higher Self.

1. Close your eyes and think about moments in your life when you felt a deep sense of inspiration and elevation. Allow your body to physiologically relive these feelings and sensations.
2. Now think about moments when you felt a deep sense of purpose, energy, and fulfillment. Let these moments guide you to an image of your aspirational Higher Self. What are the qualities that stand out most about your Higher Self?
3. Imagine a point in time when your future self becomes your aspirational Higher Self. Visualize your future self fully embodying these

qualities and fulfilling your aspirations. Savor the feelings of accomplishment, joy, and fulfillment.

4. Now try to visualize your life at this time.
   a. What gives your life meaning?
   b. Describe the relationship you have with your family and loved ones. What does your family life look like?
   c. Describe the impact you have on the wider community around you. How are you engaged with your community?
   d. What are your character traits and strengths?
   e. How do you handle setbacks?
   f. What are you passionate about? What do you do because you are not afraid to fail?

Part 5. Advice from Your Future Self

1. Now imagine a conversation with your future self. Visualize your future self showing appreciation and gratitude to your current self for embarking on the journey toward realizing your aspirations, because all the hard work and effort were worth it.
2. Ask your future self to describe what the journey involved.
3. Ask your future self to share advice and suggestions on how you overcame the challenges and obstacles along the way.
4. Write this conversation down in a letter from your future self to your current self.

Post-Exercise Reflection

To further hardwire this learning experience, please take time to write your answers to the following questions:

1. How was this experience? What insights, if any, arose?
2. What stood out most to you?
3. How can you connect to, draw inspiration from, and embody your Higher Self in your everyday life?

Additional Comments on Knowing Your Higher Self

I want to clarify that expressing the qualities of your Higher Self does not result in your becoming a doormat or a martyr who takes care of everyone else but neglects yourself. This is because the first person you need to learn to unconditionally treat with love, kindness, compassion, appreciation, and respect is yourself. Self-care is critical to fostering Brain 3.0. The reason is clear: if you don't take care of yourself physically, mentally, emotionally, and spiritually, you are more susceptible to getting pulled into Brain 1.0 and Brain 2.0.

For many people whom I have guided through these exercises, visualizing the portrait of their Higher Self and writing the letter from their future self (after becoming their aspirational Higher Self) give them an experience of the overview effect, an elevating shift that happens when people see the bigger picture from a much higher vantage point.

When you become acquainted with the great Being that you are unfolding into, it becomes much easier to let go of petty trifles and to stop beating yourself up for not conforming with social expectations. It becomes easier to surrender into pure Beingness when you "know" the essence of who *You* really are and trust in *Your* innate qualities and capabilities.

Most important, once you can intuitively sense the signal and feel of your Higher Self, it becomes much easier to tune in to and reconnect with your Higher Self whenever you notice that you have slipped into Brain 1.0 and Brain 2.0. The Higher Self is always on standby to give the embodied self a lift back up into Brain 3.0.

9.

Closing Thoughts on Embodying Our Higher Selves

No man is an island entire of itself; every man is a piece of the continent, a part of the main. —John Donne, Meditation XVII

Remember Christ has no human body now upon the earth but yours; no hands but yours; no feet but yours. Yours, my brothers and sisters, are the eyes through which Christ's compassion has to look upon the world, and yours are the lips with which His love has to speak. Yours are the hands with which He is to bless men now.

—Sarah Elizabeth Rowntree,
nineteenth-century Quaker medical missionary[1]

The most beautiful and most profound emotion we can experience is the sensation of the mystical. It is the sower of all true science. . . . To know that what is impenetrable to us really exists, manifesting itself as the highest wisdom and the most radiant beauty which our dull faculties can comprehend only in their most primitive forms—this knowledge, this feeling is at the center of true religiousness. —Albert Einstein[2]

Now that you have come this far, I hope you have done the meditations and self-reflection exercises I have shared and that you have experienced for yourself what it feels like to be in Brain 3.0. At this point, to help you guide yourself in continuing your shift into Brain 3.0, perhaps the most useful thing I can share is the understanding that guides me in my own continuing journey.

Ever since I began earnestly seeking to understand the purpose of my life and life in general, I have been slowly and steadily gathering little jigsaw puzzle pieces of the cosmic bigger picture. The jigsaw puzzle pieces and how they fit together have mostly arisen during my meditations and from what I have seen, felt, and sensed during mystical experiences. I have also found many puzzle pieces in texts from Buddhism, yoga, and Christianity; in texts by ancient philosophers such as Lao-Tzu, Plato, and Epictetus; in the poetry of mystics such as Rumi, Hafiz, William Blake, and Walt Whitman; and in works by more contemporary spiritual teachers, such as Paramahansa Yogananda and Master Choa Kok Sui, and the writings of the Theosophical Society, primarily by its founders Helena Blavatsky and William Q. Judge. (There are probably many more puzzle pieces out there in the religions I am not familiar with and the many sacred texts I have not yet read.)

In order to share the "bigger picture" that I have formed thus far, I have drawn an image that often emerges in bits and pieces when I meditate (see figure 10). As I describe and explain this drawing, please bear in mind that translating an intuitive knowing into a form that can be communicated is not easy. Choosing language presents a challenge because oftentimes there isn't a word that precisely captures the essence of what is being conveyed, and words tend to trigger different associations and memories in each person. Similarly, visual images like symbols and pictures can also trigger different associations in people's minds; yet images often do a better job at transcending language barriers. So please understand that words and images are more like fingers pointing to an essence. Rather than get stuck on the finger, you would learn more by turning your attention to what the fingers are pointing to.

In figure 10, the vertical axis represents gradations of consciousness, which progresses from being very gross and dense at the bottom to very subtle and refined at the top. As I mentioned earlier, higher consciousness envelops and interpermeates lower consciousness. This means that there are no boundaries or borders between "grades" of consciousness. It's like how molecules of different substances, when mixed together, tend to naturally separate into layers based on their mass and weight. A helpful analogy may be that of the Earth's atmosphere, which is made of many borderless layers. The layer closest to the surface has all the gases and is the only layer where the heaviest gases (which happen to include the ones critical to supporting life, such as oxygen for animals and carbon dioxide for plants) are found, while the outermost layer is composed only of the lightest gases, hydrogen and helium. Similarly, denser

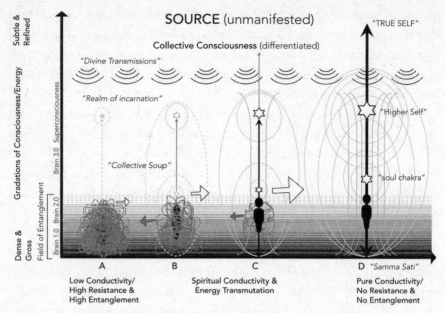

Figure 10: Bigger Picture Perspective on Brain 1.0, 2.0, and 3.0

and grosser "grades" of consciousness are restricted in how far up they can "travel," whereas subtle and refined grades of consciousness have greater "range of motion" and interpermeate the denser layers.

Consciousness is energy in a very subtle form that has characteristics different from the energy represented by physical matter. It is hard to put into words how I see that consciousness precedes or is a precursor to physical matter, and that all physical matter is a form of concretized energy *and* a form of simple, dense consciousness that is further animated by progressively "higher and more sophisticated" forms of consciousness into what we consider lifeforms (microbes, plants, animals, and humans).

The highest grade of consciousness is beyond the understanding of the human mind. So far, the most accurate word I have come across to describe something so ineffable is "Source," because, as I "understand" it, this is the formless "center" from which all consciousness somehow mysteriously proceeds, becomes differentiated, and manifests as form. Source is also called the "Supreme Being," the "Supreme Spirit," or the "Supreme Self" ("Paramatma" in Sanskrit). Because Source represents the "true nature" of consciousness, some traditions call it the "Oversoul" or "True Self." I believe this is also what Buddhists mean by the term "Buddha Nature." At the next-highest levels of consciousness, at the "place" where the unmanifested somehow becomes manifested and differentiated, what I call the "Collective Consciousness" emerges. My sense is that when we experience "oneness," we tap into this plane of consciousness, where we can experience spiritual unity and integration and yet retain our individuality. This is a bit like being a molecule of water in a vast ocean and like being a neuron in a brain.

I call the realm in which an individuated unit of the Collective Consciousness, often called a soul, evolves by incarnating a smaller portion of itself into physical form, the "Realm of Incarnation." Since the full energy of a soul cannot be contained by a physical body, only a portion is incarnated. Therefore, souls can be considered to have two manifestations. The full soul is what I call the "Higher Self" ("Atma" in Sanskrit), and the smaller portion that incarnates and animates the life of an embodied human being ("Jivatma" in Sanskrit) I prefer to call the "embodied self" rather than the "lower self," and "embodied soul" when I specifically mean the portion of the soul animating the body apart from the ego. The Collective Consciousness is continuously transmitting wisdom and new ideas into this Realm of Incarnation to anyone who is capable of receiving these transmissions. Many of these transmissions correspond to what the ancient Greeks called the Nine Muses, whom they credited for divinely inspiring creative works of art, philosophical and spiritual insights, and scientific discoveries and technological breakthroughs.

The Nine Muses are an illustration of the tendency of the human brain to humanize the formless (which is very hard to understand) by constructing an anthropomorphic conception of divine beings and projecting qualities onto these beings that align with our cultural values and iconography. Even for me, on the rare occasions I receive a transmission, the Higher Self or the Collective Consciousness often becomes personified in my mind's eye as a figure I am familiar with, such as a Christlike being or a Buddha or bodhisattva. My hunch is that how "the divine" appears to each person depends very much on his or her spiritual and cultural background, so that a Christian would likely see Christ, Mary, or an archangel; a Buddhist would likely see a Buddha or bodhisattva; a Hindu may see Brahma, Krishna, Shiva, Saraswasti, Lakshmi, Parvati, or another Hindu deity.

Within the Realm of Incarnation, the densest plane that corresponds to consciousness embodied in physical matter is what I call the "Collective Soup." It is like the innermost layer of the Earth's atmosphere, where the psychic outputs (emotions and thoughts) of all of humanity are exchanged with one another, similar to the way air inside a closed room is breathed in and out by everyone in the room. The densest, heaviest layers of the Collective Soup make up what I call the "Field of Entanglement."

Within this field, the lines that flow across horizontally signify that "lower" thought forms and emotional energies that float around in the Collective Soup affect all people the same way that the flu virus during flu season affects people all around the world. The lines correspond to heavier forms of consciousness that are picked up and transmitted by Brain 1.0 and Brain 2.0. Brain 1.0 picks up the "heaviest" thought forms and emotional energies that trigger freeze-flight-fight instincts, such as those related to aversion, fear, threat, anger, rejection, shame, worry, sadness, loneliness, and helplessness. Brain 2.0 picks up the thought forms and energies that trigger cravings, addictions, restlessness, and obsessiveness such as those related to pleasure, ego, pride, power, status, influence, control, wealth, competition, fame, and dominance. When Brain 1.0 and Brain 2.0 interact with these grosser and denser thought forms and emotional energies, people tend to resist, suppress, attach, or hold on to them. When this happens, they get entangled.

The horizontal axis represents degrees of "spiritual conductivity and energy transmutation," which progresses from being very low on the left to purely conductive at the right. "Spiritual conductivity" refers to the quality and strength of the spiritual connection between the embodied self and the Higher Self, and the ability of the embodied self to conduct and transmit the energies of the Higher Self into the physical realm. The brightness of the "soul chakra" or "soul star," which I see above people's heads, is a dynamic reflection of the degree of connection between the embodied self and the Higher Self, and the spiritual conductivity of the embodied self: the greater the conductivity of the embodied self, the stronger the connection, the brighter the soul chakra. The soul star may serve as a special type of energy transformer that converts energy from higher spiritual realms into a form that nourishes, integrates, animates, and sustains a physical human body, such that all the components that make up a body continue to cohere together as a unified life-form.

Otherwise, the natural tendency of all the particles that make up a physical life-form is to break apart.*

Spiritual conductivity is reduced by resistance, resistance is created by entanglement, and entanglement arises from our tendencies in Brain 2.0 to get attached to the things and outcomes we want and in Brain 1.0 to avoid, fight, or resist the things and outcomes we don't want. So long as we wish the present moment to be different from what is unfolding, and so long as we need something else outside of ourselves to feel fulfilled, we are entangled. The degree of entanglement varies depending on how much time we spend in Brain 1.0 and Brain 2.0 rather than in Brain 3.0.

The bidirectional vertical arrows that run through each figure signify that spiritual conductivity moves energy between higher and lower levels of consciousness in both directions. By embodying more and more of the Higher Self, a person grounds higher consciousness into his or her body and into this Earth and thus serves as a channel for higher consciousness to flow into the Collective Soup, which helps transform and clear the Field of Entanglement. As we become more spiritually conductive, more of these higher spiritual energies enter our cells. Then, once a certain degree of microcellular change has taken place, a special type of energy transmutation occurs that produces a very refined energy that India's spiritual traditions refer to as "kundalini." When kundalini is unleashed, it "rises" upward through the soul star and flows through the Higher Self into Source. My hunch is that this happens naturally because the nature of kundalini energy is similar to that of Source and its nature is to return to Source. The more kundalini our cells "naturally" generate, the stronger our conscious integration with our Higher Self and with Source. In hindsight, I intuitively sense that my vision of a portal opening was a symbolic message from my Higher Self to let me know that this process had begun for me.

In this drawing, the figures represented by A, B, C, and D present a visual

---

* The scientific name for this tendency is "entropy," a word whose general meaning, according to the *Merriam-Webster Dictionary*, is "a process of degradation or running down or a trend to disorder" or "chaos, disorganization, randomness." If you studied chemistry, you may have come across the concept of entropy when you learned about the Second Law of Thermodynamics, which states that the entropy of any isolated system always increases. In that context, the entropy of a system is a measure of its disorder and of the unavailability of energy to do work.

portrayal of my own journey. Person A depicts the period of my earlier life in which I got trapped in Brain 1.0, struggling with PTSD symptoms and barely able to contain the impulse to self-destruct. I was so full of anger and resentment that I didn't care if what I said or did hurt other people or myself. Although I didn't participate in any violent crime, a part of me admired rather than condemned people who used force to take care of their needs. I thought, "The system is so fucked up, why should anyone play by rules that exploit and oppress them?" During this period of time, I was lost in a fog of darkness, and my energy field was so dense and gross that my Higher Self couldn't "connect" or "transmit" to me, no matter how hard it may have tried to. In hindsight, I now see that my becoming suicidally depressed was a result of being spiritually disconnected. My embodied soul was so starved of spiritual sustenance, like a plant deprived of water, that I was on the verge of losing my will to live. My embodied soul resuscitated only after I learned to use neuroscience to come out of Brain 1.0 and stop choking my soul with the poisonous entanglement of anger, bitterness, and resentment.

Person B depicts me in my twenties and early thirties, when I spent most of my life in Brain 2.0, swept up in an obsession with career success and achievement. My world centered on getting ahead in life and getting what I wanted as quickly as possible. A "good" day was when things went the way I wanted them to and I got what I wanted. A "bad" day was when something didn't go the way I wanted or I didn't get something I wanted fast enough, and I spent quite a bit of time in Brain 1.0 fuming about it. Achievement required playing the game at work, which meant getting embroiled in messy politics and office melodramas. During that period, my energy was definitely not as dense, gross, or toxic as it had been when I was at point A, but it was hardly refined. The signal-to-noise ratio for my Higher Self was stronger, but I was too caught up in my tunnel vision for chasing achievement to pay any attention to my soul. Only when I grew tired of how unsatisfying the trophies were, did it begin to dawn on me that I wasn't happy and that my embodied soul felt empty and burned out. With that realization, I made the choice to come out of the maze of Brain 2.0. At that point, I finally became "spiritually conductive" enough for my Higher Self to somehow lock a tractor beam on me and pull me into a soul-searching journey that I had never before imagined.

Person C depicts me in the present moment on a "good" day. For the most part, I'm able to be in Brain 3.0 and detach from the triggers that activate my

Brain 1.0 or my Brain 2.0. Nonetheless, I still create entanglement because I can still get attached to outcomes in Inner Teen Wolf mode and am still misled by self-limiting beliefs and trauma-associated triggers that bring me into Inner Godzilla mode. However, I don't create as much entanglement as I used to because as soon as I become mindful of not being in Brain 3.0, I try to let go of what I am getting entangled with and recenter. The signal-to-noise ratio for my Higher Self is very strong in that every day, I feel myself enveloped by a divine presence and know that I am never alone. As for the Collective Soup, through Brain 1.0 and Brain 2.0, I still pick up denser thoughts and emotional energies in the Field of Entanglement, but I let them arise and pass away in my mind and body rather than struggle with them; through Brain 3.0, I also pick up divine transmissions from the Collective Consciousness and receive guidance from my Higher Self.

These days, I am probably "on average" somewhere between B and C. On a bad day (usually it has something to do with being sleep-deprived, "hangry," jet-lagged, or stressed out by family challenges, or having trauma-related memories triggered by similar incidents), I get bogged down energetically and temporarily move closer to B, as shown by the little arrow pointing to the left. Fortunately, this usually doesn't last for long, because there is also a much more powerful yet gentle force from my Higher Self pulling me toward point D.

Person D on the far right depicts what I believe happens when a person stops creating entanglement altogether. Such a person becomes so spiritually conductive that the embodied self becomes a full embodiment of the Higher Self. When this happens, the person's energy field transforms into something like a radiant golden spiritual sun—a state that is popularly described as "Enlightenment." I'm not confident that I can speak from firsthand experience about what this feels like, but from the few very brief and partial glimpses I have had of what I am depicting with Person D, I can only infer that it must be even more elevating, expansive, and energizing than any state of peak consciousness I have experienced so far.

One thing I want to point out is that the image shows that Persons C and D still navigate the Field of Entanglement. The lines moving across the image show that the same dense and gross thought forms and emotions that affect all of humanity in the Collective Soup still enter their energy fields. I consider them psychic viruses because, like common viruses, they feed on hosts to multiply and then spread even more into the Collective Soup. In Persons A and B, these thought forms and emotional energies trigger Brain 1.0 and Brain 2.0, which get entangled with them, and then these psychic viruses settle in like long-term squatters in their energy fields for as long as they remain entangled.

In contrast, Persons C and D are different in that they are able to let these thought forms and emotions pass through them without getting entangled with them. This is because Persons C and D have cultivated Brain 3.0 and Samma Sati, so they are able to mindfully allow these thought forms and emotional energies to pass through their energy fields without suppressing, resisting, or attaching to them. Thus, they have a strong "psychic immune system" that protects them from psychic viruses taking up long-term residence in their fields. They are able to stop contributing to the Field of Entanglement by not feeding and energizing psychic viruses.

The difference between Person C and Person D is the speed by which this happens. For Person C, it may take a few minutes, hours, or sometimes days for these thought forms and emotional energies to run their course through his or her field because these visitors can still trigger Brain 1.0 and Brain 2.0 to interact with them, sometimes in subtle, subconscious ways. In Person C, these thought forms and emotional energies are like visitors that check out the scenery and maybe find somewhere to crash for a night or two and then leave. In contrast, for Person D, the practice of Samma Sati is so advanced that neither Brain 1.0 nor 2.0 gets triggered or activated by these visitors entering his or her energy field. Thus these visitors find nothing of interest to interact with, so they pass straight through without making any stops.

More and more scientists are also coming to the conclusion that all life is interconnected based on evidence coming out of their fields of research. For instance, in her latest book, Nobel Prize–winning biologist Elizabeth Blackburn writes, "We are connected to one another and to all living beings at all levels from the macro to the micro, from the societal to the cellular. The separation we all feel, as if we are each on a path alone, is an illusion. The reality is that we all share much more than we can ever comprehend, both in mind and body. We are deeply interconnected with each other and nature in phenomenal ways."[3] She then goes on to explain that the eukaryotic cells that make up our bodies may have evolved 1.5 billion years ago by swallowing up bacteria, and that the mitochondria (cellular structures that supply energy) in our cells are the legacy of these bacteria and this 1.5-billion-year-old symbiosis. She also explains that roughly two to three pounds of our body weight is made of microbes living in complex communities in our gut and skin, which help keep our body and immune system balanced and healthy. We are also interconnected by our dependence on the Earth, because pollution on one side of the world can easily travel to another and affect our health and well-being. Furthermore, we are connected from generation to generation by the transmission of genetic and epigenetic information in our DNA.

Similarly, as a result of my metamorphosis, I now "see" and "know" that "the body is a temple of the Holy Spirit." Our souls are given human bodies to experience life in this physical world for relatively brief amounts of time. My embodied consciousness as "Due" arises from the symbiotic engagement between my Higher Self and the aggregate consciousness of all the cells, tissues, and organs that make up my physical body. As a human being, "Due" cannot thrive unless the cells of my body also thrive, and my cells need a continuous flow of spiritual energy and nourishment to flourish. Thus, for me to feel integrated and whole, my body has to be physiologically integrated, harmonized, and aligned with Spirit.

Whenever my ego pushes away aspects of my experience into the shadow, I can sense the dissociation process begin with my mind becoming "deaf" to the signals coming in from my body. When I am no longer tuned in and grounded in my body, I am more likely to get swept away by stories inside my head and fall out of alignment with my Higher Self. When this disconnection happens, I feel a dramatic reduction in the flow of spiritual energy nourishing my body. Since I started to tangibly sense the difference, I became much more motivated to change and create new Brain 3.0–enhancing habits so I can naturally be more and more aligned with my Higher Self and avoid ever compromising this spiritual connection.

It is very unfortunate that many people treat their body like a garbage disposal unit by feeding it unhealthy junk food and artificially flavored drinks and by feeding their mind with toxic ideas, thoughts, images, and beliefs that fuel the Inner Godzilla and Inner Teen Wolf rather than the Inner Sage. This way of living perpetuates disharmony, fragmentation, and misalignment within us and outside of us. These patterns perpetuate disconnection from the Higher Self, thus depriving the body of the spiritual energy and nourishment needed to thrive, and creating increased vulnerability to disease and depression.

A very powerful remedy to become reintegrated inside is to cultivate Samma Sati to tune in and become centered and grounded within our bodies and to stop feeding the patterns that cause entanglement. When we tune in, the wisdom of our bodies guides us to naturally and intuitively redesign our lives to foster greater connection and alignment with the Higher Self. To do this, we have to develop interoceptive awareness and listen to the wisdom contained within our bodies. When we reintegrate mind and body in this manner, we naturally take better care of ourselves. We are more selective about what we feed ourselves, not only in terms of food and drink, but also in terms of what we feed our minds. In addition, we become more selective about what thoughts and emotions we choose to energize. The more we can discern and listen to guidance transmitted from the Higher Self, the more aligned and integrated we become spiritually. This positive spiral between embodiment, discernment, alignment, and integration is where the Noble Eightfold Path and the Eight Limbs of Yoga ultimately lead.

Self-actualization involves alignment and integration at many levels and layers. Neurobiological integration is foundational to emotional, mental, and spiritual alignment and unification. It cannot be a coincidence that physio-logically, Brain 3.0 encapsulates the neural structures and pathways that in-tegrate the brain and enable us to see a bigger-picture perspective, to appreciate the whole as greater than the sum of its parts, to notice intercon-nections between divergent subjects, and to construct a unifying harmony that transcends conflicting points of view. Brain 3.0 is both a physical and a metaphysical bridge that enables us to access and embody our Higher Self while we are in our bodies. The stronger Brain 3.0 is, the more we can experi-ence the transcendence and wholeness of oneness, and the more we open our-selves to oneness, the stronger Brain 3.0 becomes. I also "see" that from a macro perspective, as more and more people shift into Brain 3.0 (moving from A and B to C), the Field of Entanglement will weaken and fewer "psychic vi-ruses" will be able to spread through the Collective Soup.

At present, as I look at the path toward point D, my logical mind continues to be skeptical toward ideological and mythical portrayals of Enlightenment as the ultimate liberation from the cycle of reincarnation—to be frank, I re-main unconvinced that the cycle of reincarnation is a horrible wheel of suf-fering one needs to seek liberation from. Nevertheless, I am still progressively shifting to the right, not because I have any goal to become fully Enlightened, but because I don't see any point in consciously energizing Brain 1.0 and Brain 2.0 and adding psychic garbage to the Field of Entanglement. I simply prefer to be in Brain 3.0 as much as possible and to continue to integrate and become more whole. I also feel that I am being continuously pulled in the direction of D by my Higher Self. Regardless of whether "Due" wants to experience Enlightenment, my Higher Self wants the embodied self to further align and unite with the Higher Self and with Source.

Although my skepticism and prior atheistic conditioning still make spirituality feel uncomfortable and awkward to me, I can't deny that I am drawn to experience what it feels like to embody more and more of my Higher Self. Similar to how birds migrate seasonally, how a moth flies to a light, or how a plant grows toward the sun, my embodied self seems "preprogrammed" to merge more and more with the Higher Self and to serve as a channel for Source to flow.

I get the sense that this is not unique to me—all the Higher Selves want to experience physical embodiment in order to ground more and more higher consciousness into our physical realm through the progressive evolution of humanity. Rather than indulge in thinking that I am in any way special, I suspect that a capacity for profound spiritual connection is an intrinsic, universal aspect of human biology.

Some time ago, as I began sharing the Calm Clarity Mindful Leadership Training with large corporations, I prudently decided to create a manifesto as a tool that would help safeguard my ability to stay anchored in Brain 3.0 in environments in which I had been previously conditioned to run around in Brain 2.0. The initial purpose of the manifesto was to trigger, activate, and prime Brain 3.0, in case I got swept up in Brain 2.0 or Brain 1.0, so I could quickly recenter and foster Brain 3.0 in my clients. By going through this exercise, I realized that there are twelve distinct elements to bring into conscious awareness on a regular basis to support a more complete shift into Brain 3.0. This manifesto now guides the Calm Clarity team to foster Brain 3.0 in ourselves, in each other, and in the people and organizations with whom we interact. I would like to share it here so you can also refer to it as a guide for your own journey.

In closing, I hope the insights I have shared from my own journey are helpful to you. I also thank you for accompanying me on this collective spiritual journey.

Calm Clarity Manifesto: 12 Elements of Fostering Brain 3.0

1	Vision, Purpose, and Intention	Every decision I make and every action I take is guided by my vision for my best self and connecting to a greater purpose. I take time to set intentions in meetings, interactions, and communications to activate Brain 3.0.
2	Integrity and Commitment	I commit to serving as a role model for Brain 3.0 by practicing what I preach and honoring my word. I refrain from overpromising and from manipulatively triggering Brain 1.0 or Brain 2.0.
3	Openness, Curiosity, and Humility	Wisdom means acknowledging I don't know everything, so whenever I encounter different perspectives or reach the edge of my comfort zone, rather than go on the offensive or defensive, I seek to learn and understand.
4	Appreciation and Gratitude	I savor positive experiences and appreciate and celebrate life. Whenever there are setbacks, I look for the positive in the negative. I am thankful for each and every person who helps me.
5	Compassion and Forgiveness	I see every person as a human being and wish for the well-being of all. I aim to do no harm in my actions, words, and thoughts to anyone (including myself). If any harm is done, I forgive and/or seek forgiveness (including from myself) and then move on.
6	Mindfulness and Authenticity	I continuously cultivate my capacity for metacognition, self-understanding, and self-acceptance so I can be more fully present, integrated, and whole. I aim to shift my autopilot into Brain 3.0.

7	Interdependence and Collaboration	Knowing that the whole world is interdependent and all humanity is interconnected, I aim to create win-win outcomes in working together internally and externally. I do not take part in the exploitation of others or of myself.
8	Flexibility, Creativity, and Innovation	Knowing that the world is dynamically evolving, I accept uncertainty and embrace change. I see challenges as opportunities for growth and invention and for letting go of inertia, resistance, and arrogance.
9	Health, Energy, and Regeneration	Knowing that self-care is critical to sustaining Brain 3.0, I proactively nurture my well-being and do not let myself burn out. I regularly tune in to monitor my energy levels and build in activities to regenerate and rejuvenate.
10	Advocacy and Moral Courage	I do not let fear or greed keep me from doing what I know in my heart is right. I stand up for my values and speak up when I see cause for concern, using skillful means so the message resonates and lands.
11	Self-Efficacy and Self-Actualization	I take ownership of my personal journey and my impact on others. I nurture my confidence in my capacity to grow, evolve, and make a positive impact on the world. I proactively ask for guidance from mentors and advisers and seek sources of wisdom, inspiration, and truth.
12	Discipline and Self-Mastery	I commit to exercising Brain 3.0 on a daily basis to access my inner wisdom and master the triggers that bring me into Brain 1.0 and Brain 2.0. When I encounter darkness, I look inside and shine light into my own shadow.

Acknowledgments

IN THE SPIRIT of transparency, I can't write an acknowledgments section without acknowledging two important things regarding this book. First, it came into being in a way that was as mysterious as how the rest of Calm Clarity has unfolded and continues to unfold. In a rather unusual sequence, as I was still exploring whether to write a book, I met my agent, George Greenfield, who coached me on how to build a proposal, and before I even wrote the proposal, I met Sara Carder at TarcherPerigee, the publisher that would later give me a book deal. In that manner, I can't thank George, Sara, and her assistant editor Joanna Ng, enough for taking a chance on a first-time author with a history of PTSD and providing the guidance, structure, and platform that enabled me to draw this book out from deep inside me onto paper.

Second, similar to how I consider myself the biggest beneficiary of the Calm Clarity Program, I consider myself the biggest beneficiary of this book because writing it was profoundly healing. One of the lingering effects of developmental trauma I still experienced long after resolving the clinical symptoms is a subclinical condition called alexithymia, which Bessel van der Kolk describes as "not having words for feelings." In *The Body Keeps the Score*, he explained that many traumatized children and adults are unable to describe what they are feeling because they cannot interpret what their interoceptive sensations mean. This is compounded by the fact that reliving traumatic memories tends to shut down the area of the brain that processes language.

Nevertheless, I have to credit alexithymia for making me turn to neuroscience to find a rational and reliable approach to deconstruct and understand emotions without being weighed down by the judgment embedded in our culture that it is "wrong" or "bad" to have certain emotions. Yet even after I developed and applied the emotional states of the brain framework to my own life, my mind and mouth were often still muzzled by alexithymia. Writing this book provided the opportunity to finally overcome it by creating intense workouts for my neural circuits for language processing and interoception.

The truth is, because of the alexithymia, it was with great reluctance and hesitation that I started writing this book. I first received inner guidance that a book was part of my life path in Kolkata in May 2012, when I reviewed the "Achieving Oneness with the Higher Soul" course with Shailesh. I found it to be a very peculiar "download" because it didn't provide any information on details I would need to move forward, such as what the book would be about. Moreover, I really didn't have much interest in being a writer. As a result of experiencing dissociation and alexithymia while writing papers during college, I had associated writing with extreme unpleasantness. After I returned to the United States, many people I met were wowed by my story and told me to write a book about my experiences. The idea seeemed so daunting that I decided to warm up by starting an anonymous blog, an experience that taught me that after years of working in business, I was much more comfortable communicating ideas in a PowerPoint presentation than in prose. Once I started creating the Calm Clarity Program as a training workshop, I gladly dropped the idea of writing a book in order to focus on building a social enterprise.

But the idea kept resurfacing. As I recruited people to join the Calm Clarity board of advisers, many of them suggested that the best way to share the Calm Clarity concepts with a wider audience was to write a book. Since I found the writing process overwhelming, I decided that if the universe wanted this book to happen, then the universe would have to intervene in a way that would make it happen and connect me to people who could help facilitate the process. Then to my amazement, the stars aligned and a path emerged.

The unfolding of this book started in spring 2015 when Chris Kohl became an adviser and introduced me to Ed, a former colleague who writes business self-help books. Ed got so excited about the content of the Calm Clarity

Program that he convinced me that it had to be captured in a book and immediately connected me with his publisher, who eagerly invited me to submit a proposal. Around that time, I was already scheduled to meet with Howard Blumenthal, then CEO of Independence Media, who had recently attended the Calm Clarity pilot for the Wharton and Penn alumni clubs in Philadelphia and offered to provide advice on my business plan. When I shared the news with Howard, he connected me with his former agent and friend George Greenfield, because, as an author and media executive, Howard knew I needed help navigating the publishing world to find the right home for the book. After George attended a Calm Clarity workshop I ran for Wharton MBA students, he guided me to develop a proposal for a book that would genuinely capture the "soul" of Calm Clarity. Meanwhile, Howard told me about Joseph Campbell's concept of the "hero's journey," pointing out that I had actually lived it, and encouraged me to share in the book what I really experienced and learned.

In parallel, the Wharton MBA students who organized the workshop George attended connected me to Adam Grant. A true giver, Adam altruistically took the time to share lessons he learned from publishing his first book, *Give and Take*. Getting Adam's overview of the publishing process from an author's perspective gave me a much better idea of what I was getting myself into and a belief that I could handle it.

For inspiration in summer 2015, I decided to attend a retreat led by Dan Siegel at the Garrison Institute called "Soul and Synapse" with a close friend. Unbeknownst to me, his editor Sara Carder rode the shuttle from the train station to the Garrison Institute with us. She then sat with us during one of the lunch breaks. When I shared that I had created a program for mindhacking called Calm Clarity and that I was working on a book proposal, she gave me her business card and asked me to submit it to her when it was ready. My friend and I were awestruck: could the universe have given a clearer sign that it was time to get serious about writing the book proposal?

By the end of 2015, I had realized that my having hidden my personal story from my former classmates and colleagues because of the fear of being stigmatized for having had PTSD was getting in the way of writing the book proposal. Therefore, as a new year's resolution, I decided to write my very first post on Medium.com, which I titled "Poor and Traumatized at Harvard," because I wanted to finally make known my experiences as a low-income first-generation college student struggling with PTSD at Harvard. I didn't expect anyone except for a few friends to read it, but the article quickly went viral. Soon the *Observer* reached out to republish the article, and for a few days it was one of the top-performing articles on their site. The article had clearly struck a chord and started a conversation across the United States and in many other countries where it was being read.

As luck would have it, when the proposal was submitted in March 2016, Sara and Joanna had come across the article and thought that it could make a great book. Sara responded right away that they wanted to move forward. As I had been warned that the chances of an unknown first-time author getting a book deal on the first round of submission was very slim, I was stunned to see the book quickly find a home at TarcherPerigee when I met with Sara and Joanna in New York City. But I was also terrified. Accepting a book deal meant taking on a legal obligation to actually write this book. I was so emotionally torn about what it would involve that I honestly didn't know whether to celebrate or cry—I ended up doing both. In case you guessed it, a torrent of tears christened the first part of the book.

That said, from the beginning, I "knew" TarcherPerigee was the right home. The editorial guidance and direction I received from Sara and Joanna helped give the book a structure that enabled the writing to finally flow through me, and Joanna's edits and feedback gave the book much more coherence and focus. I also want to honor the spirit of Jeremy Tarcher, whose legacy of publishing books on spiritual topics gave me the confidence to share my mystical experiences with an openness that a business book publisher would likely have discouraged. My gratitude also goes out to the entire team at TarcherPerigee, especially Linet Huaman for the beautiful calming cover that transcends stereotypes, copy editor Sheila Moody whose attention to detail blew me away, production editor Lavina Lee and designer Katy Riegel for making the contents of this book elegantly readable, and Keely Platte and Danielle Caravella for making sure you found out about this book!

Two cherished mentors, Rick Bellingham and Tom Tritton, helped me summon the courage and confidence to write the book and gave feedback on work-in-progress drafts. Rick, a leadership coach and former executive, persistently nudged me to share rather than hide my mystical experiences and Tom, a former cancer researcher, provided the perspective of a scientist. My dear friends Fran Snyder, Gina Scarpello, and Sujata Shah proofread and commented on the draft manuscript. I have to also thank the Calm Clarity board for their support and their patience when I unplugged for weeks at a time to write. I am similarly grateful to my cofounders of the Collective Success Network for keeping the ball moving whenever I had to take time to focus on the book.

I am very grateful to all the great philosophers, teachers, humanitarians, and scientists who I have quoted, mentioned, and cited in this book. They provided the inspiration and foundation for me to build upon to understand the human condition and to create the Calm Clarity Program. There were many more thought leaders, scientists, and writers whose work contributed in some way to my growth and understanding, but I couldn't mention all of them and their work in one book without it becoming thousands of pages long. At the risk of inadvertently leaving some people out, I'd like to do my best to name some of the people who left a mark on my brain for the better. In the field of contemplative neuroscience: Richard Davidson and the team at the Center for Healthy Minds, Daniel Siegel, Andrew Newberg, Sara Lazar and her team (Britta Hölzel and Diane Yan), Norman Farb, Wendy Hasenkamp, Tania Singer, and Emma Seppälä, as well as authors Daniel Goleman, Sharon Begley, Kelly McGonigal, and Rick Hanson. In the field of trauma: Bessel van der Kolk, Sandra Bloom, and Peter Levine. More general neuroscience and biology: Antonio Damasio, Lisa Feldman Barrett, Norman Doidge, Joseph LeDoux, Matthew Lieberman, Laurence Steinberg, Robert Sapolsky, Dacher Keltner, David Cox, John Gabrieli, and Allyson Mackey. In the field of behavioral economics: Daniel Kahneman, Dan Ariely, Richard Thaler, Abhijit Banerjee, Esther Duflo, Sendhil Mullainathan, and Eldar Shafir. In the field of management: Daniel Goleman (again), Richard Boyatzis, Annie McKee, David Rock, Adam Grant, Eric Ries, Charles Duhigg, Susan Cain, Stephen Covey, Marshall Rosenberg, Jim Collins, Chris Argyris, Peter Senge, Otto Scharmer, Daniel Pink, Frederic Laloux, Shirzad Chamine, Peter Drucker, Robert Cialdini, Simon Sinek, Dan Heath, and Chip Heath. In the field of psychology: William James, Mihaly Csikszentmihalyi, Martin Seligman, Barbara Fredrickson, Daniel Gilbert, Albert Ellis, Albert Bandura, Shawn Achor, Robert Emmons, Everett Worthington, Richard Wiseman, Jonathan Haidt, Sonja Lyubomirsky, Scott Barry Kaufman, Stuart Shanker, Michael McCullough, Shelly Gable, and David Yaden. In the field of education: Paul Tough, Carol S. Dweck, and Annette Lareau. For teachings and writings on mindfulness, meditation, and spirituality: Siddhartha Gautama, the Dalai Lama, Thich Nhat Hanh, S. N. Goenka and the Vipassana Research Institute, Matthieu Ricard, Yongey Mingyur Rinpoche, Jon Kabat-Zinn, Jack Kornfield, Sharon Salzberg, Ellen Langer, Bhante Henepola Gunaratana, Janice Marturano, Krishna Pendyala, Congressman Tim Ryan, Tara Brach, Dan Harris, David Gelles, Chade-Meng Tan, Arianna Huffington, Russell Simmons, Desmond Tutu, Michael Singer, Eckhart Tolle, and Krista Tippett.

I also feel blessed for everyone I met and all the places I visited during my spiritual explorations. To name but a few of them: Lama Zopa Rinpoche, Venerable Namgyel, Venerable Sarah, and all the other FPMT teachers, staff, and fellow students I met at the Amitabha Buddhist Centre, the Tushita Meditation Centre, and the Root Institute; Thosamling and the wonderful friends I met there, including Mary Reed; the spirit of Master Choa Kok Sui, Shailesh, the staff and volunteers at the Centre for Inner Studies, and all the pranic healers and Arhatic yogis who spent time with me; the spirit of Paramahansa Yogananda and Shibendu Lahiri for the Kriya Yoga teachings and initiation; Paalu, Satya, and the team at Tirisula Yoga for the yoga teacher training and certification; James and the team at the Yoga Shala for being the most awesome neighbors in Singapore; the Church of the Holy Child and Our Lady of Hope Catholic Parish in Philadelphia; the community of Quakers at the Arch Street Friends Meeting House and at the Friends Center (the Quaker Hub in Philadelphia); the United Lodge of Theosophists in Philadelphia for doing their best to explain theosophy to me; the Garrison Institute and my fellow retreat participants there; and, last but not least, Lama Losang Samten and the sangha at the Tibetan Buddhist Center of Philadelphia. I also want to thank Jackie Lesser and Carol Capizzi Velez for helping me embrace and own my mystical gifts.

I am very grateful for all the people who supported me to build Calm Clarity since 2013, especially all the Calm Clarity board members and advisers (many of whom have been mentioned in the book), my cousin Jo Tran, who is an incredibly talented graphic designer, the team at CultureWorks Greater Philadelphia, which has been a great fiscal sponsor and home base, and all our donors and clients. Everyone who collaborated with me to organize the initial programs (Start-up Corps [now Schoolyard Ventures] and Carver High School, Masterman High School, the University Scholars Program, the Netter Center for Community Partnerships, Central High School, KIPP DuBois

Collegiate Academy, the Cabrini Mission Corps, Cabrini University, the Wharton Club of New York, the Wharton and Penn alumni clubs of Philadelphia, Leadership Philadelphia, and Irina and Sujata, who hosted sessions in their homes) and everyone who participated in our workshops and retreats (especially those who provided invaluable feedback to improve the program and my facilitation of it), thank you for believing in the shift into Brain 3.0 and engaging with me to better explain how to make the shift. I also want to thank all our training clients for sharing Calm Clarity with their teams and making it possible for us to collaborate with social impact partners to benefit low-income students and other disadvantaged groups.

I am forever grateful to my parents for giving birth to me, supporting and taking care of me, keeping me grounded, and making sure I was always well-fed throughout the adventure of building Calm Clarity and the Collective Success Network, and writing this book. I also give never-ending thanks to the Inner Sage/Higher Self in all people, the True Self, and all its manifestations for the great cosmic mystery we call the universe and the myriad mechanisms by which higher consciousness communicates with and nurtures humanity. I didn't believe in grace until 2012 and now that I have experienced its wonders, I learned that when we embrace grace and flow with it with gratitude and humility, we create space for incredible things to unfold and manifest. Finally, I will end by thanking you, the reader, for wondering: "Is this all there is?" and "What else is possible?"

Appendix: Tools to Apply These Insights in Your Life

I would like to provide a few simple tools that can help you put the meditations and exercises provided into practice in your daily life. These are:

A. Reference Guide for the Calm Clarity Meditations and Mindfulness Practices
B. Template to Create a Practice Plan for Fostering Brain 3.0
C. Template to Plan a Daily Routine to Activate Brain 3.0
D. Recommended Reading List

A. Reference Guide for the Calm Clarity Meditations and Mindfulness Practices

The following table provides a summary of the underlying mechanisms and the benefits of the practices in the Calm Clarity Mindful Leadership Program.

Practice	Mechanisms	Purpose and Benefits
Deep Breathing (6-3-6-3)	▪ Stimulates the vagus nerve ▪ Activates the parasympathetic nervous system (PNS)	▪ Reduce stress (calm sympathetic nervous system's freeze-flight-fight cascade) ▪ Center and ground yourself in your body through breathing
Compassion (Metta) Meditation	▪ Increases activity of left prefrontal cortex (PFC) ▪ Stimulates the vagus nerve and PNS and increases oxytocin ▪ Primes compassion and empathy neural circuits ▪ Induces gamma synchrony	▪ Strengthen emotional immune system and increase positive emotions by "refilling the Brain 3.0 gas tank" ▪ Improve autopilot's capacity for compassion, empathy, and connection ▪ Reduce brain fog ▪ Steer mind wandering toward positive thoughts and images (reduce rumination)
Focused Attention and Open Awareness Meditation	▪ Exercises and strengthens neural pathways for focused attention and metacognition ▪ Enhances sensory and interoceptive awareness	▪ Learn about your autopilot through self-observation and unpacking process of sensory perception ▪ Increase ability to focus and concentrate without getting distracted ▪ Learn to stay grounded in the body and in the present moment (reduce mind wandering) ▪ Develop equanimity and beginner's mind (let go of preconceptions)

Practice	Mechanisms	Purpose and Benefits
Mindful Breathing Meditation	▪ Exercises and strengthens neural pathways for metacognition and interoceptive awareness ▪ Shifts default mode network (DMN) out of rumination mode	▪ Center and ground yourself in your body ▪ Develop space to detach from the mind-track and get out of tunnel vision ▪ Become aware of your preconditioning and neutrally watch the associative activation of neural circuits without getting swept into Brain 1.0 or Brain 2.0 ▪ Develop equanimity and beginner's mind (let go of preconceptions)
Mindful Walking (and bringing mindfulness to other daily activities)	▪ Exercises and strengthens Brain 3.0 neural pathways for focused attention and metacognition ▪ Enhances proprioceptive and interoceptive awareness	▪ Develop presence by getting grounded in your body and tuning in to your senses ▪ Appreciate daily activities like walking as opportunities to center and to learn more about your autopilot and build new habits ▪ Get out of autopilot mode and tunnel vision and create space to widen your perspective and open up to creative insights

Practice	Mechanisms	Purpose and Benefits
Mindful Eating	▪ Enhances sensory, interoceptive, and proprioceptive awareness ▪ Activates the vagus nerve to tune in to signals from your enteric nervous system and rev up the PNS rest-and-digest functions	▪ Reduce mindless overeating ▪ Develop presence by getting grounded in your body and tuning in to your senses in the process of eating: the sight, smell, texture, taste, and crunch of foods ▪ Cultivate gratitude and appreciation for food and the entire food supply chain ▪ Tap in to the body's wisdom to guide what to eat and how much to eat

B. Template to Create a Practice Plan for Fostering Brain 3.0

This Practice Plan template is designed as a tool to help you integrate the Calm Clarity practices and insights into your life. Your plan will be unique to your needs and interests. I suggest choosing three main goals to focus on at any given time. As you make progress toward these goals, you may want to choose new goals. Alternatively, if you run into obstacles that prevent you from meeting these goals, you may want to revise the goal or plan to more effectively account for and address these obstacles. Therefore, I recommend that you make a weekly or monthly appointment with yourself to review and update your Practice Plan. Please keep your plan in a convenient place where you can refer to it regularly to review and/or update it.

My Practice Plan for Fostering Brain 3.0

What I will apply	Potential challenges	Plan: When and How

C. Template to Plan a Daily Routine to Activate Brain 3.0

This Daily Routine template is designed as a tool to help you see how you can turn things that you already do into opportunities to activate Brain 3.0 neural circuits. Again, how you do something—whether you do it mindlessly or mindfully—can make a big difference. For instance, waking up can be a time to set an intention for the day, a ride on the subway or bus can become an opportunity to practice compassion meditation, a walk to the water fountain or restroom can become a mindful walking exercise, and an afternoon coffee, tea, or snack break can be an opportunity to practice mindful eating. Ideally, your Daily Routine would be aligned with the three main goals in your Practice Plan.

My Daily Routine to Activate Brain 3.0

Waking Up	
Early Morning	
Breakfast	
Commute to Work or School	
Morning	
Lunch	
Afternoon	
Commute Home	
Evening	
Dinner	
Night	
Going to Bed	

D. Recommended Reading List

To learn more about neuroscience and neuroplasticity:

- Sharon Begley, *Train Your Mind, Change Your Brain: How a New Science Reveals Our Extraordinary Potential to Transform Ourselves*
- Richard Davidson and Sharon Begley, *The Emotional Life of Your Brain: How Its Unique Patterns Affect the Way You Think, Feel, and Live—and How You Can Change Them*
- Norman Doidge, *The Brain That Changes Itself: Stories of Personal Triumph from the Frontiers of Brain Science*

To learn more about trauma:

- Bessel van der Kolk, *The Body Keeps the Score: Brain, Mind, and Body in the Healing of Trauma*
- Peter A. Levine, *In an Unspoken Voice: How the Body Releases Trauma and Restores Goodness*

To learn more about stress:

- Robert Sapolsky, *Why Zebras Don't Get Ulcers: The Acclaimed Guide to Stress, Stress-Related Diseases, and Coping*
- Kelly McGonigal, *The Upside of Stress: Why Stress Is Good for You, and How to Get Good at It*

To learn more about the autopilot and habits:

- Daniel Kahneman, *Thinking, Fast and Slow*
- Dan Ariely, *Predictably Irrational: The Hidden Forces That Shape Our Decisions*
- Charles Duhigg, *The Power of Habit: Why We Do What We Do in Life and Business*

To learn more about meditation:

- A beginner's guide to the Buddhist Vipassana meditation tradition: Bhante Henepola Gunaratana, *Mindfulness in Plain English*

- A secularized approach: Jon Kabat-Zinn, *Wherever You Go, There You Are: Mindfulness Meditation in Everyday Life*
- A psychological and neurobiological perspective: Daniel Siegel, *Mindsight: The New Science of Personal Transformation*
- A humorous autobiographical account: Dan Harris, *10% Happier: How I Tamed the Voice in My Head, Reduced Stress Without Losing My Edge, and Found Self-Help That Actually Works—A True Story*

To learn more about spiritual traditions:

My personal preference is to read the original ancient texts to gain a more direct understanding of the messages that the authors or spiritual founders of these traditions shared in their own words. Furthermore, I make it a point to read multiple translations so that I am not misled by the interpretation, biases, or agenda of any particular translator. I also find it helpful to look at translations from different eras to see how the interpretation of ancient words has evolved over time. An online search will quickly bring up multiple versions and commentaries on these texts that are freely available.

The texts that I find very insightful to read and meditate on:
- The Bhagavad Gita
- The Yoga Sutras
- The Dhammapada
- The Mahasatipatthana Sutta ("The Great Discourse on the Establishing of Mindfulness")
- The Bible (using the Interlinear Bible available at http://biblehub.com)
- The Gnostic Gospels
- The Tao Te Ching ("Dao De Jing" in Pinyin)
- *The Enchiridion* and *Discourses* of Epictetus (*The Golden Sayings of Epictetus* is available free on Kindle)

Notes

Introduction

1 Bessel van der Kolk, *The Body Keeps the Score: Brain, Mind, and Body in the Healing of Trauma* (New York: Viking, 2014).

2 Sharon Begley, *Train Your Mind, Change Your Brain: How a New Science Reveals Our Extraordinary Potential to Transform Ourselves* (New York: Ballantine Books, 2008); Richard J. Davidson and Sharon Begley, *The Emotional Life of Your Brain: How Its Unique Patterns Affect the Way You Think, Feel, and Live—and How You Can Change Them* (New York: Plume, 2013), 164–171.

3 Davidson and Begley, *The Emotional Life of Your Brain*, 161.

4 Eleanor A. Maguire, Katherine Woollett, Hugo J. Spiers, "London taxi drivers and bus drivers: A structural MRI and neuropsychological analysis," *Hippocampus* 16 (2006): 1091–1101, doi:10.1002/hipo.20233.

5 I got an in-depth understanding of the self-organizing properties of the brain by attending Dan Siegel's Soul and Synapse retreat at the Garrison Institute in May 2015. I also heard him present these ideas again at the 2016 Mindful Leadership Summit where we were both speakers. A large part of the content he shared in the retreat and talk is now available in his latest book: Daniel J. Siegel, *Mind: A Journey to the Heart of Being Human* (New York: W. W. Norton, 2017).

6 Lisa Feldman Barrett, *How Emotions Are Made: The Secret Life of the Brain* (Boston: Houghton Mifflin Harcourt, 2017).

7 Ibid., 79–80.

8 Olaf Sporns, Giulio Tononi, and Rolf Kötter, "The human connectome: A structural description of the human brain," *PLoS Computational Biology* 1.4 (2005): e42, doi:10.1371/journal.pcbi.0010042.

9 For more information on the Human Connectome Project, please refer to the two main websites: http://www.humanconnectomeproject.org and http://humanconnectome.org.

10 Carl Zimmer, "Updated Brain Map Identifies Nearly 100 New Regions," *New York Times*, July 20, 2016. https://www.nytimes.com/2016/07/21/science/human-connectome -brain-map.html?_r=0.

11 Joseph LeDoux, *Anxious: Using the Brain to Understand and Treat Fear and Anxiety* (New York: Viking Penguin, 2015), 54–55.

12 Barrett, *How Emotions Are Made*.

13 Elizabeth A. Phelps, "Human emotion and memory: Interactions of the amygdala and hippocampal complex," *Current Opinion in Neurobiology* 14 (2004): 198–202, doi:10.1016 /j.conb.2004.03.015.

14 Charles Duhigg, *The Power of Habit: Why We Do What We Do in Life and Business* (New York: Random House, 2012), 13–21; Norman Doidge, *The Brain's Way of Healing: Remarkable Discoveries and Recoveries from the Frontiers of Neural Plasticity* (New York: Viking, 2015), 38–40, 59–61, 90–92.

Part 1: My Journey Through Brains 1.0 and Brain 2.0 to Brain 3.0

Chapter 1: A Traumatic Start

1 "The Sanctuary Model," Sanctuary Institute, accessed July 17, 2017, http://thesanctuary institute.org/about-us/the-sanctuary-model.

2 "DSM-5 Fact Sheet: Posttraumatic Stress Disorder," American Psychiatric Association, accessed July 17, 2017, https://www.psychiatry.org/psychiatrists/practice/dsm/educa tional-resources/dsm-5-fact-sheets.

3 Van der Kolk, *The Body Keeps the Score*, 96–97.

4 Judith Shulevitz, "The Science of Suffering," *New Republic*, November 16, 2014, https:// newrepublic.com/article/120144/trauma-genetic-scientists-say-parents-are-passing -ptsd-kids.

5 "About the CDC-Kaiser ACE Study," Centers for Disease Control and Prevention, accessed July 17, 2017, https://www.cdc.gov/violenceprevention/acestudy/about.html.

6 "Got Your Aces Score?," ACEs Too High, accessed October 14, 2017, https://acestoohigh .com/got-your-ace-score.

7 "ACEs Science 101," ACEs Too High, accessed October 14, 2017, https://acestoohigh.com /aces-101.

8 Research and Evaluation Group at Public Health Management Corporation, "Findings from the Philadelphia Urban ACE Survey," September 18, 2013. This report can be downloaded at http://www.instituteforsafefamilies.org/philadelphia-urban-ace-study.

9 Ebonie D. Hazle, "Survey Finds Depression Pervasive in College," *Harvard Crimson*, March 31, 2003, http://www.thecrimson.com/article/2003/3/31/survey-finds-depression-pervasive -in-college/?page=single.

10 Mariel A. Klein, "The Harvard Condition," *Harvard Crimson*, October 8, 2015, http:// www.thecrimson.com/article/2015/10/8/scrutiny-harvard-condition.

11 American College Health Association, *American College Health Association–National College Health Assessment II: Undergraduate Student Reference Group Executive Summary, Fall 2016* (Hanover, MD: American College Health Association, 2017), http://www.acha -ncha.org/reports_ACHA-NCHAIIc.html.

12 B. A. Strange, P. C. Fletcher, R. N. A. Henson, K. J. Friston, and R. J. Dolan, "Segregating the functions of human hippocampus," *Proceedings of the National Academy of Sciences of the United States of America* 96.7 (1999): 4034–39, doi: 10.1073/pnas.96.7.4034.

13 Omer Bonne, Meena Vythilingam, Masatoshi Inagaki, et al., "Reduced posterior hippocampal volume in posttraumatic stress disorder," *Journal of Clinical Psychiatry* 69.7 (2008): 1087–91, https://www.ncbi.nlm.nih.gov/pmc/articles/PMC2684983.

14 Claudia Lieberwirth, Yongliang Pan, Yan Liu, Zhibin Zhang, Zuoxin Wang, "Hippocampal adult neurogenesis: Its regulation and potential role in spatial learning and memory," *Brain Research* 1644 (2016): 127–40, doi:10.1016/j.brainres.2016.05.015.

15 Bonne, Vythilingam, Inagaki, et al., "Reduced posterior hippocampal volume in posttraumatic stress disorder"; Linghui Meng, Jing Jiang, Changfeng Jin, Jia Liu, Youjin Zhao, Weina Wang, Kaiming Li, and Qiyong Gong, "Trauma-specific grey matter alterations in PTSD," *Scientific Reports* 6 (2016): 33748, doi:10.1038/srep33748.

Chapter 2: Becoming a Mind-Hacker

1 Davidson and Begley, *The Emotional Life of Your Brain*, 23–27.

2 Ibid.

3 Ibid., 30–31.

4 Antonio Damasio, *Looking for Spinoza: Joy, Sorrow, and the Feeling Brain* (Orlando, FL: Harcourt, 2003), 62; Begley, *Train Your Mind, Change Your Brain*, 225–26.

5 Van der Kolk, *The Body Keeps the Score*, 60–62.

6 Davidson and Begley, *The Emotional Life of Your Brain*, 71.

7 Olga Khazan, "The Upside of Pessimism," *The Atlantic*, September 12, 2014, https://www.theatlantic.com/health/archive/2014/09/dont-think-positively/379993.

8 Heidi Grant, "Be an Optimist Without Being a Fool," *Harvard Business Review*, May 2, 2011, https://hbr.org/2011/05/be-an-optimist-without-being-a.

9 Van der Kolk, *The Body Keeps the Score*, 142–43.

Chapter 3: An Unhappy Pursuit of Happiness

1 Bruce Lee, *Striking Thoughts: Bruce Lee's Wisdom for Daily Living*, ed. John Little (Rutland, VT: Tuttle Publishing, 2000).

2 Daniel Kahneman, *Thinking, Fast and Slow* (New York: Farrar, Straus and Giroux, 2011), 364–66.

3 Simon Moss, "Approach and Avoidance Motivation," Sicotests, accessed July 17, 2017, http://www.sicotests.com/psyarticle.asp?id=60.

4 Ibid.

5 Daniel C. Molden and David B. Miele, "The origins and influences of promotion-focused and prevention-focused achievement motivations," in *Advances in Motivation and Achievement: Social Psychological Perspectives on Motivation and Achievement*, vol. 15, ed. Martin L. Maehr, Stuart A. Karabenick, and Timothy C. Urdan (Bingley, Wales: Emerald Group, 2008), 81–118.

6 Duhigg, *The Power of Habit*, 13–21; Robert S. Marin and Patricia A. Wilkosz, "Disorders of diminished motivation," *Journal of Head Trauma Rehabilitation* 20.4 (2005): 377–88.

7 Tim Layden, "After Rehabilitation, the Best of Michael Phelps May Lie Ahead," *Sports Illustrated*, November 9, 2015, https://www.si.com/olympics/2015/11/09/michael-phelps-rehabilitation-rio-2016.

8 Doidge, *The Brain's Way of Healing*, 38–40, 59–61, 90–92.

9 Ibid.

10 I learned this from Rick Bellingham, one of my mentors, through a powerful piece he wrote about his experience living with Tourette syndrome: Rick Bellingham, "Taming Tourette," Perspectives and Possibilities (blog), January 3, 2017, https://rickbellingham.com/2017/01/03/taming-tourette.

11 James Olds and Peter Milner, "Positive reinforcement produced by electrical stimulation of the septal area and other regions of rat brain," *Journal of Comparative and Physiological Psychology* 47 (1954): 419–27, doi: 10.1037/h0058775.

12 "The Science of Willpower: Kelly McGonigal at TEDxBayArea," YouTube video, 15:43, posted by TEDx Talks, May 18, 2012, https://youtu.be/W_fQvcBCNbA.

13 Ibid.

14 Nora D. Volkow, Joanna S. Fowler, and Gene-Jack Wang, "The addicted human brain: Insights from imaging studies," *Journal of Clinical Investigation* 111.10 (2003): 1444–51, doi: 10.1172/jci18533.

15 Laurence Steinberg, *Age of Opportunity: Lessons from the New Science of Adolescence* (Boston: Eamon Dolan, Houghton Mifflin Harcourt, 2014), 73–74.

16 David Rock, *Your Brain at Work* (New York: Harper Collins, 2009), 192.

17 "Basic Principles of Classical Conditioning: Pavlov," Boundless.com, September 20, 2016, accessed August 31, 2017, https://www.boundless.com/psychology/textbooks/boundless-psychology-textbook/learning-7/classical-conditioning-46/basic-principles-of-classical-conditioning-pavlov-192-12727; "Ivan Pavlov," Wikipedia, accessed August 31, 2017, https://en.wikipedia.org/wiki/Ivan_Pavlov.

Chapter 4. A Skeptic Stumbles on "Enlightenment"

1 This is an authenticated quote from Albert Einstein: "Religious and Philosophical Views of Albert Einstein," Wikipedia, accessed July 21, 2017, https://en.wikipedia.org/wiki/Religious_and_philosophical_views_of_Albert_Einstein.

2 René Descartes, "Discourse on Method, Part V (1637)" in *Discourse on Method and Meditations*, trans. Elizabeth S. Haldane and G. R. T. Ross (New York: Dover Publications, 2003).

3 "And India's 7 Wonders Are," *Times of India*, August 5, 2007, http://timesofindia.india times.com/And-Indias-7-wonders-are-/articleshow/2256323.cms?.

4 This is an authenticated quote from Albert Einstein: Albert Einstein, *The World as I See It*, trans. Alan Harris (New York: Philosophical Library, 1949), 21.

5 Tenzin Gyatso, "Our Faith in Science," *New York Times*, November 12, 2005, http://www.nytimes.com/2005/11/12/opinion/our-faith-in-science.html.

6 *Farlex Partner Medical Dictionary*, s.v. "phosphenes," accessed July 18, 2017, http://medical-dictionary.thefreedictionary.com/Phosphenes.

7 "What Do You 'See' When You Close Your Eyes?," Vision Eye Institute, accessed July 18, 2017, http://www.visioneyeinstitute.com.au/article/see-close-eyes.

8 David Bryce Yaden, Jonathan Iwry, Andrew Newberg, et al., "The overview effect: Awe and self-transcendent experience in space flight," *Psychology of Consciousness: Theory, Research, and Practice* 3.1 (2016): 1–11, doi: 10.1037/cns0000086; Julia Calderone,

"Something Profound Happens When Astronauts See Earth from Space for the First Time," *Business Insider*, August 31, 2015, http://www.businessinsider.com/overview -effect-nasa-apollo8-perspective-awareness-space-2015-8.

9 *Oxford English Dictionary*, s.v. "mystical," accessed July 18, 2017, https://en.oxforddictio naries.com/definition/mystical.

10 *Merriam-Webster Dictionary*, s.v. "mystical," accessed July 18, 2017, https://www.merriam -webster.com/dictionary/mystical.

11 Drake Baer, "Awe Is the Everyperson's Spiritual Experience," *New York Magazine*, Science of Us, March 9, 2017, http://nymag.com/scienceofus/2017/03/why-awe-is-so-good-for -you.html.

12 David Bryce Yaden, Jonathan Haidt, Ralph W. Hood, David R. Vago, and Andrew Newberg, "The varieties of self-transcendent experience," *Review of General Psychology* (2017), doi: 10.1037/gpr0000102.

13 Jack Kornfield, *After the Ecstasy, the Laundry: How the Heart Grows Wise on the Spiritual Path* (New York: Bantam Books, 2000), 93.

Chapter 5. Guided by an Inner Compass

1 To verify that this quote attributed to the Buddha is authentic, I have traced it to the Jnanasara-Samuccaya. This particular English rendering of these lines was shared by the Dalai Lama in John F. Avedon, *An Interview with The Dalai Lama* (New York: Littlebird Publications, 1980), 51–53. These lines are also provided on this website: "Our Teachers," the White Lotus Buddhist Center, accessed July 18, 2017, http://www.whitelotusdharma .org/teachers.

2 Immanuel Kant, "What Is Enlightenment?," trans. Mary C. Smith, accessed July 18, 2017, http://www.columbia.edu/acis/ets/CCREAD/etscc/kant.html.

3 Mary provides a very moving account of her struggles to accept her mystical awakening and a detailed description of her visions in her memoir: Mary Reed, *Unwitting Mystic: Evolution of the Message of Love* (n.p.: CreateSpace, 2014).

4 William James, *Talks to Teachers on Psychology: And to Students on Some of Life's Ideals* (New York: Henry Holt, 1899), 64–66.

5 Ibid., 76–77.

6 "Introduction to the Technique," under "The Code of Discipline," Vipassana Meditation as taught by S. N. Goenke, accessed July 17, 2017, https://www.dhamma.org/en-US /about/code.

7 Kahneman, *Thinking, Fast and Slow*, 51.

8 According to Wikipedia this term was first introduced by William James in this publication: William James, *The Principles of Psychology* (New York: Henry Holt, 1890). Source: "Stream of Consciousness (Narrative Mode)," Wikipedia, accessed July 18, 2017, https:// en.wikipedia.org/wiki/Stream_of_consciousness_(narrative_mode).

9 A. D. Craig, "How do you feel? Interoception: The sense of the physiological condition of the body," *Nature Reviews Neuroscience* 3 (2002): 655-66, doi:10.1038/nrn894; A. D. Craig, "Interoception: The sense of the physiological condition of the body," *Current Opinion in Neurobiology* 13 (2003): 500–505, doi:10.1016/S0959-4388(03)00090-4; Norman A. S. Farb, Jennifer Daubenmier, Cynthia J. Price, Tim Gard, Catherine Kerr, Barnaby D. Dunn, Anne Carolyn Klein, Martin P. Paulus, and Wolf E. Mehling, "Interoception, contemplative practice, and health," *Frontiers in Psychology* 6 (2015): 763, doi:10.3389/fpsyg .2015.00763.

10 Drake Baer, "Wait, What's That? The Science behind Why Your Mind Keeps Wandering," *Fast Company*, August 26, 2013, https://www.fastcompany.com/3016114/wait-whats -that-the-science-behind-why-your-mind-keeps-wandering.

11 Matthew A. Killingsworth and Daniel T. Gilbert, "A wandering mind is an unhappy mind," *Science* 330 (2010): 932, doi:10.1126/science.1192439; Matt Killingsworth, "Does Mind-Wandering Make You Unhappy?," *Greater Good Magazine*, July 16, 2013, http://greater good.berkeley.edu/article/item/does_mind_wandering_make_you_unhappy.

12 Killingsworth and Gilbert, "A wandering mind is an unhappy mind."

13 "Equanimity," Wikipedia, accessed July 18, 2017, https://en.wikipedia.org/wiki/Equa nimity.

14 "The Eight-fold Path," BuddhaNet, accessed July 18, 2017, http://www.buddhanet.net /e-learning/8foldpath.htm.

15 The translations I propose are partly derived from examining the Pali words using the free online glossary provided by Wisdom Library at http://www.wisdomlib.org/index .php.

16 "The SRF Line of Gurus," Self-Realization Fellowship, accessed July 18, 2017, http:// www.yogananda-srf.org/lineageandleadership/The_SRF_Line_of_Gurus.aspx#.WIeN 8PkrIdU.

17 "Traditional Knowledge Digital Library," Wikipedia, accessed September 2, 2017, https:// en.wikipedia.org/wiki/Traditional_Knowledge_Digital_Library.

18 *MedicineNet.com Medical Dictionary*, s.v. "proprioception," accessed July 18, 2017, http:// www.medicinenet.com/script/main/art.asp?articlekey=6393.

19 Artur Struzik, Bogdan Pietraszewski, Adam Kawczynski, Sławomir Winiarski, Grzegorz Juras, and Andrzej Rokita, "Manifestations of proprioception during vertical jumps to specific heights," *Journal of Strength and Conditioning Research* 31.6 (2017): 1694–1701, doi: 10.1519/JSC.0000000000001868.

20 Uwe Proske and Simon C. Gandevia, "The proprioceptive senses: Their roles in signaling body shape, body position and movement, and muscle force," *Physiological Reviews* 92.4 (2012): 1651–97, doi: 10.1152/physrev.00048.2011.

21 Van der Kolk, *The Body Keeps the Score*, 207.

22 Robert M. Sapolsky, *Why Zebras Don't Get Ulcers: The Acclaimed Guide to Stress, Stress-Related Diseases, and Coping* (New York: Henry Holt, 2004), 37–48.

23 Ibid., 48–49.

24 Sapolsky, *Why Zebras Don't Get Ulcers*.

25 Grethen Cuda, "Just Breathe: Body Has a Built-In Stress Reliever," NPR, December 6, 2010, http://www.npr.org/2010/12/06/131734718/just-breathe-body-has-a-built-in-stress -reliever.

26 Matthew 11:28–30, English Standard Version.

27 Antonio Damasio, *Descartes' Error: Emotion, Reason and the Human Brain* (New York: Penguin Books, 2005), 249, citing: René Descartes, "Discourse on Method, Part IV (1637)" in *The Philosophical Works of Descartes, Volume I*, trans. Elizabeth S. Haldane and G. R. T. Ross (New York: Cambridge University Press, 1970), 101.

Chapter 6. Creating and Sharing Calm Clarity

1 This is an authenticated quote from Albert Einstein: Albert Einstein, *The Einstein Reader* (New York: Citadel Press, 2006), 26. Please note, this book was previously published under the title *Out of My Later Years* in 1950.

2 Trip Gabriel, "Budget Cuts Reach Bone for Philadelphia Schools," *New York Times,* June 16, 2013, http://www.nytimes.com/2013/06/17/education/budget-cuts-reach-bone-for -philadelphia-schools.html.

3 David J. Kearney, Carol A. Malte, Carolyn McManus, Michelle E. Martinez, Ben Felleman, and Tracy L. Simpson, "Loving-kindness meditation for posttraumatic stress disorder: A pilot study," *Journal of Traumatic Stress* 26 (2013): 426–34, doi: 10.1002/jts.21832; David J. Kearney, Carol A. Malte, Carolyn McManus, Michelle E. Martinez, Ben Felleman, and Tracy L. Simpson, "Loving-kindness meditation and the broaden-and-build theory of positive emotions among veterans with posttraumatic stress disorder," *Medical Care* 52.12 Suppl 5 (2014): S32–38, doi: 10.1097/MLR.0000000000000221.

4 Barbara L. Fredrickson, Michael A. Cohn, Kimberly A. Coffey, Jolynn Pek, and Sandra M. Finkel, "Open hearts build lives: Positive emotions, induced through loving-kindness meditation, build consequential personal resources," *Journal of Personality and Social Psychology* 95.5 (2008): 1045–62, doi: 10.1037/a0013262; Bethany E. Kok, Kimberly A. Coffey, Michael A. Cohn, Lahnna I. Catalino, Tanya Vacharkulksemsuk, Sara B. Algoe, Mary Brantley, and Barbara L. Fredrickson, "How positive emotions build physical health: Perceived positive social connections account for the upward spiral between positive emotions and vagal tone," *Psychological Science* 24.7 (2013): 1123–32, doi: 10.1177/095679761247 0827.

5 Cendri A. Hutcherson, Emma M. Seppälä, and James J. Gross, "Loving-kindness meditation increases social connectedness," *Emotion* 8 (2008), 720–24, doi: 10.1037/a0013237.

6 Barbara Fredrickson, *Love 2.0: Creating Happiness and Health in Moments of Connection* (New York: Plume, 2014), 47–53; Kelly McGonigal, *The Upside of Stress: Why Stress Is Good for You, and How to Get Good at It* (New York: Avery, 2015), 52–53; Bethany E. Kok and Barbara L. Fredrickson, "Upward spirals of the heart: Autonomic flexibility, as indexed by vagal tone, reciprocally and prospectively predicts positive emotions and social connectedness," *Biological Psychology* 85.3 (2010): 432–36, doi: 10.1016/j.biopsycho.2010.09.005.

7 There are slight variations in wording between translations: "The Gospel of Thomas," trans. Stevan Davies, Gnostic Society Library, accessed July 18, 2017, http://www.gnosis.org /naghamm/gosthom-davies.html, and "The Gospel of Thomas," trans. Stephen Patterson and Marvin Meyer, Gnostic Society Library, accessed July 18, 2017, http://www.gnosis .org/naghamm/gosthom.html.

8 "Ellen Langer: Science of Mindlessness and Mindfulness," *On Being with Krista Tippett,* September 10, 2015, accessed July 18, 2017, https://onbeing.org/programs/ellen-langer -science-of-mindlessness-and-mindfulness.

9 Sendhil Mullainathan and Eldar Shafir, *Scarcity: Why Having Too Little Means So Much* (New York: Times Books, Henry Holt, 2013), 48–52.

Part 2: A Mind-Hacker's Guide to Shifting into Brain 3.0

1 Jonathan Haidt, "Religion, Evolution, and the Ecstasy of Self-Transcendence," filmed February 2012, TED video, 18:09, posted March 2012, https://www.ted.com/talks/jona than_haidt_humanity_s_stairway_to_self_transcendence.

Chapter 7. Overview of the Journey

1 Stephen Cope, *The Great Work of Your Life: A Guide for the Journey to Your True Calling* (New York: Bantam Books, 2012), 168.

2 Marion Woodman and Jill Mellick, *Coming Home to Myself: Reflections for Nurturing a Woman's Body and Soul* (Boston: Conari Press, 2000), 259.

3 Van der Kolk, *The Body Keeps the Score*, 233.

4 Ibid., 206.

5 Thich Nhat Hanh, *The Sun My Heart: Reflections on Mindfulness, Concentration, and Insight* (Berkeley, CA: Parallax Press, 2010),12–13.

6 Kelly McGonigal explains the body and brain's energy budgeting mechanisms and the effects of low energy levels, sleep deprivation, and tiredness on willpower, one of the key networks of Brain 3.0 in: Kelly McGonigal, *The Willpower Instinct: How Self-Control Works, Why It Matters, and What You Can Do to Get More of It* (New York: Avery, 2012), 45–48, 51–53, 60–80.

7 Guy Claxton, *Intelligence in the Flesh: Why Your Mind Needs Your Body Much More Than It Thinks* (New Haven: Yale University Press, 2015), 103.

8 Damasio, *Looking for Spinoza*, 6.

9 Barrett, *How Emotions Are Made*, 104, 27–28.

10 Ibid., 78, 134–39.

11 Martin Seligman and John Tierney, "We Aren't Built to Live in the Moment," *New York Times*, May 9, 2017, https://www.nytimes.com/2017/05/19/opinion/sunday/why-the-future-is-always-on-your-mind.html.

12 Thorsten Barnhofer, Tobias Chittka, Helen Nightingale, Claire Visser, and Catherine Crane, "State effects of two forms of meditation on prefrontal EEG asymmetry in previously depressed individuals," *Mindfulness* 1.1 (2010), 21–27, doi: 10.1007/s12671-010-0004-7; Richard J. Davidson, "What does the prefrontal cortex 'do' in affect: Perspectives on frontal EEG asymmetry research," *Biological Psychology* 67 (2004): 219–33.

13 Begley, *Train Your Mind, Change Your Brain*, 225–29.

14 Elizabeth Blackburn and Elissa Epel, *The Telomere Effect: A Revolutionary Approach to Living Younger, Healthier, Longer* (New York: Grand Central Publishing, 2017), 37; Hal E. Hershfield, Susanne Scheibe, Tamara L. Sims, and Laura L. Carstensen, "When feeling bad can be good: Mixed emotions benefit physical health across adulthood," *Social Psychological and Personality Science* 4.1 (2013), 54–61, doi: 10.1177/1948550612444616.

15 David Rock, "SCARF: A brain-based model for collaborating with and influencing others," *Neuroleadership Journal* 1 (2008), accessed July 15, 2017, http://web.archive.org/web/20100705024057/http://www.your-brain-at-work.com:80/files/NLJ_SCARFUS.pdf.

16 Rick Hanson and Richard Mendius, *Buddha's Brain: The Practical Neuroscience of Happiness, Love, and Wisdom* (Oakland, CA: New Harbinger Publications, 2009), 45–46.

17 I heard Marc Brackett talk when we were both speakers at the Mindful Leadership Summit in September 2016: Marc Brackett, "Emotionally Intelligent Leadership," 2016 Mindful Leadership Summit Talk.

18 Matthew D. Lieberman, Naomi I. Eisenberger, Molly J. Crockett, Sabrina M. Tom, Jennifer H. Pfeifer, and Baldwin M. Way, "Putting feelings into words: Affect labeling disrupts amygdala activity to affective stimuli," *Psychological Science* 18.5 (2007), 421–28, doi: 10.1111/j.1467-9280.2007.01916.x; Lisa J. Burklund, J. David Creswell, Michael R. Irwin, and Matthew D. Lieberman, "The common and distinct neural bases of affect labeling and reappraisal in healthy adults," *Frontiers in Psychology* 5 (2014): 221.

19 Stuart Wolpert, "Putting Feelings into Words Produces Therapeutic Effects in the Brain; UCLA Neuroimaging Study Supports Ancient Buddhist Teachings," UCLA Newsroom, June 21, 2007, http://newsroom.ucla.edu/releases/Putting-Feelings-Into-Words-Produces-8047.

20 Duhigg, *The Power of Habit*, 13–21.

21 Ibid.

22 Tim Gard, Maxime Taquet, Rohan Dixit, Britta K. Hölzel, Bradford C. Dickerson, and Sara W. Lazar, "Greater widespread functional connectivity of the caudate in older adults who practice kripalu yoga and vipassana meditation than in controls," *Frontiers in Human Neuroscience* 9 (2015): 137, doi:10.3389/fnhum.2015.00137; Sanne de Wit, Poppy Watson, Helga A. Harsay, Michael X. Cohen, Irene van de Vijver, and K. Richard Ridderinkhof, "Corticostriatal connectivity underlies individual differences in the balance between habitual and goal-directed action control," *Journal of Neuroscience* 32.35 (2012): 12066–75, doi: 10.1523/jneurosci.1088-12.2012.

23 Daniel Siegel expounded upon the integrative functions of the prefrontal cortex in his talk at the 2016 Mindful Leadership Summit: Daniel Siegel, "Mind: A Journey to the Heart of Being Human by Dan Siegel," 2016 Mindful Leadership Summit Talk.

24 Claxton, *Intelligence in the Flesh*, 66.

25 Damasio, *Descartes' Error.*

26 Damasio, *Looking for Spinoza*, 62; Begley, *Train Your Mind, Change Your Brain*, 65.

27 Britta K. Hölzel, Sara W. Lazar, Tim Gard, Zev Schuman-Olivier, David R. Vago, Ulrich Ott, "How does mindfulness meditation work? Proposing mechanisms of action from a conceptual and neural perspective," *Perspectives on Psychological Science* 6.6 (2011): 537–59, doi: 10.1177/1745691611419671; Andrew Newberg and Mark Robert Waldman, *How God Changes Your Brain: Breakthrough Findings from a Leading Neuroscientist* (New York: Ballantine Books, 2009), 19.

28 Sara W. Lazar, Catherine E. Kerr, Rachel H. Wasserman, et al., "Meditation experience is associated with increased cortical thickness," *Neuroreport* 16.17 (2005): 1893–97, https://www.ncbi.nlm.nih.gov/pmc/articles/PMC1361002/; Norman A. S. Farb, Zindel V. Segal, and Adam K. Anderson, "Mindfulness meditation training alters cortical representations of interoceptive attention," *Social Cognitive and Affective Neuroscience* 8.1 (2013): 15–26, doi: 10.1093/scan/nss066; Véronique A. Taylor, Véronique Daneault, Joshua Grant, et al., "Impact of meditation training on the default mode network during a restful state," *Social Cognitive and Affective Neuroscience* 8.1 (2013), 4–14, doi: 10.1093/scan/nsr087.

29 Damasio, *Descartes' Error*, 62–63.

30 John M. Allman, Atiya Hakeem, Joseph M. Erwin, et al., "The anterior cingulate cortex: The evolution of an interface between emotion and cognition," *Annals of the New York Academy of Sciences* 935 (2001): 107–17, doi: 10.1111/j.1749-6632.2001.tb03476.x.

31 Matthew D. Lieberman, *Social: Why Our Brains Are Wired to Connect* (New York: Crown Publishers, 2013); Mary Helen Immordino-Yang, Andrea McColl, Hanna Damasioa, and Antonio Damasio, "Neural correlates of admiration and compassion," *Proceedings of the National Academy of Sciences of the United States of America* 106.19 (2009): 8021–26, doi: 10.1073/pnas.0810363106.

32 Newberg and Waldman, *How God Changes Your Brain*, 124–26, 215; Allman, Hakeem, Erwin, et al., "The anterior cingulate cortex"; Regina Bailey, "Cingulate Gyrus and the Limbic System," ThoughtCo, last modified August 29, 2016, accessed July 20, 2017, https://www.thoughtco.com/cingulate-gyrus-and-the-limbic-system-4078935; Claudio Lavin, Camilo Melis, Ezequiel Mikulan, et al., "The anterior cingulate cortex: An integrative hub for human socially-driven interactions," *Frontiers in Neuroscience* 7.64 (2013), doi: 10.3389/fnins.2013.00064.

33 Wendy Hasenkamp, Christine D. Wilson-Mendenhall, Erica Duncan, and Lawrence W. Barsalou, "Mind wandering and attention during focused meditation: A fine-grained temporal analysis of fluctuating cognitive states," *NeuroImage* 59.1 (2012): 750–60, doi:

10.1016/j.neuroimage.2011.07.008; Tania Singer, Hugo D. Critchley, and Kerstin Preus-
choff, "A common role of insula in feelings, empathy and uncertainty," *Trends in Cognitive
Sciences* 13.8 (2009): 334–40, doi: 10.1016/j.tics.2009.05.001.

34 Andrew Newberg and Mark Robert Waldman, *How Enlightenment Changes Your Brain:
The New Science of Transformation* (New York: Avery, 2016), 133.

35 Ingfei Chen, "Brain Cells for Socializing," *Smithsonian Magazine*, June 2009, http://www
.smithsonianmag.com/science-nature/brain-cells-for-socializing-133855450.

36 Camilla Butti, Chet C. Sherwood, Atiya Y. Hakeem, John M. Allman, and Patrick R. Hof,
"Total number and volume of Von Economo neurons in the cerebral cortex of cetaceans,"
Journal of Comparative Neurology 515.2 (2009): 243–59, doi: 10.1002/cne.22055.

37 Chen, "Brain Cells for Socializing"; K. K. Watson, T. K. Jones, and J. M. Allman, "Dendritic
architecture of the von Economo neurons," *Neuroscience* 141.3 (2006): 1107–12, doi:
10.1016/j.neuroscience.2006.04.084.

38 Chen, "Brain Cells for Socializing."

39 Yaden, Iwry, Newberg, et al., "The overview effect"; Olivia Goldhill, "Astronauts Report an
'Overview Effect' from the Awe of Space Travel—and You Can Replicate It Here on Earth,"
Quartz, September 6, 2015, https://qz.com/496201/astronauts-report-an-overview-effect
-from-the-awe-of-space-travel-and-you-can-replicate-it-here-on-earth.

40 Haiying Shao and Ming-Sheng Zhou, "Cardiovascular action of oxytocin," *Journal of Au-
tacoids and Hormones* 3.1 (2014): e124, doi: 10.4172/2161-0479.1000e124.

41 Sapolsky, *Why Zebras Don't Get Ulcers*, 33–34.

42 McGonigal, *The Upside of Stress*, 135–80 (chapter 5).

43 Ibid., 52–53.

44 Ibid., 50–53.

45 Ibid., 109–12.

46 Ibid., 139.

47 Ibid., 140–41.

48 Van der Kolk, *The Body Keeps the Score*, 207; Christopher Bergland, "The Neurobiology of
Grace under Pressure," *Psychology Today* website, February 2, 2013, https://www.psychol
ogytoday.com/blog/the-athletes-way/201302/the-neurobiology-grace-under-pressure.

49 Bergland, "The Neurobiology of Grace under Pressure"; Kok and Fredrickson, "Upward
spirals of the heart."

50 Dacher Keltner, "The Compassionate Species," *Greater Good Magazine*, July 31, 2012,
http://greatergood.berkeley.edu/article/item/the_compassionate_species.

51 Fredrickson, *Love 2.0*, 54.

Chapter 8. Nurturing and Strengthening Brain 3.0

1 *Merriam-Webster Dictionary*, s.v. "meditate," accessed July 20, 2017, https://www.merriam
-webster.com/dictionary/meditate.

2 Jon Kabat-Zinn, *Full Catastrophe Living: Using the Wisdom of Your Body and Mind to Face
Stress, Pain, and Illness* (New York: Bantam Books, 2013), 188. I confirmed Kabat-Zinn's
etymology using: Walter W. Skeat, *A Concise Etymological Dictionary of the English Language* (Oxford: Clarendon Press, 1885); *Online Etymology Dictionary*, s.v. "meditation,"
accessed July 20, 2017, http://www.etymonline.com/index.php?allowed_in_frame=0&
search=meditation; and *Online Etymology Dictionary*, s.v. "medicine," accessed July 20,
2017, http://www.etymonline.com/index.php?allowed_in_frame=0&search=medicine.

3 Kabat-Zinn, *Full Catastrophe Living*, 188.

4 This is an authenticated quote from Albert Einstein: Albert Einstein, letter to Norman Salit, March 4, 1950, quoted in Walt Martin and Magda Ott, eds., *The Cosmic View of Albert Einstein: Writings on Art, Science, and Peace* (New York: Sterling, 2013), 2.

5 Kearney, McManus, Malte, Martinez, Felleman, and Simpson, "Loving-kindness meditation for posttraumatic stress disorder"; Kearney, Malte, McManus, Martinez, Felleman, and Simpson, "Loving-kindness meditation and the broaden-and-build theory of positive emotions"; Kok, Coffey, Cohn, Catalino, Vacharkulksemsuk, Algoe, Brantley, and Fredrickson, "How positive emotions build physical health."

6 Marc Kaufman, "Meditation Gives Brain a Charge, Study Finds," *Washington Post*, January 3, 2005, A05, http://www.washingtonpost.com/wp-dyn/articles/A43006-2005Jan2 .html.

7 Begley, *Train Your Mind, Change Your Brain*, 234–35.

8 Antoine Lutz, Lawrence L. Greischar, Nancy B. Rawlings, Matthieu Ricard, and Richard J. Davidson, "Long-term meditators self-induce high-amplitude gamma synchrony during mental practice," *Proceedings of the National Academy of Sciences of the United States of America* 101.46 (2004): 16369–73, doi: 10.1073/pnas.0407401101.

9 Richard Davidson, phone conversation with author, September 3, 2016.

10 Barnhofer, Chittka, Nightingale, Visser, and Crane, "State effects of two forms of meditation on prefrontal EEG asymmetry."

11 Matthew 22:37 and 22:39, King James Version.

12 Kahneman, *Thinking, Fast and Slow*, 58.

13 Timothy Wilson, *Strangers to Ourselves: Discovering the Adaptive Unconscious* (Cambridge, MA: Belknap Press, 2002).

14 "Priming," *Psychology Today*, accessed July 20, 2017, https://www.psychologytoday.com /basics/priming.

15 Kahneman, *Thinking, Fast and Slow*, 52.

16 Kahneman, *Thinking, Fast and Slow*, 54; Fritz Strack, Leonard Martin, Sabine Stepper, "Inhibiting and facilitating conditions of the human smile: A nonobtrusive test of the facial feedback hypothesis," *Journal of Personality and Social Psychology* 54 (1988): 768–77, doi: 10.1037/0022-3514.54.5.768.

17 Ralf Rummer, Judith Schweppe, René Schlegelmilch, and Martine Grice, "Mood is linked to vowel type: The role of articulatory movements," *Emotion* 14.2 (2014): 246–50, doi: 10.1037/a0035752; University of Cologne—Universität zu Köln, "We speak as we feel, we feel as we speak," *ScienceDaily*, June 26, 2014, https://www.sciencedaily.com/releases /2014/06/140626095717.htm.

18 To learn more about the frustrations of a badly designed door, check out this article by 99% Invisible, which features a video by Joe Posner of Vox on this topic: "Norman Doors: Don't Know Whether to Push or Pull? Blame Design," 99% Invisible, February 26, 2016, http://99percentinvisible.org/article/norman-doors-dont-know-whether-push-pull -blame-design.

19 This quote is popularly attributed to Viktor Frankl, yet there is no evidence he actually said or wrote it. According to the Viktor Frankl Institute Vienna, these lines by an unknown author were used by Stephen Covey to describe Viktor Frankl's essential teachings: Franz Vesely, "Alleged Quote: 'Between Stimulus and Response . . . ,'" Victor Frankl Insitute, accessed October 21, 2017, http://www.viktorfrankl.org/e/quote_stim ulus.html.

20 My understanding of the Mahasatipatthana Sutta comes from studying multiple translations and searching online for various interpretations of Pali words. The version I relied

on most provided the Pali version and English translation side by side: "Mahāsatipatthāna Sutta," Vipassana Research Institute, trans., Pali Tipitaka, accessed July 20, 2017, http://www.tipitaka.org/stp-pali-eng-parallel.

21 A good overview of the Five Aggregates is contained in: Thich Nhat Hanh, *The Heart of the Buddha's Teaching: Transforming Suffering into Peace, Joy, and Liberation* (New York: Broadway Books, 1999), 176–83.

22 Barrett, *How Emotions Are Made*, 84, 145–46.

23 Ibid., 65–66.

24 Baer, "Wait, What's That?"

25 Scott Barry Kaufman, "The Real Neuroscience of Creativity," *Scientific American*, August 19, 2013, https://blogs.scientificamerican.com/beautiful-minds/the-real-neuroscience-of-creativity.

26 Lieberman, *Social*, 15–19.

27 Barrett, *How Emotions Are Made*.

28 Seligman and Tierney, "We Aren't Built to Live in the Moment."

29 Lieberman, *Social*.

30 Kahneman, *Thinking, Fast and Slow*, 79–88.

31 Yongey Mingyur Rinpoche and Eric Swanson, *Joyful Wisdom: Embracing Change and Finding Freedom* (New York: Three Rivers Press, 2010), 15–16.

32 Lazar, Kerr, Wasserman, et al. "Meditation experience is associated with increased cortical thickness."

33 Norman A. S. Farb, Zindel V. Segal, and Adam K. Anderson, "Attending to the present: Mindfulness meditation reveals distinct neural modes of self-reference," *Social Cognitive and Affective Neuroscience* 2.4 (2007): 313–22, doi: 10.1093/scan/nsm030; Rock, *Your Brain at Work*, 93–95.

34 Judson A. Brewer, Patrick D. Worhunsky, Jeremy R. Gray, Yi-Yuan Tang, Jochen Weber, and Hedy Kober, "Meditation experience is associated with differences in default mode network activity and connectivity," *Proceedings of the National Academy of Sciences of the United States of America* 108.50 (2011): 20254–59, doi: 10.1073/pnas.1112029108.

35 Wendy Hasenkamp and Lawrence W. Barsalou, "Effects of meditation experience on functional connectivity of distributed brain networks," *Frontiers in Human Neuroscience* 6 (2012): 38, doi: 10.3389/fnhum.2012.00038.

36 Wendy Hasenkamp, "How to Focus a Wandering Mind," *Greater Good Magazine*, July 17, 2013, https://greatergood.berkeley.edu/article/item/how_to_focus_a_wandering_mind.

37 Victoria Singh-Curry and Masud Husain, "The functional role of the inferior parietal lobe in the dorsal and ventral stream dichotomy," *Neuropsychologia* 47.6 (2009): 1434–48, doi: 10.1016/j.neuropsychologia.2008.11.033.

38 Taylor, Daneault, Grant, et al., "Impact of meditation training on the default mode network during a restful state."

39 Aurelie Fontan, Fabien Cignetti, Bruno Nazarian, Jean-Luc Anton, Marianne Vaugoyeau, and Christine Assaiante, "How does the body representation system develop in the human brain?," *Developmental Cognitive Neuroscience* 24 (2017): 118–28, doi: 10.1016/j.dcn.2017.02.010.

40 Chantal Villemure, Marta Čeko, Valerie A. Cotton, and M. Catherine Bushnell, "Insular cortex mediates increased pain tolerance in yoga practitioners," *Cerebral Cortex* 24.10 (2014): 2732–40, doi: 10.1093/cercor/bht124; Laura Schmalzl, Chivon Powers, and Eva Henje Blom, "Neurophysiological and neurocognitive mechanisms underlying the effects of yoga-based practices: Toward a comprehensive theoretical framework," *Frontiers in*

Human Neuroscience 9 (2015): 235, doi: 10.3389/fnhum.2015.00235; Stephani Sutherland, "How Yoga Changes the Brain," *Scientific American*, March 1, 2014, https://www.scientificamerican.com/article/how-yoga-changes-the-brain.

41 Henepola Gunaratana, *The Four Foundations of Mindfulness in Plain English* (Somerville, MA: Wisdom Publications, 2012), 8.

42 Henepola Gunaratana, *Beyond Mindfulness in Plain English* (Somerville, MA: Wisdom Publications, 2009), 43.

43 Rock, *Your Brain at Work*.

44 John 14:6–7, New International Version.

45 Megan M. Filkowski, R. Nick Cochran, and Brian W. Haas, "Altruistic behavior: Mapping responses in the brain," *Neuroscience and Neuroeconomics* 5 (2016): 65–75, doi: 10.2147/NAN.S87718; Jill Suttie and Jason Marsh, "5 Ways Giving Is Good for You," *Greater Good Magazine*, December 13, 2010, https://greatergood.berkeley.edu/article/item/5_ways_giving_is_good_for_you; Scott Bea, "Wanna Give? This Is Your Brain on a 'Helper's High,'" Cleveland Clinic, November 15, 2016, https://health.clevelandclinic.org/2016/11/why-giving-is-good-for-your-health.

46 Stephanie L. Brown, Randolph M. Nesse, Amiram D. Vinokur, and Dylan M. Smith, "Providing social support may be more beneficial than receiving it: Results from a prospective study of mortality," *Psychological Science* 14.4 (2003): 320–27, doi: 10.1111/1467-9280.14461.

47 K. I. Hunter and Margaret W. Linn, "Psychosocial differences between elderly volunteers and non-volunteers," *International Journal of Aging and Human Development* 12.3 (1981), 205–13, doi: 10.2190/0H6V-QPPP-7JK4-LR38.

Chapter 9. Closing Thoughts on Embodying Our Higher Selves

1 Sarah Elizabeth Rowntree, quoted in "Joining Hands with You: Equipping You to Become His Hands and Feet to the World," *Today's Christian Doctor*, Summer 2014, Christian Medical and Dental Associations, accessed July 23, 2017, https://cmda.org/resources/publication/tcd-summer-2014-joining-hands-with-you. These lines were originally published in a British Quaker monthly journal in 1892: *The British Friend*, volume 1, number 1, 1892, 15.

2 This is an authenticated quote from Albert Einstein: Martin and Ott, eds., *The Cosmic View of Albert Einstein*, 2.

3 Blackburn and Epel, *The Telomere Effect*, 322.

Index

Brain 2.0, *294 (cont.)*
  overactivation of, 46–47
  physical cues of, 210
  and priming, 245
  regret for behavior in, 114–15
  renunciation of, 71–74
  rewards orientation of, xiii–xiv, 210–13
  rewiring of, 213
  and right PFC (prefrontal cortex), 46
  and SCARF framework, 205
  and shadow, 194
  and social environment, 50–51
  and souls, 189
  and spiritual connectivity, 155–56
  and spiritual teachings, 189–90
  and strengthening the brain, 200
  and System 1 (fast-thinking autopilot)/System
      2 (slow-thinking pilot) framework, *116*
  triggering of, 198–99, 225, *248*, 250, 300
  and two-arrows metaphor, 120
  See also Inner Teen Wolf
Brain 3.0, xxx–xxxi, 214–20, *294*
  approach/avoid balance in, *206*
  Brain 1.0's connection to, 35, 197
  Brain 2.0's connection to, 52, 197
  brain activation pattern of, xxix
  and "Buddha Nature," 120
  building, 200
  in business settings, *178*, 178, *181*, 181–82
  Calm Clarity associated with, xiv
  connectome of, xxx
  contagious activation of, 179
  and creating positive impact on world, 74
  and cues to prompt shift into, 213
  and discomfort, 199
  expectations for, 198–201
  and gratitude practices, 164
  and Higher Self, 189, 197–98
  hijack of, *207*
  impaired by Brains 1.0 and 2.0, 46
  and impulse control, 211–12
  and interpersonal interactions, *180*
  and intrinsic motivation, 211
  lifelong process of cultivating, 200–201
  and maturation of frontal lobes, 52, 57, 212
  and meditation practices, 115–16, 164–65,
      228–29, 260
  and mind-track, 273–74
  and mystical experiences, 155
  as natural state of being, 115, 118
  and Noble Eightfold Path, 253–54
  and observer function, 209
  physical cues of, 210
  and priming, 245, 250
  and problem solving, 19

  recharging as necessity of, 199–200
  and SCARF framework, 205
  and self-actualization, xv
  and self-care, 291
  and spiritual connectivity, 156
  and System 1 (fast-thinking autopilot)/System 2
      (slow-thinking pilot) framework, *116*, 251
  and tend-and-befriend response, 223
  and transcendent experiences, 285
  triggering of, *248*, 250
  and triggering of Brains 1.0/2.0, 199
  and two-arrows metaphor, 120
  underdeveloped state of, 208–9, 210–11
  and Vipassana practices, 105–6, 107
  See also Inner Sage
The Brain That Changes Itself (Doidge), xxvi
breathing exercises
  in Kriya Yoga, 139
  and meditation practices, 232
  mindful breathing meditation, 277–80
  ujjayi breathing, 147–48
  and vagus nerve, 147, 225
Brewer, Judson, 260
Brief Resilience Scale, 175
Bruno, Giordano, 95n
Buddha
  in redemption dream, 187
  Reed's encounters with, 101
  statues of, 85–86, 89
  and Theravada teachings, 77
  and Vipassana practices, 104
  See also Gautama, Siddhartha
Buddha's Brain (Hanson), 208
Buddhism
  and archaeological discoveries, 123, 124
  and "Buddha Nature," 92, 120, 254, 284n, 295
      (*see also* Higher Self)
  in Burma, 77, 84–86
  and dukkha (suffering), 119–20
  "emptiness" concept of, 91
  and fasting, 106
  Five Aggregates of, 253
  Five Precepts of, 106, 117–18
  Four Immeasurables prayer of, 239–40
  Four Noble Truths of, 118–19, 120, 253
  and Hinduism, 70
  introductory course on, 87
  and karma, 90–91
  Mahayana branch of, 77–78
  monuments of, 123–24
  mythology surrounding, 122, 125
  Noble Eightfold Path of, 119, 120, *121*, 251,
      253–54, 284, 302
  and personifications of Collective Conscious-
      ness, 295

About the Author

Due Quach (pronounced "Zway Kwok") is the founder and CEO of Calm Clarity, a social enterprise that uses science to help people master their mind and be their best self. A refugee from Vietnam and a graduate of Harvard College and the Wharton MBA program, Quach overcame the long-term effects of poverty and trauma by turning to neuroscience and meditation. After building a successful international business career in management consulting and private equity investments, Quach created the Calm Clarity Program to make mindful leadership accessible to people of all backgrounds. She now leads Calm Clarity workshops in inner-city high schools, university lecture halls, and corporate executive board rooms alike. She is also the founding chair and executive director of the Collective Success Network, a nonprofit that supports low-income, first-generation college students in achieving their academic, personal, and professional aspirations. The Collective Success Network collaborates with the wider business community to create innovative approaches to foster socioeconomic diversity and inclusion. After living and traveling all around the world, Quach is once again a proud resident of Philadelphia, her hometown.